METHODOLOGIES AND APPLICATIONS FOR ANALYTICAL AND PHYSICAL CHEMISTRY

Innovations in Physical Chemistry: Monograph Series

METHODOLOGIES AND APPLICATIONS FOR ANALYTICAL AND PHYSICAL CHEMISTRY

Edited by

A. K. Haghi, PhD
Sabu Thomas, PhD
Sukanchan Palit, MChE
Priyanka Main, MTech

APPLE
ACADEMIC
PRESS

Apple Academic Press Inc.	Apple Academic Press Inc.
3333 Mistwell Crescent	9 Spinnaker Way
Oakville, ON L6L 0A2	Waretown, NJ 08758
Canada	USA

© 2018 by Apple Academic Press, Inc.

First issued in paperback 2021

Exclusive worldwide distribution by CRC Press, a member of Taylor & Francis Group

No claim to original U.S. Government works

ISBN-13: 978-1-77463-059-4 (pbk)

ISBN-13: 978-1-77188-621-5 (hbk)

Library and Archives Canada Cataloguing in Publication

Methodologies and applications for analytical and physical chemistry / edited by A.K. Haghi, PhD, Sabu Thomas, PhD, Sukanchan Palit, M.ChE, Priyanka Main, M.Tech.

(Innovations in physical chemistry : monograph series)
Includes bibliographical references and index.
Issued in print and electronic formats.
ISBN 978-1-77188-621-5 (hardcover).--ISBN 978-1-315-15953-9 (PDF)

1. Chemistry, Physical and theoretical. 2. Chemistry, Analytic. 3. Chemical engineering.
I. Haghi, A. K., editor II. Series: Innovations in physical chemistry. Monograph series

| QD453.3.M48 2018 | 541 | C2018-901673-6 | C2018-901674-4 |

Library of Congress Cataloging-in-Publication Data

Names: Haghi, A. K., editor.
Title: Methodologies and applications for analytical and physical chemistry / editors, A.K. Haghi, PhD [and three others].
Description: Toronto : Apple Academic Press, 2018. | Series: Innovations in physical chemistry. Monograph series | Includes bibliographical references and index.
Identifiers: LCCN 2018012389 (print) | LCCN 2018016769 (ebook) | ISBN 9781315159539 (ebook) | ISBN 9781771886215 (hardcover : alk. paper)
Subjects: LCSH: Chemistry, Analytic. | Chemistry, Physical and theoretical.
Classification: LCC QD75.22 (ebook) | LCC QD75.22 .M48 2018 (print) | DDC 543--dc23
LC record available at https://lccn.loc.gov/2018012389

Apple Academic Press also publishes its books in a variety of electronic formats. Some content that appears in print may not be available in electronic format. For information about Apple Academic Press products, visit our website at **www.appleacademicpress.com** and the CRC Press website at **www.crcpress.com**

ABOUT THE EDITORS

A. K. Haghi, PhD

A. K. Haghi, PhD, is the author and editor of 165 books, as well as 1000 published papers in various journals and conference proceedings. Dr. Haghi has received several grants, consulted for a number of major corporations, and is a frequent speaker to national and international audiences. Since 1983, he served as professor at several universities. He is currently Editor-in-Chief of the *International Journal of Chemoinformatics and Chemical Engineering* and *Polymers Research Journal* and on the editorial boards of many international journals. He is also a member of the Canadian Research and Development Center of Sciences and Cultures (CRDCSC), Montreal, Quebec, Canada. He holds a BSc in urban and environmental engineering from the University of North Carolina (USA), an MSc in mechanical engineering from North Carolina A&T State University (USA), a DEA in applied mechanics, acoustics and materials from the Université de Technologie de Compiègne (France), and a PhD in engineering sciences from the Université de Franche-Comté (France).

Sabu Thomas, PhD

Sabu Thomas, PhD, is a Professor of Polymer Science and Engineering at the School of Chemical Sciences and Director of the International and Inter University Centre for Nanoscience and Nanotechnology at Mahatma Gandhi University, Kottayam, Kerala, India. The research activities of Professor Thomas include surfaces and interfaces in multiphase polymer blend and composite systems; phase separation in polymer blends; compatibilization of immiscible polymer blends; thermoplastic elastomers; phase transitions in polymers; nanostructured polymer blends; macro-, micro- and nanocomposites; polymer rheology; recycling; reactive extrusion; processing–morphology–property relationships in multiphase polymer systems; double networking of elastomers; natural fibers and green composites; rubber vulcanization; interpenetrating polymer networks; diffusion and transport; and polymer scaffolds for tissue engineering. He has supervised 68 PhD theses, 40 MPhil theses, and 45 Masters thesis. He has three patents to his credit. He also received the coveted Sukumar Maithy Award for the best polymer researcher in the country for the year 2008. Very recently, Professor

Thomas received the MRSI and CRSI medals for his excellent work. With over 600 publications to his credit and over 23,683 citations, with an h-index of 75, Dr. Thomas has been ranked fifth in India as one of the most productive scientists.

Sukanchan Palit, MChE

Sukanchan Palit, MChE is a Chemical Engineer by training and profession. He did B.ChE in Chemical Engineering and M.ChE in Chemical Engineering from Jadavpur University, Kolkata, India. He has 23 years of experience in industry, teaching, and research. His research areas are environmental engineering, advanced oxidation processes, ozonation, multiobjective optimization, genetic algorithm, and other branches of Chemical Engineering. He did research at The Queen's University of Belfast, Northern Ireland, United Kingdom and taught at Salalah College of Technology, Salalah, Oman. He has 41 journal papers including 5 chapters in internationally renowned journals.

Priyanka Main, MTech

Priyanka Main, MTech is a postgraduate in Polymer Science and Technology from IIT Delhi, India (2016). Her graduation was from Cochin University of Science and Technology, (CUSAT), Kerala in Polymer Science and Engineering. She was born in Kerala in 1992 but was raised in Muscat, Oman where her parents still are working and her schooling was completed from Indian School Muscat, Sultanate of Oman. She has done projects in VSSC, Trivandrum, and JNU, New Delhi, and has undergone training in RRII, Kottayam and also Primus Gloves, CSEZ, Cochin. Currently, she has joined M.G. University under the guidance of Prof. Sabu Thomas as Project Associate. When she is not pursuing Science, she loves to read, travel, and meet new people.

INNOVATIONS IN PHYSICAL CHEMISTRY: MONOGRAPH SERIES

This new book series, Innovations in Physical Chemistry: Monograph Series, offers a comprehensive collection of books on physical principles and mathematical techniques for majors, non-majors, and chemical engineers. Because there are many exciting new areas of research involving computational chemistry, nanomaterials, smart materials, high-performance materials, and applications of the recently discovered graphene, there can be no doubt that physical chemistry is a vitally important field. Physical chemistry is considered a daunting branch of chemistry—it is grounded in physics and mathematics and draws on quantum mechanics, thermodynamics, and statistical thermodynamics.

Innovations in Physical Chemistry has been carefully developed to help readers increase their confidence when using physics and mathematics to answer fundamental questions about the structure of molecules, how chemical reactions take place, and why materials behave the way they do. Modern research is featured throughout also, along with new developments in the field.

Editors-in-Chief

A. K. Haghi, PhD
Editor-in-Chief, *International Journal of Chemoinformatics* and *Chemical Engineering and Polymers Research Journal*; Member, Canadian Research and Development Center of Sciences and Cultures (CRDCSC), Montreal, Quebec, Canada
E-mail: AKHaghi@Yahoo.com

Lionello Pogliani, PhD
University of Valencia-Burjassot, Spain
E-mail: lionello.pogliani@uv.es

Ana Cristina Faria Ribeiro, PhD
Researcher, Department of Chemistry, University of Coimbra, Portugal
E-mail: anacfrib@ci.uc.pt

BOOKS IN THE SERIES

- High-Performance Materials and Engineered Chemistry
- Applied Physical Chemistry with Multidisciplinary Approaches
- Methodologies and Applications for Analytical
  and Physical Chemistry
- Physical Chemistry for Engineering and Applied Sciences:
  Theoretical and Methodological Implication
- Theoretical Models and Experimental Approaches in Physical
  Chemistry: Research Methodology and Practical Methods
- Engineering Technology and Industrial Chemistry with Applications
- Modern Physical Chemistry: Engineering Models, Materials,
  and Methods with Applications
- Engineering Technologies for Renewable and Recyclable Materials:
  Physical-Chemical Properties and Functional Aspects
- Physical Chemistry for Chemists and Chemical Engineers:
  Multidisciplinary Research Perspectives
- Chemical Technology and Informatics in Chemistry with
  Applications

CONTENTS

LIST OF CONTRIBUTORS

Marat Belotserkovskiy
JIME NSA Belarus, Academic Street 12, Minsk 220072, Republic of Belarus

Karim Bouchmella
CMOS/Institut Charles Gerhardt Montpellier, Université de Montpellier, 34095 Montpellier, France

Bruno Boury
CMOS/Institut Charles Gerhardt Montpellier, Université de Montpellier, 34095 Montpellier, France.
E-mail: bruno.boury@univ-montp2.fr

Gabriela Sanchez Brambila
Russell Research Center-ARS, Quality and Safety Assessment Research Unit USDA,
950 College Station Road, Athens, GA 30605, USA

Antonio Francisco Aguilera Carbo
Department of Animal Nutrition, Universidad Autonoma Agraria Antonio Narro,
Calzada Antonio Narro 1923, Colonia Buenavista, Saltillo 25315, Coahuila, Mexico

Gloria Castellano
Departamento de Ciencias Experimentales y Matemáticas, Facultad de Veterinaria y Ciencias
Experimentales, Universidad Católica de Valencia San Vicente Mártir, Guillem de Castro-94,
E-46001 València, Spain

Xochitl Ruelas Chacon
Department of Food Research, Faculty of Chemistry, Universidad Autonoma de Coahuila,
Blvd. V. Carranza, Colonia Republica Oriente, Saltillo 25280, Coahuila, Mexico.
E-mail: xochitl.ruelas@uaaan.mx; xruelas@yahoo.com

Tanmoy Chakraborty
Department of Chemistry, Manipal University Jaipur, Dehmi-Kalan, Jaipur 303007, India.
E-mail: tanmoychem@gmail.com/tanmoy.chakraborty@jaipur.manipal.edu

A. K. Chaudhary
Advanced Centre of Research in High Energy Materials, University of Hyderabad, Hyderabad 500046,
India. E-mail: anilphys@yahoo.com; akcphys@gmail.com

Juan Carlos Contreras Esquivel
Department of Food Research, Faculty of Chemistry, Universidad Autonoma de Coahuila,
Blvd. V. Carranza, Colonia Republica Oriente, Saltillo 25280, Coahuila, Mexico

Indrayudh Ganguly
School of Electrical Engineering, VIT University, Vellore, India

K. G. Gopchandran
University of Kerala, Thiruvananthapuram, India. E-mail:gopchandran@yahoo.com

G. Jyothi
University of Kerala, Thiruvananthapuram, India

Yury Korobov
UrFU, Mira St. 19, Ekaterinburg 620002, Russia

Ajay Kumar
Department of Mechatronics Engineering, Manipal University Jaipur, Dehmi-Kalan,
Jaipur 303007, India

Surendra Kumar
Department of Bioengineering, Birla Institute of Technology, Mesra, Ranchi, India.
E-mail: Sikuranchi@gmail.com

L. Sandhya Kumari
Oregon State University, Corvallis, OR 97330, USA

Sharanya M.
Department of Bioinformatics, Vels University, Chennai 621117, India.
E-mail: sharanya.bioinfo@gmail.com

Udit Mishra
School of Electrical Engineering, VIT University, Vellore, India

Julio Montañez
Department of Chemical Engineering, Faculty of Chemistry, Universidad Autonoma de Coahuila,
Blvd. V. Carranza, Colonia Republica Oriente, Saltillo 25280, Coahuila, Mexico

Emmanuel Mousset
Université Paris-Est, Laboratoire Géomatériaux et Environnement (LGE), EA 4508, UPEM, 5 bd
Descartes, 77454 Marne-la-Vallée Cedex 2, France. E-mail: emmanuel.mousset@univ-lorraine.fr
Laboratoire Réactions et Génie des Procédés, UMR CNRS 7274, Université de Lorraine,
1 rue Grandville BP 20451, 54001 Nancy Cedex, France

Ranjith G. Nair
Department of Physics, National Institute of Technology Silchar, Silchar 788010, Assam, India.
E-mail: rgnair2007@gmail.com

R. Nivedita
School of Electrical Engineering, VIT University, Vellore, India

Mehmet A. Oturan
Université Paris-Est, Laboratoire Géomatériaux et Environnement (LGE), EA 4508, UPEM,
5 bd Descartes, 77454 Marne-la-Vallée Cedex 2, France

Nihal Oturan
Université Paris-Est, Laboratoire Géomatériaux et Environnement (LGE), EA 4508,
UPEM, 5 bd Descartes, 77454 Marne-la-Vallée Cedex 2, France

Manoj Kumar P
PSG Institute of Technology and Applied Research, Coimbatore, Tamil Nadu, India

Vindhya P. S.
Department of Physics, Bishop Moore College, Mavelikara, Alappuzha 690110, Kerala, India Center for
Advanced Materials Research, Department of Physics, Govt. College for Women, Trivandrum 695014,
Kerala, India

Sukanchan Palit
Department of Chemical Engineering, University of Petroleum and Energy Studies, Post-Office Bidholi
via Premnagar, Dehradun 248007, India. E-mail: sukanchan68@gmail.com; sukanchan92@gmail.com

Vidya Raj
PSG Institute of Technology and Applied Research, Coimbatore, Tamil Nadu, India

K. Raju
St. Joseph Engineering College, Mangalore 575028, India

Prabhat Ranjan
Department of Mechatronics Engineering, Manipal University Jaipur, Dehmi-Kalan, Jaipur 303007, India

G. Ravichandran
St. Joseph Engineering College, Mangalore 575028, India. E-mail: ravig_s@rediffmail.com

Rene Dario Peralta Rodriguez
Research Center for Applied Chemistry, Blvd. Enrique Reyna Hermosillo No. 140, Saltillo 25253, Coahuila, Mexico

Sarat Kumar Sahoo
School of Electrical Engineering, VIT University, Vellore, India

D. Sajan
Department of Physics, Bishop Moore College, Mavelikara, Alappuzha 690110, Kerala, India.
E-mail: dsajand@gmail.com
Department of Food Science and Technology, Universidad Autonoma Agraria Antonio Narro, Calzada Antonio Narro 1923, Colonia Buenavista, Saltillo 25315, Coahuila, Mexico

R. T. Sapkal
Nano-Materials Laboratory, Department of Physics, Tuljaram Chaturchand College, Baramati 413103, India. E-mail: rt_sapkal@yahoo.co.in

B. Suresha
The National Institute of Engineering, Mysore 570008, India

Francisco Torrens
Institut Universitari de Ciència Molecular, Universitat de València, Edifici d'Instituts de Paterna, P.O. Box 22085, E-46071 València, Spain

R. R. Usmanova
Ufa State Technical University of Aviation, Ufa 450000, Bashkortostan, Russia.
E-mail: Usmanovarr@mail.ru, chembio@chph.ras.ru

Y. S. Varadarajan
The National Institute of Engineering, Mysore 570008, India

Maria de la Luz Reyes Vega
Department of Food Research, Faculty of Chemistry, Universidad Autonoma de Coahuila, Blvd. V. Carranza, Colonia Republica Oriente, Saltillo 25280, Coahuila, Mexico. E-mail: mlrv20@yahoo.com

M. Venkatesh
Advanced Centre of Research in High Energy Materials, University of Hyderabad, Hyderabad 500046, India

G. E. Zaikov
Institute of Biochemical Physics, Russian Academy of Sciences, Moscow 119991, Russia

LIST OF ABBREVIATIONS

2D	two-dimensional
2DL	2D-layered
3D	three-dimensional
AAS	atomic absorbance spectrometric
AOP	advanced oxidation process
As	answers
BCS	Bardeen–Cooper–Schrieffer
BDD	boron-doped diamond
BHs	black holes
BPD	balanced photo diode
BS	beam splitter
BTEX	benzene, toluene, ethylbenzene, and xylene
BWO	backward wave oscillator
CCT	correlated color temperature
CDFT	conceptual density functional theory
CNT	carbon nanotube
COD	chemical oxygen demand
CS	chitin synthase
CSP	chemical spray pyrolysis
CW	continuous
DDM	diaminodiphenylmethane
DFT	density functional theory
DPHE	Department of Public Health Engineering
DW	dry weight
EAOP	electrochemical AOP
EBP	ergosterol biosynthesis pathway
ED	electrodialysis
EDAX	energy dispersive X-ray analysis
EDS	energy dispersive spectroscopy
EF	electro-Fenton
EM	electromagnetic
EMR	electromagnetic radiation
EO	electrooptic
EOS	electrooptic sampling

FCM	fuzzy C-mean
FDA	Food and Drug Administration
FEL	free-electron laser
FL	fuzzy logic
FTIR	Fourier transform infrared
FTO	fluorine-doped tin oxide
GDL	gas-diffusion layer
GR	graphene
GRO	GR oxide
GS	glucan synthases
HDPE	high-density polyethylene
HF	fluorhydric acid
HM	heavy metal
HOC	hydrophobic organic compound
HOMO	highest occupied molecular orbital
HPCD	hydroxypropyl-beta-cyclodextrin
HRTEM	high-resolution transmission electron microscope
Hs	hypotheses
IAEA	International Atomic Energy Agency
IH	industrial hygiene
IMF	intrinsic mode function
inverse IH	inverse IH
IR	infrared
LDA	linear discriminate analysis
LFL	Laisez-Faire
LRP	labor risk prevention
LSDA	Local Spin Density Approximation
LT-GaAs	low-temperature gallium arsenide
LTs	low temperatures
LUMO	lowest unoccupied molecular orbital
MC	moisture content
MEA	membrane electrode assembly
MED	multieffect distillation
MF	magnetic field
MIT	Massachusetts Institute of Technology
MMs	molecular magnets
MSF	multistage flash
MW	molar weight
MWCNT	multiwalled carbon nanotube
NF	nanofiltration

NL	nonlinear
NM	nanomaterial
NOM	natural organic matter
NP	nanoparticle
NPr	nanoproduct
NT	nanotube
OM	organic matter
OS	oxidation state
PAH	polycyclic aromatic hydrocarbon
PC	photoconductive
PCA	photoconductive antenna
PCS	photoconductive sampling
PDB	Protein Data Bank
PE	polyethylene
PEM	proton exchange membrane
PEMFC	proton exchange membrane fuel cell
PET	polyethylene terephthalate
PL	photoluminescence
POM	polyoxometalate
PP	periodic property
PPE	personal protective equipment
PTE	periodic table of the elements
PVC	photovoltaic cell
Q/A	questions/answer
QCL	quantum cascade laser
Qs	questions
QT	quantum theory
RA	risk assessment
RD-SOS	radiation damaged silicon on sapphire
RO	reverse osmosis
RPE	respiratory protective equipment
S/N ratio	signal-to-noise ratio
SCE	saturated calomel electrode
SE	squalene epoxidase
SEM	scanning electron microscope
SF	soil flushing
SI-GaAs	semiinsulating gallium arsenide
SNR	signal-to-noise ratio
SOM	soil organic matter
SPR	surface plasmons resonance

SVM	support vector machine
SW	soil washing
SWCNT	single-walled carbon nanotubes
TEM	transmission electron microscope
THz	terahertz
THz-ES	THz emission spectroscopy
THz-TDS	THz time domain spectroscopy
TL	transformational
TMAP	1-2,4,5-tri methoxy acetophenone
TMD	transition metal dichalcogenide
TMPMP	1-2,4,5-tri methoxy phenyl-1¢-methoxy propionaldehyde
TNS	2-(p-toluidino)naphthalene-6-sulfonic acid sodium
TRTS	time resolved THz spectroscopy
TS	tensile strength
TTS	time-temperature superposition
TTT	time-temperature-transformation
USEPA	Environmental Protection Agency of United States
UV	ultraviolet
VA	variable attenuator
VIS	visible
WP	Wollaston prism
WVP	water vapor permeability
WVTR	water vapor transmission rate
XRD	X-ray diffraction
ZnO	zinc oxide

PREFACE

This volume generates understanding through numerous examples and practical applications drawn from research and development chemistry. The authors allow a greater understanding of problems more quickly and easily than purely intuitive methods.

At the same time, each topic is framed within the context of a broader more interdisciplinary approach, demonstrating its relationship and interconnectedness to other areas. The premise of this work, therefore, is to offer both a comprehensive understanding of applied science and engineering as a whole and a thorough knowledge of individual subjects. This approach appropriately conveys the basic fundamentals, state-of-the-art technology, and applications of the involved disciplines and further encourages scientific collaboration among researchers.

The book presents an up-to-date review of modern materials and physical chemistry concepts, issues, and recent advances in the field. Distinguished scientists and engineers from key institutions worldwide have contributed chapters that provide a deep analysis of their particular subjects.

This volume emphasizes the intersection of chemistry, math, physics, and the resulting applications across many disciplines of science and explores applied physical chemistry principles in specific areas.

PART I
Nanoscience

NANOSCIENCE: FROM A TWO-DIMENSIONAL TO A THREE-DIMENSIONAL PERIODIC TABLE OF THE ELEMENTS

FRANCISCO TORRENS[1,*] and GLORIA CASTELLANO[2]

[1]*Institut Universitari de Ciència Molecular, Universitat de València, Edifici d'Instituts de Paterna, P.O. Box 22085, E-46071 València, Spain*

[2]*Departamento de Ciencias Experimentales y Matemáticas, Facultad de Veterinaria y Ciencias Experimentales, Universidad Católica de Valencia San Vicente Mártir, Guillem de Castro-94, E-46001 València, Spain*

Corresponding author. E-mail: torrens@uv.es

CONTENTS

ABSTRACT

This chapter presents the general principles that are at the basis of the construction of artificial molecular devices and machines, and the main characteristics of the systems, with a special focus for the kind of energy inputs needed to make them work. Human progress was always determined by the design and construction of novel devices and machines, beginning with the wheel to the most sophisticated items. Nowadays, the new trend is to reduce the dimensions and weight of the component parts as much as possible. The most important and common examples are in the information technology domain, but some other fields exist that will be the beneficiary of the trend (e.g., medicine, energy, materials, and environment). Miniaturization is a must for the actual progress, and the scientific achievements of the last few years show that the concepts of *device* and *machine* can be extended at the molecular level, that is, nanometer scale, that, at present, seems to be the ultimate frontier to miniaturization. The extension at the molecular level of device and machine concepts is important not only for the practical applications of nanoscience and nanotechnology but also for fundamental research. By the chemical molecule-by-molecule *bottom-up* approach, science and technology move from the microworld to the nanoworld and, because of the nature of inputs (light, chemical), they move from electronics to photonics and chemionics. The bottom-up approach offers unlimited opportunities for the design and construction of nanoscale supramolecular structures, combining the high precision of the chemical synthesis with scientists having a device-driven ingenuity.

1.1   INTRODUCTION

Industrial hygiene (IH), understood as an activity destined to prevent the diseases of labor origin, goes back to antiquity, where the first knowledge of adverse effects caused by the labor activity appeared.[1-4] To remember them is not only an exercise of anthropological interest but also reflects that situations known way back are already of total actuality. No better way exists to beginning to talk about the present and the future than remembering the past. One must revise some particular aspects of IH day to day, which those devoted to this should have always present, if not that one cannot see the wood for the trees. Once these questions (Qs) are treated, a reflection is made of the present situation of IH activities and expert opinions with regard to its future. The themes that are considered of most actuality are revised:

working with nanomaterials (NMs), the study of which will indicate what could be the future IH activities.

The present report presents inverse IH (IIH), the simplified methods of evaluation of the exposure to NMs and the limitations for its application. The aim is the interchange of experiences in the utilization of these simplified methods of evaluation. The first objective is to expose our experience and, over all, our proposal of action on the focus and development of the evaluation model of NMs exposure, in order to be a quality process, which would give people trust in that the environmental conditions, by NMs exposure, are going to imply a risk for the health of workers in neither the short nor long term. The second purpose is to propose a new model of action, based on the introduction of improvements and practical measures of control of NMs-exposition conditions, and environmental-evaluation programs, as a tool to demonstrate conditions acceptability on working with NMs. The main and first IH objective is to achieve that hurts for health would not be produced by causes derived from the conditions of the working environment. It shares the same objective that labor medicine, but it is a *more primary* preventive discipline. It exerts an anticipated prevention: it evaluates the working conditions before the signs and symptoms of any professional disease appear. Labor medicine acts on the person analyzing those alterations of any biological, biochemical, or physiological parameter not to be produced. In the new model, IIH, the order of preventive actions is inverted. (1) The most evident improvements and corrective measures are adapted, based on inspections and observations that allow defining them, without having to apply to samplings. (2) Programs are established for residual-IH risk assessment (RA) and confirming the acceptability in the long-term exposition. A comparative analysis of IIH methods can be carried out with tools [e.g., COSHH Essentials (UK), International Chemical Control Toolkit (OIT), Easy-to-Use Workplace Control Scheme for Hazardous Substances (Germany), and Méthodologie d'Évaluation Simplifiée du Risque Chimique (France)]. As an example of all ideas above, the determination of the group of danger of working with NMs is analyzed.

In earlier publications, the periodic table of the elements (PTE),[5,6] molecular simulators,[7–9] labor risk prevention (LRP) and preventive healthcare at work with NMs[10] were reviewed. In the present report, the main objective is to initiate a debate suggesting a number of Qs that can arise on LRP when handling NMs and providing, when possible, answers (As) and/or hypotheses (Hs). When managing NMs, one must apply the *principle of caution*. Nanoparticles (NPs) must be considered as least as toxic as their corresponding bulk materials. The concept of *elementarity* is revised. The PTE is discussed. In PTE framework, NMs are presented as an extension from a two-dimensional (2D) to a three-dimensional (3D) PTE. Some approaches of the Research Project *Nanotechnology and Labour Risk Prevention* are analyzed. The discussion includes the philosophical conflict between chemistry and physics, and the chemistry–culture relationship.

1.2   LABOR RISKS PREVENTION: NANOMATERIALS

Generalitat Valenciana/INVASSAT organized a day on sensitizing and spreading as regards LRP, nanotechnology, NPs as an emergent risk to workers safety and health, a step forward as regards their exposition to workers and their possible consequences.[11] A Q appeared.

Q1. How can NPs affect people?
   Bergamaschi raised Qs on human biomonitoring and epidemiological studies.

Q2. Human toxicity of NMs: what disease endpoint?

Q3. How safe are NMs?

Q4. What is the estimated number of workers *actually* exposed to engineered NPs (ENPs)?[12,13]
   He presented the main objectives of biological monitoring:[14] (1) individual or group exposure RA; (2) identification of early (preferably specific) effects, which are indicative of actual or potential health effects; (3) health RA to exposed subjects. He discussed the layout of biomarkers research, as condition of the responsible development of nanotechnologies and safety of workers exposed to engineered NMs (ENMs, cf. Fig. 1.1).
   Fito proposed Qs and problem (P) on legal framework and limitations for NMs and nanoproducts (NPrs) RA.

Q5. What part of sprayed does it present particles under 100 nm?

P1. Particles aggregate and long-time studies cannot be made.

Q6 What do people know about the legal framework and present limitations for RA of NMs/NPrs?
   Gálvez proposed the following Qs and H on limit values and exposition RA.

Q7. What NMs are there in the work environment?

H1. *Principle of caution: An NM is toxic unless the opposed be demonstrated.*

Q8. The problem of quantitative RA, what do people measure?

Q9. What do people measure with?

Q10. Limit values: mass/volume?

Q11. Limit values: number of particles/volume?

Q12. Limit values: surface area/volume?

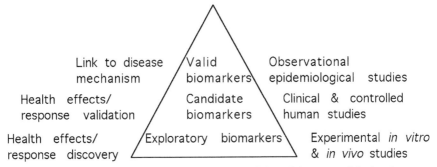

FIGURE 1.1   Layout of biomarkers research as condition of the responsible development of nanotechnologies and safety of workers exposed to ENMs.

López de Ipiña presented collective protection measures (collective protective equipment) versus NMs. He classified safety: (1) safe product; (2) safe production; (2.1) safe machine; (2.2) safe exposition; and (3) safe environment. He emphasized the prevention importance of safe machine. He examined the hierarchy of controls: (1) elimination; (2) substitution; (3) engineering controls; (4) administration controls; (5) protect the worker with proper equipment. He analyzed the order of NMs hazard riskiness: slurry/suspension < agglomerated < highly disperse. He raised a Q on collective protection measures versus NMs.

Q13. Has the risk been adequately reduced?

He discussed three approaches: (1) approach based on danger; (2) approach based on control banding; (3) approach based on the state of the art. He differentiated: (1) safety of machines (emission) and (2) health and safety at work (exposition).

Carlos proposed Q/As on efficiency of respiratory protective equipments (RPEs) versus NMs.

Q14. What are people breathing now?

A14. The number of NPs rises: laboratory (100) < foundry (1000) < solder (10,000 NPs/m^3).

Q15. Risk of NPs exposition, how must one act?

Q16. On the basis of RA, how to select proper RPE?

Q17. Isolated exposition?

Q18. Exposition extended during the day?

Q19. Will other RPEs be needed?

Q20. Is the personal protective equipment (PPE) correct?

Q21. Is the PPE appropriate?

Q22. Will the PPE be used well?

Q23. Why to validate factory's PPEs?

A23. It is motivating the workers to use them.

  Caballero proposed Qs and an A on preventing measures in the use of NMs.

Q24. Beard and cosmetics when using masks?

A24. Masks should be used without beard and cosmetics; beard affects the adjustment of mask.

Q25. How to protect oneself?

Q26. How to make the most of NPs?

  Llorca raised a Q on the revision of the labor risks in the hotel sector.

Q27. What are the psychosocial factors of greatest risk for health in the hotel sector?

  Lobato raised Qs on application of nanotechnology in medicine.

Q28. What has nanotechnology in medicine?

Q29. Why is nanotechnology important in medicine?

  Martínez-Jiménez raised two Qs on healthy program and psycho-social health.

Q30. Factors of work and psychosocials, what?

Q31. Indicators?

  López-Vilchez analyzed transformational (TL)–Laisez-Faire (LFL) leadership-styles relationship, proposing Hs.

H2. The TL and LFL styles affect burnout.

H3. The TL dimensions strong and negatively correlate with burnout.

H4. The LFL dimensions strong and positively correlate with burnout.

Carrasco Báez raised some Qs on PYMESA, and healthy persons and enterprises.

Q32. What is exactly your objective?

Q33. What is the present situation?

Q34. What resources does it need?

Q35. What do you want?

Q36. What do you think when you feel so?

Q37. Where is the evidence that this is as you want?

Q38. Workplace bullying, in the face of this situation, what can I do?

Q39. What resources have I?

She presented the wheels of life of the professional and personal scopes. In the *wheel of life of the professional scope* (cf. Fig. 1.2), some possible areas are: relationships with the boss/es, relationship with the workmates, relationships with the collaborators, communication, personal development, professional development, working team, career road map, labor climate, future project, private-life–work equilibrium, physical environment, professional skills, finances/pay, and so on. The *wheel of life of the personal scope* includes Qs, for example,

Q40. How do you get on with your couple?

She presented affective communication: describe, explain, ask, and thank. She raised an additional Q.

Q41. Is your enterprise an excellent place to work?

Macías raised a Q on labor suitability after sudden cardiac arrest in the health environment.

Q42. What to assess?

Piñaga raised Qs on the action of the prevention services versus chronic renal failure.

Q43. Are physicians adequately diagnosing all cases?

Q44. Are the cases that physicians have well handled, attended, or laborly incorporated?

Rueda de la Vida

Tienes ante ti la Rueda de la Vida Profesional. El conjunto de sus 10 secciones o partes representan el equilibrio. Busca cuales son respecto a tu trabajo los pilares básicos en que se debe apoyar para que obtengas una plena satisfacción y coloca cada uno de ellos al lado de uno de los números 10 de la circunferencia exterior. Una vez elegidos todos puntúa de 0 a 10 tu grado de satisfacción actual en cada una de las áreas. El centro de la rueda representa el nivel más bajo, cero y la circunferencia externa el más alto, 10.

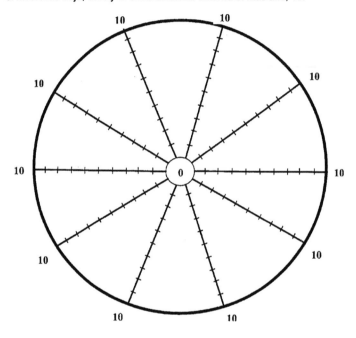

ALGUNAS ÁREAS POSIBLES: Relaciones con el/los jefes, Relación con los compañeros, Relaciones con los colaboradores, Comunicación, Desarrollo Personal, Desarrollo profesional, Equipo de trabajo, Plan de carrera, Clima laboral, Proyecto futuro, Equilibrio vida privada-trabajo, Ambiente físico, Habilidades profesionales, Finanzas/ Retribución, etc

FIGURE 1.2   The wheel of life. Source: N. Carrasco Báez.[32]

1.3   EVOLUTION OF THE CONCEPT OF *ELEMENTARITY*

Starting from the idea of physical particle of quantum mechanics, García Canal discussed that corresponding to a quantum object, for which the name of particle is already maintained but that is conceptually different.[15] After getting past this step in the evolution of the concept of elementarity, he gave the next pass that conduces to that of confined quantum objects, the level of quarks. He completed with the presentation of theoretical ideas that includes the present knowledge of the microworld, those that are mainly based on symmetry considerations. He proposed Qs, H, and A on *elementarity* evolution.

Q1.   What is elemental?

Q2.   Under the macrocosmos, is the microcosmos simple?

H1.   The wave–particle duality is like the Necker (1832) cube (cf. Fig. 1.3).[16]

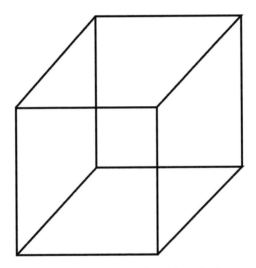

FIGURE 1.3   The Necker cube: a wire frame cube with no depth cues.

Q3.   In addition, what does it happen to the theory?

A3.   It is based on *symmetry* notions.

Q4.   In addition, what does it happen if one wants to define α in different situations?

Q5.   How to reobtain invariance?

Q6.   How to compensate for the variations of $d\alpha/dt$?

Q7.   Symmetries signs, of what?

Q8.   Quality?

Q9.   Chirality, why?

Q10. Choice of the good hand?

1.4 PERIODIC TABLE OF THE ELEMENTS

Sheehan represented a PTE according to elements relative abundance on Earth's surface (cf. Fig. 1.4).[17]

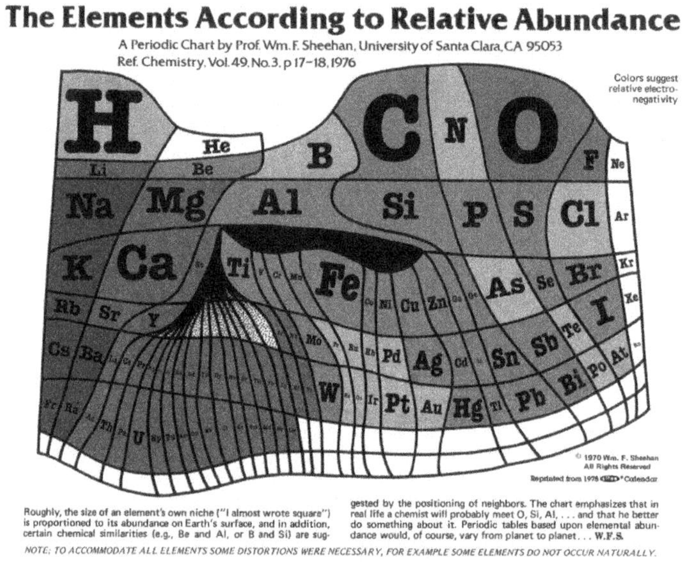

FIGURE 1.4 The elements according to their relative abundance.

The PTE regularly incorporated into living matter (H, B–O, Na, Mg, P–Cl, K, Ca, Mn–Zn, cf. Fig. 1.5) shows oxidation states (OSs) 2, 1, 3, and so on.[18,19]

```
   1A  2A                                                  3A  4A  5A  6A  7A  8A
  -----                                                                        -----
1 | H |                                                                        |   |
  |---+----                                                -----------------------+---|
2 |   |   |                                                | B | C | N | O |   |   |
  |---+---|                                                |---+---+---+---+---+---|
3 |Na |Mg |3B  4B  5B  6B  7B |      8B      |1B  2B |   |   | P | S |Cl |   |
  |---+---+---------------------------------------------+---+---+---+---+---+---|
4 | K |Ca |   |   |   |   |Mn |Fe |Co |Ni |Cu |Zn |   |   |   |   |   |   |
  |---+---+---+---+---+---+---+---+---+---+---+---+---+---+---+---+---+---|
5 |   |   |   |   |   |   |   |   |   |   |   |   |   |   |   |   |   |   |
  |---+---+---+---+---+---+---+---+---+---+---+---+---+---+---+---+---+---|
6 |   |   |LAN|   |   |   |   |   |   |   |   |   |   |   |   |   |   |   |
  |---+---+---+-----------------------------------------------------------------
7 |   |   |ACT|
  -------------
              -----------------------------------------------------------------
   Lanthanide |   |   |   |   |   |   |   |   |   |   |   |   |   |   |   |
              |---+---+---+---+---+---+---+---+---+---+---+---+---+---+---|
   Actinide   |   |   |   |   |   |   |   |   |   |   |   |   |   |   |   |
              -----------------------------------------------------------------
```

FIGURE 1.5 PTE showing those regularly incorporated into living matter.

Thims reported interactive human molecular PTE with 26-element standard (cf. Fig. 1.6).[20]

1	2	3	4	5	6	7	8	9	10	11	12	13	14	15	16	17	18
													Intellect (animation) ↓	Power (energy) ↓	Body (fluidity) ↓		
1 H 1.007																	2 He 4.003
3 Li 6.941	4 Be 9.012											5 B 10.81	6 C 12.01	7 N 14.01	8 O 16.00	9 F 19.00	10 Ne 20.18
11 Na 22.99	12 Mg 24.31											13 Al 26.98	14 Si 28.09	15 P 30.97	16 S 32.07	17 Cl 35.45	18 Ar 39.95
19 K 39.10	20 Ca 40.08	21 Sc 44.96	22 Ti 47.88	23 V 50.94	24 Cr 52.00	25 Mn 54.94	26 Fe 55.85	27 Co 58.93	28 Ni 58.69	29 Cu 63.55	30 Zn 65.39	31 Ga 69.72	32 Ge 72.59	33 As 74.92	34 Se 78.96	35 Br 79.90	36 Kr 83.80
37 Rb 85.47	38 Sr 87.62	39 Y 88.62	40 Zr 91.22	41 Nb 92.91	42 Mo 95.94	43 Tc (98)	44 Ru 101.1	45 Rh 102.9	46 Pd 106.4	47 Ag 107.9	48 Cd 112.4	49 In 114.8	50 Sn 118.7	51 Sb 121.8	52 Te 127.6	53 I 126.0	54 Xe 131.3
55 Cs 132.9	56 Ba 137.3	57 La 138.9	72 Hf 178.5	73 Ta 180.9	74 W 183.9	75 Re 183.9	76 Os 190.2	77 Ir 192.2	78 Pt 195.1	79 Au 197.0	80 Hg 200.6	81 Tl 204.4	82 Pb 207.2	83 Bi 209.0	84 Po (210)	85 At (210)	86 Rn (222)
87 Fr (223)	88 Ra (226)	89 Ac (227)	104 Rf (257)	105 Ha (260)	106 Sg (263)	107 Ns (262)	108 Hs (262)	109 Mt (266)	110	111	112						
				58 Ce 140.1	59 Pr 140.9	60 Nd 144.2	61 Pm (147)	62 Sm 150.4	63 Eu 152.0	64 Gd 157.3	65 Tb 158.9	66 Dy 162.5	67 Ho 164.9	68 Er 167.3	69 Tm 168.9	70 Yb 173.0	71 Lu 175.0
				90 Th 232.0	91 Pa (231)	92 U 238.0	93 Np (237)	94 Pu (242)	95 Am (243)	96 Cm (247)	97 Bk (247)	98 Cf (249)	99 Es (254)	100 Fm (253)	101 Md (256)	102 No (254)	103 Lr (257)

FIGURE 1.6 Human molecular periodic table of the elements: 26-elements standard.

Thims informed human molecular PTE with relative abundance by percent mass (cf. Fig. 1.7).

Some transition metals NPs are used in paints (cf. Fig. 1.8).

A Ph. D. thesis dissertation was presented in this laboratory on Au NMs decorated with DNA bases: design, synthesis, and applications; Au NPs; and nanoclusters (cf. Fig. 1.9).

1.5   FROM A TWO- TO A THREE-DIMENSIONAL PERIODIC TABLE OF THE ELEMENTS

Before nanoscience, people played chemistry as *to sink the fleet*: D9, cobalt.[21] It was because the elements were placed and their periodic properties (PPs) determined by their position in PTE: *melting point* (MP), *boiling point*, *density*, and so on. Now, it is all over [e.g., C has not a unique MP because its properties are different if it is the case of a graphene (GR), nanotube (NT), and fullerene]. The GR is a C-atoms single layer in the form of a honeycomb lattice. Its properties suggest applications. Many GR layers placed on top of each other become ordinary graphite. A C-NT is a GR lattice curled up to make a tube. Fullerenes are roughly a spherical form of the structure. The simplest fullerene is made of 60 atoms that lie at points described by the geometry of a soccer ball of the age of the black-and-white television. Now, one has a heap of different materials, which share the same nature and atomic composition but completely different properties.[22–24]

Nanotechnology adds a 3D to PTE; for example, graphite-like monolayer MoS_2 is the most known and stable 2D-layered (2DL) transition metal dichalcogenide (TMD). Monolayer MoS_2 has exceptional charge-carrier mobility and is a contender for the next electronics generation, as Si chip reaches its fundamental limits. It is an *n*-type semiconductor, useful for photovoltaic cells (PVCs).

1.6   CALCULATION RESULTS

The chemistry of 2DL TMD nanosheets allowed using them as solid lubricants.[25] In PVCs, a need exists to collect all electromagnetic radiation (EMR) spectrum [e.g., infrared (IR), visible (VIS)]. Some NPs of certain size and shape collect IR; some other NPs of different size and shape, VIS. Intercalating both types of NPs, PVCs collect a greater portion of EMR spectrum. Monolayers proved to be stable under ambient conditions [room

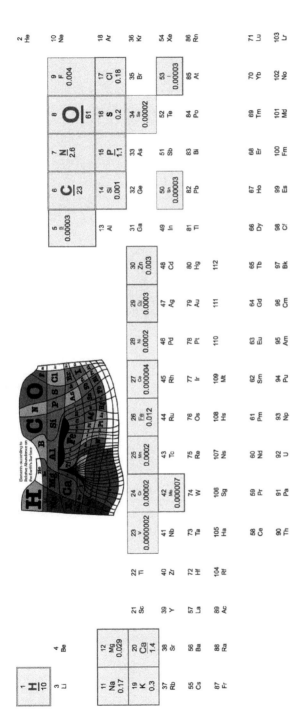

FIGURE 1.7   Human molecular periodic table of the elements: relative abundance by percent mass.

FIGURE 1.8   Some transition metals nanoparticles are used in paints.

FIGURE 1.9 A thesis dissertation presented in this laboratory on Au nanomaterials.

temperature (RT) in air, GR; hexagonal BN, *white GR*; boron carbon nitride; fluorographene; GR oxide; MoS_2, WS_2, $MoSe_2$, WSe_2; micas, Bi/Sr/Ca/Cu oxide; H, B–F, Si, S, Ca, Cu, Se, Sr, Mo, W, Bi, cf. Fig. 1.10], those probably stable in air (semiconducting dichalcogenides: $MoTe_2$, WTe_2, ZrS_2, $ZrSe_2$, etc.; MoO_3, WO_3; layered Cu oxides; Zr, Te) and those unstable in air but that may be stable in inert atmosphere (metallic dichalcogenides: $NbSe_2$, NbS_2, TaS_2, TiS_2, $NiSe_2$, etc.; layered semiconductors: GaSe, GaTe, InSe, Bi_2Se_3, etc.; Ti, Ni, Ga, Nb, In, Ta); 3D compounds that were successfully exfoliated to monolayers [TiO_2, MnO_2, V_2O_5, TaO_3, RuO_2, etc.; perovskite-type: $LaNb_2O_7$, $(Ca,Sr)_2Nb_3O_{10}$, $Bi_4Ti_3O_{12}$, $Ca_2Ta_2TiO_{10}$, etc.; hydroxides: $Ni(OH)_2$, $Eu(OH)_2$, etc.; V, Mn, Ru, La, Eu] with OSs 3, 4, 2, 5, 6, and so on.

```
    1A  2A                                           3A  4A  5A  6A  7A  8A
   -----                                            ----------------------
1 | H |                                                                | |
  |---+----                                    ----------------------- |-+-|
2 |   |   |                                   | B | C | N | O | F |     | |
  |---+---|                                   |---+---+---+---+---+---- |-+-|
3 |   |   |3B  4B  5B  6B  7B |    8B    |1B  2B |   |Si |   | S |      | |
  |---+---+----------------------------------------+---+---+---+---+--- |-+-|
4 |   |Ca |   |Ti | V |   |Mn |   |   |Ni |Cu |   |Ga |   |   |Se |     | |
  |---+---+---+---+---+---+---+---+---+---+---+---+---+---+---+---+---+- |-+-|
5 |   |Sr |   |Zr |Nb |Mo |   |Ru |   |   |   |   |In |   |   |Te |     | |
  |---+---+---+---+---+---+---+---+---+---+---+---+---+---+---+---+---+- |-+-|
6 |   |   |LAN|   |Ta | W |   |   |   |   |   |   |   |Bi |   |   |     | |
  |---+---+---+---+---+---+---+---+---+---+---+---+---+---+---+---+---+- |-+-|
7 |   |   |ACT|
   -----------

  Lanthanide |La |   |   |   |   |Eu |   |   |   |   |   |   |   |
             |---+---+---+---+---+---+---+---+---+---+---+---+---+---|
  Actinide   |   |   |   |   |   |   |   |   |   |   |   |   |   |   |
```

FIGURE 1.10   Monolayers PTE: stable, probably, unstable, and 3D compounds exfoliated to monolayers.

Materials obtained by the incorporation of metals, for example, Ti–Co, Cu, Zr, Mo, Sn, and W, in the frame or at the surface of nanoporous oxides or hydroxides (cf. Fig. 1.11) with OSs 2, 4, 3, 6, and so on are able to catalyze selective reactions between organic compounds and hydroperoxides or H_2O_2.[26]

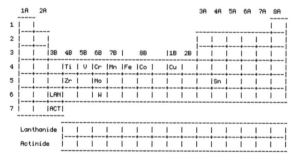

FIGURE 1.11   PTE of metals incorporated in the frame or at surface of nanoporous oxides or hydroxides.

Research Project *Nanotechnology and Labour Risks Prevention* surveys of practices in preventive health matters, with Qs on the NMs to which workers are exposed: $CaCO_3$, dendrimers, SiO_2, TiO_2, fullerene, GR, nano-clays, single/multiple-walled carbon nanotubes, ferrous, organic, C-black, Au, Al_2O_3, Ce_2O_3, ZnO, Ag, polystyrene, polymers, quantum dots, and $BaTiO_3$.[27] Matching PTE (cf. Fig. 1.12) shows different groups, periods, and OSs 2, 0, 3, 4, and so on.

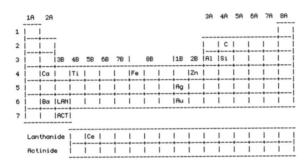

FIGURE 1.12   PTE of nanomaterials in a survey of practices in preventive matters.

The PTE of 2DL TMDs MX_2 (M = Ti–Cr, Zr–Tc, Hf–Re; X = S, Se, Te, cf. Fig. 1.13) shows OSs 4, 5, 6, 7, and 3.[28]

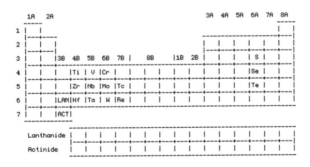

FIGURE 1.13   Periodic table of the elements of two-dimensional-layered metal dichalco-genides MX_2.

The PTE of metal oxide NPs ZnO, CuO, V_2O_3, Y_2O_3, Bi_2O_3, In_2O_3, Sb_2O_3, Al_2O_3, Fe_2O_3, SiO_2, ZrO_2, SnO_2, TiO_2, CoO, NiO, Cr_2O_3, and La_2O_3 (of Al, Si, Ti–Cr, Fe–Zn, Y, Zr, In–Sb, La, and Bi, cf. Fig. 1.14) shows OSs 3, 2, and 4.[29]

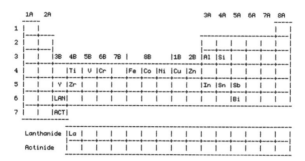

FIGURE 1.14   Periodic table of the elements of metal oxide nanoparticles.

1.7   ETHICS IN THE TIMES OF INTERNET, DIGITALIZATION, AND SOCIAL MEDIA

The following question was raised on ethics.

Q1. *British Medical Journal* introduced open peer review for *ethical reasons*, does not it make one wonder why peer review was ever anything else?
    Plasencia proposed some Qs and an A on ethics, digitalization, and social media.

Q2. People live in exponential times, are they times to ethics?[30]

Q3. What is the global framework?

Q4. How to live with this gigantic information explosion that encircles people?

Q5. Is there an ethics of digitalization?

A5. No, there is a general ethics.

Q6. Is there an ethics of cyberspace?

Q7. Is there an ethics of Internet?

Q8. Is *ethical* the rising of social media?

Q9. Is *ethical* Moore law?

Q10. How do people react to Internet risks for younger?

Q11. What must be the ethical rules for this field of the Era of the Net in which people are leaving?

Q12. Which direction must people orientate them towards?

Q13. What is cybercrime?

Q14. Crime in Internet?

He reviewed some applications in education, raising additional Qs.

Q15. Did you know…?

Q16. What does this all mean?

Q17. Are you doing this in school?

Q18. As a father/mother, do you demand this to school?

Q19. As information is duplicated each two years, how do you prevent that what pupils learn in first does not expire in third year?

1.8   DISCUSSION

1.8.1   PHILOSOPHICAL CONFLICT BETWEEN CHEMISTRY AND PHYSICS

Chemistry presents an important idea ever elucidated in 400 years of experimental science: PTE (the other is bioevolution).

Confrontation, expressed in terms of elements versus atoms, leads one to the consideration of Mendeleev's element concept, which is a philosophical and historical concept in chemical history.[31] Mendeleev's viewpoint allows raising the *reductionism* subject, which became important after quantum mechanics. The PTE reflects a delicate abstract philosophical understanding of the element, which was questioned after the isotopes and atomic physics. Chemists were misunderstood as naïf *positivists* when they refused to accept the existence of the atoms in the 19th century. After Compte and Mach, chemistry is positivist and not positivist at once in its approximation. Ostwald and Duhem showed the limit of positivism in chemistry. Atomism discussion generated by positivism allows exploring atomisms variety that existed and exist. The PTE represents chemist's distinctive atomism, which focus on the atom as a node of chemical relationships.

Does nanotechnology mark chemistry end as a subject? A strong continuity exists in the science–society relationship, which will survive the transition from chemistry to nanotechnology. A further examination of the natural entities and a new interest in these at the nanoscale produced a revival of the Faustian ambitions associated with chemistry. Nanotechnology does not only search for imitating nature but also surpassing it, with an increasing number of scientific visionaries that announce artificial life and self-propagating nanomachines as precursors of the control of life by the humanity. A general philosophical guide could help to construct an ethics adequate for contemporary research in the nanorevolution context.

1.8.2   CHEMISTRY AND CULTURE

Chemical science is embedded in cultural values, which matter for the public acceptance of scientific and technological innovations. The ethics of chemistry needs to negotiate priorities via a democratic collective deliberation. Comparing physics and chemistry, is chemistry the future? New trends (nanotechnology, etc.) show decaying differences between physics and chemistry, science and technology, economy and capitalism, and so on. It is interesting to study the possible generalization of the ideas above to PPs, periodic law, and so on.

The necessity of communicating science is a secondary effect of scientific creativity.

The natural/artificial dilemma comes from the times of alchemy. The cause is the lack of spreading.

1.9   FINAL REMARKS

From the present calculation results and discussion, the following final remarks can be drawn:

1. In NMs, one has a heap of different materials, which share the same nature and atomic composition but completely different properties.
2. On handling NMs, one must apply the *principle of caution.*
3. The NPs must be considered as least as toxic as their corresponding bulk materials.
4. Miniaturization is a must for the actual progress.
5. It is important whether one's research field becomes fashionable.
6. Nature is not classical.
7. Physical properties of NMs differ significantly from those of conventional ones because of large surface-to-volume ratios and quantum effects.
8. Feynman foresaw that (1) when one gets to the small world, one has a lot of new things that would happen that represent completely new opportunities for design;[22] (2) other way to simulate the probabilistic nature is by a computer, which itself be probabilistic.[23]

9. Some advances are expected in quantum chemistry, for example, superconductivity, and so on.

10. The outcome and implications of this report are that there should be no indication of prejudice and enmity in a literary work, and that, researchers should not be afraid to transcend cultural boundaries in search for the truth or present the view of the *other* objectively.

11. Something wrong exists in quantum mechanics. It predicts probabilities for experiments, not certainties. However, the formal mathematics of quantum mechanics predicts the quantum state with certainty. The mathematical conclusion is a Schrödinger's cat that is both alive and dead. People who observe Schrödinger's cat are also in a *superposition* of two states, but no way exists they can be aware of their other half. It impresses many as too exotic that people themselves are in two (actually many) states, and quantum mechanics probabilities are only artifacts of their existence in the form of many quantum states.

12. Every time, more people think that theory emergence must be search in an esthetic conception of the world, which can be difficult to understand, surprising, complex, counterintuitive, yes, but it cannot be ugly! However, in physics, perhaps, it is not so important that intuition (educated in a world) feel comfortable (in a different world inaccessible to people's senses), which is what happens with quantum mechanics. The global esthetic criterion, which compensates the unintuitive character of quantum mechanics, could be summarized in the expression: *Science cannot depend on who looks at it.*

13. Physics continues to make important contributions, which make a difference to everyone's life (e.g., new materials, progress in fluid dynamics, miniaturization techniques, and quantum manipulations with applications to computing). The future will include better batteries and solar cells, which will make a big difference in people's energy usage. Future developments include chips directly interfacing with other components via light rather than electrical signals. However, the really exciting possibilities will be those we have not yet imagined.

14. Know thyself! Although philosophers continue to be necessary, they will perform better their essential work if they would be more interested in the relevant data that scientists unveil.

ACKNOWLEDGMENTS

Francisco Torrens belongs to the Institut Universitari de Ciència Molecular, Universitat de València. Gloria Castellano belongs to the Departamento de Ciencias Experimentales y Matemáticas, Facultad de Veterinaria y Ciencias Experimentales, Universidad Católica de Valencia *San Vicente Mártir*. The authors thank support from the Spanish Ministerio de Economía y Competitividad (Project No. BFU2013-41648-P), EU ERDF, Generalitat Valenciana (Project No. PROMETEO/2016/094) and Universidad Católica de Valencia *San Vicente Mártir* (Project No. PRUCV/2015/617).

KEYWORDS

- **nanomaterials**
- **elementarity**
- **internet**
- **digitalization**
- **social media**

REFERENCES

1. Guardino, J., Ed. *Seguridad y Condiciones de Trabajo en el Laboratorio*; Institut Nacional de Seguretat i Higiene en el Teball: Barcelona, 1992.
2. OIT España. *Enciclopedia de Salud y Seguridad en el Trabajo*; Ministerio de Trabajo y Asuntos Sociales: Madrid, 1999.
3. Guardino, X. *Occupational Health for Health Care Workers*; Elsevier: Amsterdam, 1999.
4. Guardino, X. Occupational Hazards in the Laboratory Environment. In *Occupational Health for Health Care Workers: A Practical Guide*; Toomingas, A., Hasselhorn, H. M., Lagerstrom, M., Eds.; Elsevier: Amsterdam, 1999.
5. Torrens, F.; Castellano, G. Reflections on the Nature of the Periodic Table of the Elements: Implications in Chemical Education. In: *Synthetic Organic Chemistry*; Seijas, J. A., Vázquez Tato, M. P., Lin, S. K., Eds.; MDPI: Basel, Switzerland, 2015; Vol. 18, pp 1–15.
6. Torrens, F.; Castellano, G. Periodic Table. In *The Explicative Dictionary of Nanochemistry*; Putz, M. V., Ed.; Apple Academic–CRC; Waretown, NJ (in press).
7. Torrens, F.; Castellano, G. Ideas in the History of Nano/Miniaturization and (Quantum) Simulators: Feynman, Education and Research Reorientation in Translational Science.

In *Synthetic Organic Chemistry*; Seijas, J. A., Vázquez Tato, M. P., Lin, S. K., Eds.; MDPI: Basel, Switzerland, 2015; Vol. 19, pp 1–16.

8. Torrens, F.; Castellano, G. Reflections on the Cultural History of Nanominiaturization and Quantum Simulators (Computers). In *Sensors and Molecular Recognition*; Laguarda Miró, N., Masot Peris, R., Brun Sánchez, E., Eds.; Universidad Politécnica de Valencia: València, Spain, 2015; Vol. 9, pp 1–7.

9. Torrens, F.; Castellano, G. Nanominiaturization and Quantum Computing. In *Sensors and Molecular Recognition*; Costero Nieto, A. M., Parra Álvarez, M., Gaviña Costero, P., Gil Grau, S. Eds.; Universitat de València: València, Spain, 2016; Vol. 10, pp 31-1–31-5.

10. Torrens, F.; Castellano, G. *Book of Abstracts, Certamen Integral de la Prevención y el Bienestar Laboral,* València, Spain, September 28–29, 2016; Generalitat Valenciana–INVASSAT: Valencia, Spain, 2016; pp 3.

11. *Book of Abstracts, Certamen Integral de la Prevención y el Bienestar Laboral,* València, Spain, September 28–29, 2016; Generalitat Valenciana–INVASSAT: Valencia, Spain, 2016.

12. Liou, S. H.; Tsai, C. S. J.; Pelclova, D.; Schubauer-Berigan, M. K.; Schulte, P. A. Assessing the First Wave of Epidemiological Studies of Nanomaterial Workers. *J. Nanopart. Res.* **2015**, *17*, 413-1–413-19.

13. Bergamaschi, E.; Poland, C.; Canu, I. G.; Prina-Mello, A. The Role of Biological Monitoring in Nano-safety. *Nano Today* **2015**, *10*, 274–277.

14. Manno, M.; Viau, C.; Cocker, J.; Colosio, C.; Lowry, L.; Mutti, A.; Nordberg, M.; Wangh, S. Introduction: Biomonitoring for Occupational Health Risk Assessment. *Toxicol. Lett.* **2010**, *192*, 3–16.

15. García Canal, C. Personal communication.

16. Necker, L. A. Observations on Some Remarkable Optical Phaenomena Seen in Switzerland; and on an Optical Phenomenon Which Occurs on Viewing a Figure of a Crystal or Geometrical Solid. *London Edinburgh Philos. Mag. J. Sci.* **1832,** *1*(5), 329–337.

17. Sheehan, W. F. Periodic Table of Elements with Emphasis. *Chemistry* **1976**, *49*(3), 17–18.

18. Cloud, P. *Cosmos, Earth, and Man: A Short History of the Universe*; Yale University Press: Newhaven, UK, 1978.

19. Cloud, P. *Oasis in Space: Earth History from the Beginning*; W. W. Norton & Company: New York, NY, 1988.

20. Thims, L. *Human Thermodynamics*; Institute of Human Thermodynamics: Chicago, IL, 2002; Vol. 1.

21. Cerdà, H. Entrevista: Javier García Martínez. *Téc. Ind.* **2012**, *2012*(300), 82–85.

22. Feynman, R. P. There is Plenty of Room at the Bottom. *Caltech Eng. Sci.* **1960**, *23*, 22–36.

23. Feynman, R. P. Simulating Physics with Computers. *Int. J. Theor. Phys.* **1982**, *21*, 467–488.

24. Feynman, R. P. Feynman's Lectures on Computation; Hey, A. J. G., Allen, R. W., Eds.; Addison-Wesley: Reading, MA, 1996.

25. Chhowalla, M.; Shin, H. S.; Eda, G.; Li, L. J.; Loh, K. P.; Zhang, H. The Chemistry of Two-dimensional Layered Transition Metal Dichalcogenide Nanosheets. *Nat. Chem.* **2013**, *5*, 263–275.

26. Kholdeeva, O. A. Recent Developments in Liquid-phase Selective Oxidation Using Environmentally Benign Oxidants and Mesoporous Metal Silicates. *Catal. Sci. Technol.* **2014**, *4*, 1869–1889.

27. Díaz Soler, B. M. *Nanotecnología y Prevención de Riesgos Laborales*; Universidad de Granada: Granada, Spain, 2015.

28. Geim, A. K.; Grigorieva, I. V. Van der Waals heterostructures. *Nature (London)* **2013**, *499*, 419–425.

29. Puzyn, T.; Rasulev, B.; Gajewicz, A.; Hu, X.; Dasari, T. P.; Michalkova, A.; Hwang, H. M.; Toropov, A.; Leszczynska, D.; Leszczynski, J. Using Nano-QSAR to Predict the Cytotoxicity of Metal Oxide Nanoparticles. *Nat. Nanotechnol.* **2011**, *6*, 175–178.

30. Plasencia, A. Personal Communication.

31. Bensuade-Vincent, B.; Simon, J. *Chemistry: The Impure Science*; Imperial College Press: London, UK, 2012.

32. Carrasco Báez, N. *Book of Abstracts, Certamen Integral de la Prevención y el Bienestar Laboral,* València, Spain, September 28–29, 2016; Generalitat Valenciana–INVASSAT: Valencia, Spain, 2016; PRLCL5.

CHAPTER 2

NANOMINIATURIZATION, CLASSICAL/QUANTUM COMPUTERS/ SIMULATORS, SUPERCONDUCTIVITY, AND UNIVERSE

FRANCISCO TORRENS[1,*] and GLORIA CASTELLANO[2]

[1]*Institut Universitari de Ciència Molecular, Universitat de València, Edifici d'Instituts de Paterna, P. O. Box 22085, E-46071 València, Spain*

[2]*Departamento de Ciencias Experimentales y Matemáticas, Facultad de Veterinaria y Ciencias Experimentales, Universidad Católica de Valencia San Vicente Mártir, Guillem de Castro-94, E-46001 València, Spain*

*Corresponding author. E-mail: torrens@uv.es

CONTENTS

ABSTRACT

A reorientation of research in translational (marketable) science is necessary. The reorientation can be carried out by culture-guided reverse engineering. In this laboratory, Coronado group developed a strategy to improve the efficiency of powerful quantum bits, which consists in making these nanomagnets invisible to the magnetic field. The finding results another step forward toward one of the Holy Grails of modern applied physics: the building of quantum computers.

2.1   INTRODUCTION

After Feynman's conference *Simulating physics with computers* (1982), a series of hypotheses (Hs) were proposed to introduce miniaturization on the nanoscale and quantum simulators, for example[1–5]:

H1. (Feynman, 1982).[3] No classic computer can simulate quantum physics in an efficient way.

H2. (Feynman, 1982).[3] He suggested utilizing quantum systems to simulate quantum systems.

      In earlier publications, fractal hybrid-orbital analysis,[6,7] resonance,[8] molecular diversity,[9] periodic table of the elements (PTE),[10,11] law, property, information entropy, molecular classification, simulators,[12–14] labor risk prevention, and preventive healthcare at work with nanomaterials[15,16] were reviewed. In the present report, questions were raised on the nature of nanominiaturization cultural history and provide facts. Computational simulating physics was examined. Nature is quantum mechanical and problem is quantum-physics simulation.

2.2   THE INDUSTRIAL REVOLUTION THAT NANOTECHNOLOGY WILL INVOLVE

García interviewed Cerdà proposing questions/answers (Q/A) on the industrial revolution that nanotechnology will involve.

Q1. What are the foundations of nanotechnology?

Q2. In which sense nanotechnology makes a metal *like* other?

Q3. Must people pass from a two-dimensional (2D) to a three-dimensional (3D) PTE?

Q4. Are different properties determined by the size and structure that it receives?

Q5. What is the path of the nanotechnological wave?

Q6. What position is Spain to take this wave in?

Q7. What is the inversion in nanotechnology compared by countries?

Q8. And in patents?

Q9. Is it easy that a patent office concede a copyright in nanotechnology?

Q10. (J. J. Gómez-Cadenas). Must advances got with public funds be available to all society?

Q11. As Spain is without a National Nanotechnology Initiative Strategic Plan, is even the little that people have in danger?

Q12. What is the probability for people to lose this new technological wave?

Q13. What alternative does it remain to Spain if it loses this train?

Q14. Where are these people going to?

Q15. How is an experience to create a business in the field of nanotechnology?

Q16. Is it a kind of ambition?

Q17. Where is the key of the success of an innovative business?

Q18. Is it in technology?

A18. No, it is in the team, in the persons that form it. It is also in the setting where it is generated.

Q19. Is the survival rate of these innovative new businesses in 5 years ca. 5%?

Q20. Is the acceptance of risk needed?

Q21. And of failure?

Q22. What did it happen with Massachusetts Institute of Technology with this strategy?

Q23. To look at atoms from that height of an expert in nanotechnology and play with them, does it produce the opposed sensation to humbleness of the one that contemplates Earth from space?

2.3 QUANTUM COMPUTING

The physical components of computers became increasingly complex in the course of the development of ever-newer branches of physics.[17] The mechanical gears of Pascal, Leibniz, and Babbage–Lovelace's plans were replaced by electromechanical relays. The first computers used such devices. The application of electron tubes represented progress. After the formulation of the theory of stored programs by John von Neumann, computers attained their modern form. In the 1950s came transfer variator [*transistor* (Shockley, Bardeen, and Brattain)]-based computers, those with discrete switching elements and, later, integrated circuits. Computer technology successfully discharged its debt to basic research, that is, by making it possible, via Internet, to spread ideas to the entire world in statu nascendi, and in such a way that the concepts could be built on or rejected. People are approaching a stage at which the basic computational elements consist of individual atoms. However, the laws of quantum mechanics come into play with their probabilistic statements and uncertainty relations.

Q1. Does this establish a limit on the possibilities for further development?

A1. No, it is, rather, a challenge.

　　　As a result of interpretations of the debates on the fundamentals of quantum mechanics and repetition of basic experiments, the contours of possibilities for application came to light.

Q2. (Newman) Was there a *mechanical process* (definite method), which could be applied to a mathematical statement, and which would come up with answer as to whether it was provable?

A2. Alan Turing's *a* (automatic)-*machine* (1936).

　　　Alan Turing (1936) created the general theory of *classical* computers.[18,19] John von Neumann (1945) made the general theory of *modern* computers.[20] David Deutsch (1985)[21] proposed the quantum analogue of *Turing a-machine* (1936). The difference between the two types of computers follows. While conventional computers store data in the form of bits that can take on the values 0 and 1, a quantum computer stores information in two-stage quantum states, for example, the spin of a proton. The point is that the quantum states, called quantum bits (*qubits*), become *entangled*, meaning that N qubits exist in 2^N different states. In such a quantum superposition, all the states can be processed simultaneously. However, this means that a quantum computer can operate correspondingly faster than a conventional one (Di Vincenzo and Terhal, 1998).[22] Some possibilities for the realization of a qubit (term introduced by Ben Schumacher, 1995) follow: for photons, vertical and horizontal polarization; for atoms, electrons, and atomic nuclei, the two spin alignments; and for atoms, the two energy levels.[23] The construction of a quantum computer is straightforward in principle. One plainly takes simple quantum-logical elements and combines them into a quantum net. However, with a growing number of elements, difficulties following from quantum laws arise that today still await a solution. The most serious is the *decoherence* of the quantum state, resulting in loss of information, because of the quantum system's interaction with its environment. Irrespective of whether new practical applications might be successfully implemented, the techniques and accompanying experimental facts will create momentum in the investigations of basic quantum-mechanical problems, for example, riddle of the measurement process or Einstein–Podolsky–Rosen (EPR) paradox (1935)[24] with Bell's inequalities (1964).[25,26]

2.4 NANOCOSMOS TRAVEL: MINIATURE UNIVERSE AND CHALLENGES

In this laboratory, Coronado proposed Q/As on nanocosmos travel, miniature universe, and molecular nanoscience challenges.

Q1. Why?

Q2. How?

Q3. Molecular electronics?

Q4. Nanotechnology: a superior form of either evolution or humiliation?

Q5. What is the toxicity of nanoparticles (NPs)?

A5. An NP is toxic if its corresponding bulk is.

Q6. Do you collaborate with Dr. Cirac in quantum computations with cold-trapped ions?

A6. Yes, in quantum simulations with magnetic materials; decoherence rises as: magnetic < solid < superconductor < gas.

2.5 FROM CLASSICAL TO QUANTUM COMPUTERS

In comparison with present computers and devices, which are based on transistors to process information bits in the form of binary "0" and "1," quantum computers augur an exponential rise in speed at the time of carrying out computational tasks. The high power of qubits could become leaving behind present machines and revolutionizing fields (e.g., computational chemistry, cryptography) basic for communications safety. Advances seem possible in the world of atoms and subatomic particles, where the physical laws that rule quantum-objects' behavior are different from those of the *classic* world in which people live. However, quantum states are fragile and sensitive to the environment in which they are immersed, so that the development of the advanced devices that quantum mechanics augures results extremely complicated to achieve. The main problem, which quantum computing based in magnetic qubits finds, is that these must communicate between each other in a too noisy environment, till today difficult to reduce, which prevents that quantum information be efficiently transported; that is, interaction between qubits (although theoretically feasible in the quantum world) is turned up full of magnetic noise when one works in a real environment, so interfering with the calculations.

Advance consists in designing molecular magnets (MMs), which become invisible when interacting with a magnetic field (MF). In a certain way, MM qubits are analogues to *metamaterials* (materials invisible to light), which allows them to communicate between one another without magnetic noise, generated by both environment and magnetic interactions present between them on coming near, affect them. Such molecules are based on Ho ions encapsulated by a molecular oxide, polyoxometalate (POM). In order to achieve invisibility, scientists made the most of a process similar to that used in the so-called *atomic clocks*, high-precision devices for time measuring, which use the fact that the resonance frequency between two atomic states stay constant and insensitive to external perturbations, for example, those produced by an MF. Scientists named the operations *atomic-clock transitions*. It is about only one more step but an essential step to continue moving forward in the design of more robust magnetic qubits, which allow, at short time, improving the communication between them and processing quantum information in a more efficient way and, at longer time, building quantum computers based on MMs.

In classical computing, numbers are represented by either "0"s or "1"s, and calculations are carried out according to an algorithm's *instructions*, which manipulate the 0s/1s to transform an input to an output.[27] In contrast, quantum computing relies on atomic-scale units (qubits), which can be simultaneously 0/1 (a state known as a *superposition*). In this state, a single qubit can essentially perform two separate streams of calculations in parallel, making computations far more efficient than a classical computer. Like a compass, an electron MF dictates the binary code of either "0" or "1." In a quantum system, particles can exist in two states simultaneously too (superposition). A two-qubit system can perform simultaneous operations on four values, a three-qubit system, on eight values, and so on. A quantum computer with 300 qubits could hold 2^{300}–10^{90} values (number of atoms in the universe) simultaneously, performing an incredible quantity of calculations at once.

2.6 BCS THEORY

Bardeen–Cooper–Schrieffer (BCS) theory (named after John Bardeen, Leon Cooper, and John Robert Schrieffer) is the first microscopic theory of superconductivity since its discovery in 1911.[28,29] The theory describes superconductivity as a microscopic effect caused by a condensation of Cooper pairs into a boson-like state. The theory is also used in nuclear physics to describe the pairing interaction between nucleons in an atomic nucleus. Bardeen, Cooper, and Schrieffer proposed it in 1957; they received the Nobel Prize in Physics for the theory in 1972. Theory BCS (1957) is the first microscopic theory of superconductivity. The theory was first published in April 1957 in the letter *Microscopic theory of superconductivity*.[28] The demonstration that the phase transition is second order, reproduces Meissner effect and calculations of specific heats and penetration depths appeared in the December 1957 article, *Theory of superconductivity*.[29] At low temperatures (LTs), electrons near Fermi surface become unstable versus Cooper-pairs formation. Cooper showed such binding will occur in the presence of an attractive potential, no matter how weak. In conventional superconductors, an attraction is generally attributed to an electron–lattice interaction. Theory BCS, however, requires only that the potential be attractive, regardless of its origin. In BCS framework, superconductivity is a macroscopic effect, which results from the condensation of Cooper pairs, which have some bosonic properties, and bosons, at LT, can form a large Bose–Einstein condensate. Superconductivity was simultaneously explained by Nikolay Bogolyubov via Bogoliubov transformations.

In many superconductors, the attractive interaction between electrons (necessary for pairing) is brought about, indirectly, by the interaction between the electrons and the vibrating crystal lattice (phonons). The picture follows. An electron moving via a conductor will attract nearby positive charges in the lattice, which deformation causes another electron, with opposite spin, to move into the region of higher positive charge density. The two electrons then become correlated. Because a lot of such electron pairs exist in a superconductor, the pairs overlap strongly and form a highly collective condensate. In this *condensed* state, the breaking of one pair will change the energy of the entire condensate (not just a single electron or a single pair). The energy required to break any single pair is related to the energy required to break *all* pairs (or more than just two electrons). Because the pairing increases the energy barrier, it kicks from oscillating atoms in the conductor (which are small at LTs) are not enough to affect the condensate as a whole, or any individual *member pair* within the condensate. Electrons stay paired together and resist all kicks, and the electron flow as a whole (current via superconductor) will not experience resistance. Condensate collective behavior is necessary for superconductivity.

Some key background to BCS theory follows (1) evidence of a band gap at the Fermi level (*a key piece in the puzzle*); (2) isotope effect on the critical temperature, suggesting lattice interactions; (3) an exponential rise in heat capacity near the critical temperature for some superconductors; and (4) the lessening of the measured energy gap toward the critical temperature.

2.7 COSMIC ARCHITECTURE: THE QUANTUM UNIVERSE

Ibáñez and Pérez Cañellas organized a conference cycle on the quantum universe.[30]

Pérez Cañellas proposed questions, answers, Hs, and paradox (Pa) on quantum world.[31]

Q1. Classical simulation of a quantum system?

A1. However, 40 spins 1/2 require $2^{40} \cdot 2^{40} = 2^{80} \approx 10^{24}$ matrix elements in a classic computer!

Q2. In *entangled* states, what state is the particle in?

A2. The state of particle-1 is not defined but that of both particles.

H1. After quantum mechanics, collapse is instantaneous.

    Pa1. (EPR, 1935). Either an *instantaneous action* exists or quantum mechanics is not *complete* (*local realism, hidden variables*).

H2. Bell (1964): inequalities.[25,32]

H3. After two independent experiments (2015): There is no *local realism*.

H4. Noncloning theorem. It is impossible to design a quantum device that clones arbitrary states: $|\Psi> |\Phi> = |\Psi> |\Psi>$.

H5. (Feynman, 1982).[3] No classic computer can simulate quantum physics in an efficient way.

H6. (Feynman, 1982).[3] He suggested utilizing quantum systems to simulate quantum systems.

    He drew the following conclusions (Cs) and questions.

C1. In the future, quantum computers can be built.

C2. At shorter term, one can design quantum simulators.

C3. Q3. What more does the quantum world prepare people?

Q4. To build a quantum computer that be more rapid than a classical computer?
     Roldán proposed Qs, As, and Hs on quantum light.[33]

Q5. (Samuel Johnson). What is light?

A5. Light is some times waves, some other times, particles.

Q6. What does it say quantum theory with regard to the nature of light?

Q7. What does it mean to search for an answer to the question, *what is light*?

A7. It means to find a mechanical model of light.

H7. To understand ideal monochromatic waves allows people to conceive more complicated waves.

H8. (Heisenberg, 1927).[34] Uncertainty principle: $\Delta x \Delta p \geq \hbar/2$.

Q8. Light, wave, or particle?

A8. The photon is a wave but it is also a particle.

Q9. What are the sources that can generate photons?

A9. Light amplification by stimulated emission of radiation (*laser*, coherent states); optical parametric oscillator (compressed states).

Q10. In addition, what are the number states good for?

A10. Metrology and quantum information (quantum coding, computing, and simulation).

Q11. If the vacuum is not empty, why do physicists call it *vacuum*?

A11. It should properly be called *quantum vacuum*.

Q12. What is the use of compressing vacuum?

A12. Vacuum can be compressed and presents light but with lesser fluctuations in one quadrature.

Q13. How does an interferometer work?

A13. It compares two lights, the compressed light and the light to be measured.
     Bañuls proposed questions and answers on quantum simulations.[35]

Q14. What is the power of the quantum computer?

A14. Quantum computers need algorithms working better for a quantum than for classic computer.

Q15. How is a quantum computer built?

A15. (Bennett and DiVincenzo, 2000).[36] Requirements for a quantum computer: (1) scalability; (2) preparation of initial state; (3) long decoherence time; (4) universal set of gates; and (5) reading.

Q16. What is the simplest model that describes the properties?

A16. It is the Fermi–Hubbard model.

Q17. What type of question does one want to answer?

A17. One wants to answer Hamiltonian, configuration of lowest energy, time variation, and so on.

Q18. Is simulating a quantum system in the quantum simulator easier than in quantum computer?

A18. *Dedicated* quantum computer. It needs: (1) no complete description (the whole state); (2) (local) observables; (3) tolerance to errors.

Q19. In dynamics (nonequilibrium), what is the quantum description?

Q20. How does a generic system relax?

A20. The question cannot be answered because people have no methods.

Q21. Classical simulation of quantum systems?

A21. Impossible but quantum information provides tools in cases: interesting systems are not entangled.

Q22. Can quantum simulations be applied to mathematical simulations?

A22. Chemical systems fulfill certain mathematical conditions and work well for local connections.

Q23. Is there any case of quantum simulator that be universal?

A23. Yes, some universal quantum simulators exist, for example, in 2D Ising models.

Q24. In quantum simulations with ultracold atoms, is there a limit for cooling?

A24. Magnetic order takes a lot; each technique has a limit in temperature order: laser > evaporation.

Q25. Is there a connection between mathematical fuzzy logic (FL) and quantum entanglement?

A25. An FL-quantum mechanics connection exists but not between FL and entanglement.

Q26. In the Fermi–Hubbard model, why are there superconductors?

A26. No easy cases exist but guide is doping with lack of electrons, needing Monte Carlo calculations.

Q27. Can tensor networks be used in 2D for quantum Hall effect (topological crystals)?

A27. Yes, but it computationally costs a lot; it is difficult to understand what state is obtained.

Navarrete proposed the following questions and answers on new quantum technologies.[37]

Q28. Where is the ball going to fall?

Q29. What baskets do you think that become more filled, those of the center or the ones of ends?

A29. Those of the center because in order to arrive to the center more ways exist than for the ends.

Q30. Is, so, the electron more like a *wave*?

A30. Wave–particle duality: It behaves *wave* like when it *propagates freely*, *particle* like when one *looks at it…* but it is delocalized: *superposition of different positions*.

Q31. However, what is light polarization?

Q32. Diffraction of fullerenes C_{60}, is it a macroscopic quantum phenomenon?

A32. People go to the great.

Q33. Are there action protocols for the case when a research group solves factorization?

A33. It is secret, but, yes, there are in USA, China, and so on (not Spain).

Q34. Are there quantum experiments?

A34. Experiments, for example, Schrödinger's cat, exist but the coupled macroscopic system has a *classical* probability of being in one/another state; it is not *quantum* but *classical superposition*.

Q35. Is there any way to generate energy from quantum technology?

A35. No, but ways exist to generate/transport it more efficiently, for example, quantum photosynthesis.

Olmo proposed Hs, questions, As, Pa, and Cs on quantum black holes (BHs).[38]

H9. (Misner, Thorne, Wheeler, 1973).[39] Uniqueness/no-*hair* theorem: BH solutions of Einstein–Maxwell equations of gravitation/electromagnetism in general relativity are characterized by three *externally* observable classical parameters: mass/electric charge/angular momentum.

Q36. In addition, BHs exist, do not?

A36. Advanced Laser Interferometer Gravitational-Wave (GW) Observatory (LIGO) detected GWs from two BHs collision.

Pa2. BH information (loss) Pa: BHs evaporate, which violates quantum information principles.

Q37. Really, does information get lost?

A37. There are the following three Hs.

H10. Not completely thermal evaporation (backreaction).

H11. Incomplete evaporation (remainders).

H12. (Susskind). The BH complementarity principle.

In addition, he presented four recent proposals.

H13. ('t Hooft, Susskind and Maldacena). Holographic principle.

H14. (Almheiri, Marolf, Polchinski, and Sully, 2012).[40] Firewalls.

H15. (Susskind and Maldacena, 2013).[41] ER = EPR: Entangled particles are linked by a wormhole.

H16. (Hawking, Perry, Strominger, 2016).[42] A BH may carry *soft hair* (low-energy quantum excitations that release information when BH evaporates).

He proposed additional questions.

Q38. Quantum gravity?

Q39. What is one measuring?

Q40. How to measure the Hawking radiation produced by an acoustic horizon?

He concluded the following:

C4.   The BHs are real.

C5.   The quantum properties of the horizons can be studied in laboratory.

He proposed additional questions and answers.

Q41. When two BHs collide and horizon rises, why do not they act like a vacuum cleaner absorbing whole universe?

A41. They absorb only the matter that is within their reach.

Q42. Is BH information Pa like measure in quantum mechanics where part of information gets lost?

A42. All the system (not every individual particle) must be in a pure state.

Q43. Could it be possible that BH would break some conservation principle?

A43. Correspondence (conservation principles) exists in particles/antiparticles, which must not break if either enters BH.

Q44. What have people learned from quantum theory (QT)?

A44. QT necessity; two cases where gravity/QT are related: horizon and BH inside (singularity); it can be applied to universe emergence theories.

Q45. Do wormholes function like in science fiction?

A45. Geometrically, yes; entanglement–wormhole relationship exists: if two particles were connected by a wormhole they would be two signs of the same reality.

2.8 ENHANCING COHERENCE IN MOLECULAR SPIN FOR QUANTUM COMPUTING

In this laboratory, Coronado group reviewed POM-chemistry relevance to provide model objects in molecular magnetism.[43] They presented potential applications in nanomagnetism [e.g., molecular spin electronics (*spintronics*), quantum computing]. *Decoherence* is the loss of fragile quantum-phase information because of an uncontrolled *entanglement* with its environment. It is errors in calculations caused by interference from many factors. They studied POM single-ion magnet $[GdW_{30}P_5O_{110}]^{14-}$ by generalized Rabi oscillation experiments.[44] They increased the number of coherent rotations tenfold via matching Rabi frequency with the proton one. Achieving high coherence with POM chemistry, they showed its excellent potential not only for quantum-information storage but even quantum-algorithms realization. Usually, quantum algorithms assume independent spin qubits to produce trivial $|\uparrow\rangle = |0\rangle$, $|\downarrow\rangle = |1\rangle$ mappings, which can be unrealistic in many solid-state implementations with sizeable magnetic interactions. They showed that the lower spectrum part of a molecule containing three exchange-coupled metal ions with $S = 1/2$, $I = 3/2$ is equivalent to nine electron-nuclear qubits.[45] They derived spin–qubit-states relation in reasonable parameter ranges for lanthanoid Tb^{3+}/transition metal Cu^{2+} and studied possibility to implement Shor's quantum error correction (QEC) on the molecule. They discussed developed molecular systems that are experimentally adequate.

Quantum computing is an emerging area within the information sciences revolving around qubit concept. A major obstacle is the extreme fragility of qubits because of interactions with their environment, which destroy their quantumness. The phenomenon (decoherence) is of fundamental interest. Many competing candidates exist for qubits (e.g., superconducting circuits, quantum optical cavities, ultracold atoms, and spin qubits) every one with its strengths and weaknesses. When dealing with spin qubits, the strongest source of decoherence is the magnetic dipolar interaction. In order to minimize it, spins are typically diluted in a diamagnetic matrix; for example, dilution can be taken to the extreme of a single P atom in Si, whereas in molecular matrices, a typical ratio is one magnetic molecule per 10,000 matrix molecules. However, a fundamental contradiction exists between reducing decoherence by dilution and allowing quantum operations via the interaction between spin qubits. In order to resolve the contradiction, quantum-hardware design and engineering benefit from a *bottom-up* approach, whereby the electronic structure of magnetic molecules is chemically tailored to give the desired physical behavior. They presented a way of enhancing coherence in solid-state molecular spin qubits without extreme dilution.[46] It was based on molecular-structures design with crystal field ground states possessing large tunneling gaps, which gave rise to optimal operating points (*atomic-clock transitions*), at which the quantum spin dynamics became protected versus dipolar decoherence. The approach is illustrated with a Ho molecular nanomagnet in which long coherence times (up to 8.4 μs at 5K) are obtained at unusually high concentrations. The finding opened avenues for quantum computing based on molecular spin qubits.

2.9   DISCUSSION

Like a compass, the MF of an electron dictates the binary code of either "0" or "1." In a quantum system, particles can exist in two states simultaneously too (*superposition*). A two-qubit system can perform simultaneous operations on four values, a three-qubit system, on eight values, and so on. A quantum computer with 300 qubits could hold 2^{300}–10^{90} values simultaneously (number of atoms in the universe) performing an incredible quantity of calculations at once.

Giordani proposed questions, problems (Ps), and fact (F) on quantum computing.[47]

Q1.   A quantum leap in processors?

Q2.   Beyond Si?

P1.   Qubits are prone to errors; one needs lots of extra (*ancilla*) bits for a secondary error correction.

Q3.   Will such a technology really scale in the way it needs to?

P2. There are a lot of overheads to make a photon-based quantum computer function.

P3. System of single electron trapped in semiconductor nanostructure presents data-loss problems caused by material strains.

F1. Such problems were solved when a strong external MF was applied.

2.9.1 CHEMISTRY AND ETHICS

Colborn et al. alerted to endocrine disruptors.[48] The American Chemical Society reported *The Chemical Professional's Code of Conduct*[49]; however, it does not explain what to do if the science policy is in conflict with demands related to the obligation with the boss or other employees. Designers McDonough and Braungart promoted using synthetics as raw materials.[50]

2.10 CONCLUDING REMARK

From the present questions, answers hypotheses, *etc*. the following remark can be drawn.

1. Main problem to design qubits is decoherence minimization, particularly from nuclear spins and dipolar interactions. The use of molecular spin qubits offers advantages (i.e., reduced decoherence, high reproducibility). Research is oriented to a better understanding of the basic yet extremely fragile building blocks. As the size of an object becomes increasingly smaller, quantum (entanglement) effects become increasingly important.

ACKNOWLEDGMENTS

Francisco Torrens belongs to the Institut Universitari de Ciència Molecular, Universitat de València. Gloria Castellano belongs to the Departamento de Ciencias Experimentales y Matemáticas, Facultad de Veterinaria y Ciencias Experimentales, Universidad Católica de Valencia *San Vicente Mártir*. The authors thank support from the Spanish Ministerio de Economía y Competitividad (Project No. BFU2013-41648-P), EU ERDF, Generalitat Valenciana (Project No. PROMETEO/2016/094) and Universidad Católica de Valencia *San Vicente Mártir* (Project No. PRUCV/2015/617).

KEYWORDS

- **industrial revolution**
- **quantum computing**
- **nanocosmos travel**
- **quantum computers**
- **BCS theory**

REFERENCES

1. Feynman, R. P. There is Plenty of Room at the Bottom. *Caltech Eng. Sci.* **1960**, *23*, 22–36.
2. Cerdà, H. Entrevista: Javier García Martínez. *Técnica Ind.* **2012**, *2012*(300), 82–85.
3. Feynman, R. P. Simulating Physics with Computers. *Int. J. Theor. Phys.* **1982**, *21*, 467–488.
4. Feynman, R. P. In *Feynman's Lectures on Computation*; Hey, A. J. G., Allen, R. W., Eds.; Addison-Wesley: Reading, MA, 1996.
5. Cirac, I.; Zoller, P. Quantum Computations with Cold Trapped Ions. *Phys. Rev. Lett.* **1995**, *74*, 4091–4094.
6. Torrens, F. Fractals for Hybrid Orbitals in Protein Models. *Complexity Int.* **2001**, *8*, torren01-1–torren01-13.
7. Torrens, F. Fractal Hybrid-Orbital Analysis of the Protein Tertiary Structure. *Complexity Int.* (in press).
8. Torrens, F.; Castellano, G. Resonance in Interacting Induced-Dipole Polarizing Force Fields: Application to Force-Field Derivatives. *Algorithms* **2009**, *2*, 437–447.
9. Torrens, F.; Castellano, G. Molecular Diversity Classification *via* Information Theory: A Review. *ICST Trans. Complex Syst.* **2012**, *12*(10–12), e4-1–e4-8.
10. Torrens, F.; Castellano, G. Reflections on the Nature of the Periodic Table of the Elements: Implications in Chemical Education. In *Synthetic Organic Chemistry*; Seijas, J. A., Vázquez Tato, M. P., Lin, S. K., Eds.; MDPI: Basel, Switzerland, 2015; Vol. 18, pp 8-1–8-15
11. Putz, M. V., Ed. *The Explicative Handbook of Nanochemistry*; Apple Academic–CRC: Waretown, NJ (in press).
12. Torrens, F.; Castellano, G. Reflections on the Cultural History of Nanominiaturization and Quantum Simulators (Computers). In *Sensors and Molecular Recognition*; Laguarda Miró, N., Masot Peris, R., Brun Sánchez, E., Eds.; Universidad Politécnica de Valencia: València, Spain, 2015; Vol. 9, pp. 1–7.
13. Torrens, F.; Castellano, G. Ideas in the History of Nano/Miniaturization and (Quantum) Simulators: Feynman, Education and Research Reorientation in Translational Science. In *Synthetic Organic Chemistry*; Seijas, J. A., Vázquez Tato, M. P., Lin, S. K., Eds.; MDPI: Basel, Switzerland, 2016; Vol. 19, pp. 1–16.

14. Torrens, F.; Castellano, G. Nanominiaturization and Quantum Computing. In *Sensors and Molecular Recognition*; Costero Nieto, A. M., Parra Álvarez, M., Gaviña Costero, P., Gil Grau, S., Eds.; Universitat de València: València, Spain, 2016; Vol. 10, pp. 31-1–31-5.

15. Torrens, F.; Castellano, G. In *Book of Abstracts, Certamen Integral de la Prevención y el Bienestar Laboral*, València, Spain, September 28–29, 2016; Generalitat Valenciana–INVASSAT: València, Spain, 2016; p 3.

16. Torrens, F.; Castellano, G. Nanoscience: From a Two-Dimensional to a Three-Dimensional Periodic Table of the Elements. In *Innovations in Physical Chemistry*; Haghi, A. K., Ed., Apple Academic–CRC: Waretown, NJ; Vol. 3 (in press).

17. Simonyi, K. *A Cultural History of Physics*; CRC: Boca Raton, FL, 2012.

18. Turing, A. M. On Computable Numbers, with an Application to the Entscheidungsproblem. *Proc. London Math. Soc.* **1937**, *s2–42*, 230–265.

19. Turing, A. M. On Computable Numbers, with an Application to the Entscheidungsproblem: A Correction. *Proc. London Math. Soc.* **1937**, *43*, 544–546.

20. Von Neumann, J. First draft of a report on the EDVAC. Contract No. W-670-ORD-4926 between the United States Army Ordnance and the University of Pennsylvania; University of Pennsylvania: Philadelphia, PA, 1945.

21. Deutsch, D. Quantum theory, the Church-Turing Principle and the Universal Quantum Computer. *Proc. R. Soc. London, Ser. A* **1985**, *400*, 97–117.

22. DiVincenzo, D.; Terhal, B. Decoherence: The Obstacle to Quantum Computation. *Phys. World* **1998**, *11*, 53–57.

23. Schumacher, B. Quantum coding. *Phys. Rev. Sect. A* **1995**, *51*, 2738–2747.

24. Einstein, A.; Podolsky, B.; Rosen, N. Can Quantum-Mechanical Description of Physical Reality be Considered Complete? *Phys. Rev.* **1935**, *47*, 777–780.

25. Bell, J. On the Einstein Podolsky Rosen paradox. *Physics* **1964**, *1*, 195–200.

26. Schafmeister, C. Clasp: Common Lisp Using LLVM and C++ for Molecular Metaprogramming: Towards a Matter Compiler, Google Tech Talk, June 10, 2015. https://www.youtube.com/watch?v=6X89_42Mj-g (accessed December 23, 2016).

27. Monz, T.; Nigg, D.; Martinez, E. A.; Brandl, M. F.; Schindler, P.; Rines, R.; Wang, S. X.; Chuang, I. L.; Blatt, R. Realization of a Scalable Shor Algorithm. *Science* **2016**, *351*, 1068–1070.

28. Bardeen, J.; Cooper, L. N.; Schrieffer, J. R. Microscopic Theory of Superconductivity. *Phys. Rev.* **1957**, *106*, 162–164.

29. Bardeen, J.; Cooper, L. N.; Schrieffer, J. R. Theory of Superconductivity. *Phys. Rev.* **1957**, *108*, 1175–1204.

30. Ibáñez, J. M.; Pérez Cañellas, A. Book of Abstracts. *Ciclo de Conferencias El Universo Cuántico*, València, Spain, September 22–October 27, 2016; Fundación Valenciana de Estudios Avanzados: València, Spain, 2016.

31. Pérez Cañellas, A. Book of Abstracts. *Ciclo de Conferencias El Universo Cuántico*, València, Spain, September 22–October 27, 2016; Fundación Valenciana de Estudios Avanzados: València, Spain, 2016; O-1.

32. Clauser, J. F.; Horne, M. A.; Shimony, A.; Holt, R. A. Proposed Experiment to Test Local Hidden-Variable Theories. *Phys. Rev. Lett.* **1969**, *23*, 880–884.

33. Roldán, E. Book of Abstracts. *Ciclo de Conferencias El Universo Cuántico*, València, Spain, September 22–October 27, 2016; Fundación Valenciana de Estudios Avanzados: València, Spain, 2016; O-2.

34. Heisenberg, W. Über den anschaulichen Inhalt der quantentheoretischen Kinematik und Mechanik. *Z. Phys.* **1927,** *43,* 172–198.

35. Bañuls, M. C. Book of Abstracts. *Ciclo de Conferencias El Universo Cuántico,* València, Spain, September 22–October 27, 2016; Fundación Valenciana de Estudios Avanzados: València, Spain, 2016; O-3.

36. Bennett, C. H.; DiVincenzo, D. P. Quantum Information and Computation. *Nature (London)* **2000,** *404,* 247–255.

37. Navarrete, C. Book of Abstracts. *Ciclo de Conferencias El Universo Cuántico,* València, Spain, September 22–October 27, 2016; Fundación Valenciana de Estudios Avanzados: València, Spain, 2016; O-4.

38. Olmo, G. J. Book of Abstracts. *Ciclo de Conferencias El Universo Cuántico,* València, Spain, September 22–October 27, 2016; Fundación Valenciana de Estudios Avanzados: València, Spain, 2016; O-5.

39. Misner, C. W. Thorne, K. S.; Wheeler, J. A. *Gravitation*; W. H. Freeman: Gordonsville, V. A. 1973.

40. Almheiri, A.; Marolf, D.; Polchinski, J.; Sully, J. Black Holes: Complementarity or Firewalls? *J. High Energy Phys.* **2013,** *62,* 1–19.

41. Maldacena, J.; Susskind, L. Cool Horizons for Entangled Black Holes. *Fortschr. Phys.* **2013,** *61,* 781–811.

42. Hawking, S. W.; Perry, M. J.; Strominger, A. Soft Hair on Black Holes. *Phys. Rev. Lett.* **2016,** *116,* 231301-1–9.

43. Clemente-Juan, J. M.; Coronado, E.; Gaita-Ariño, A. Magnetic Polyoxometalates: From Molecular Magnetism to Molecular Spintronics and Quantum Computing. *Chem. Soc. Rev.* **2012,** *41,* 7464–7478.

44. Baldoví, J. J.; Cardona-Serra, S.; Clemente-Juan, J. M.; Coronado, E.; Gaita-Ariño, A.; Prima-García, H. Coherent Manipulation of Spin Qubits based on Polyoxometalates: The Case of the Single Ion Magnet $[GdW_{30}P_5O_{110}]^{14-}$. *Chem. Commun.* **2013,** *49,* 8922–8924.

45. Baldoví, J. J.; Cardona-Serra, S.; Clemente-Juan, J. M.; Escalera-Moreno, L.; Gaita-Ariño, A.; Espallargas, G. M. Quantum Error Correction with Magnetic Molecules. *ArXiv.* **2014,** *1404,* 6912-1–6912-5.

46. Shiddiq, M.; Komijani, D.; Duan, Y.; Gaita-Ariño, A.; Coronado, E.; Hill, S. Enhancing Coherence in Molecular Spin Qubits *via* Atomic Clock Transitions. *Nature (London)* **2016,** *531,* 348–351.

47. Giordani, A. A Quantum Leap in Processors? *Sci. Comput. World* **2015,** *2015*(145), 14–16.

48. Colborn, T.; Dumonski, D.; Myers, J. P. *Our Stolen Future: Are We Threatening Our Fertility, Intelligence and Survival—A Scientific Detective Story*; Penguin: New York, NY, 1996.

49. American Chemical Society. *The Chemical Professionals Code of Conduct*; 2012. http://www.acs.org/content/acs/en/careers/career-services/ethics/the-chemical-professionals-code-of-conduct.html (accessed December 23, 2016).

50. McDonough, W.; Braungart, M. *Cradle to Cradle: Remarking the Way We Make Things*; North Point: New York, NY, 2002.

CHAPTER 3

COMPUTATIONAL INVESTIGATION OF Au-DOPED Ag NANOALLOY CLUSTERS: A DFT STUDY

PRABHAT RANJAN[1], TANMOY CHAKRABORTY[2,*], and AJAY KUMAR[1]

[1]Department of Mechatronics Engineering, Manipal University Jaipur, Dehmi Kalan, Jaipur 303007, India

[2]Department of Chemistry, Manipal University Jaipur, Dehmi Kalan, Jaipur 303007, India

*Corresponding author. E-mail: tanmoychem@gmail.com; tanmoy.chakraborty@jaipur.manipal.edu

CONTENTS

ABSTRACT

Bimetallic silver–gold nanoalloy clusters up to nine atoms are investigated by using conceptual density functional theory methodology. The Ag–Au nanoalloy cluster have wide range of applications in the field of nanotechnology, material science, biomedicine, and solid-state chemistry, due to their unique electronic, optical, magnetic, and catalysis properties. In this chapter, we have investigated the electronic and optical properties of Au-doped Ag Ag_nAu ($n = 1$–8) nanoalloy clusters in terms of conceptual density functional based descriptors, namely, highest occupied molecular orbital (HOMO)–lowest unoccupied molecular orbital (LUMO) energy gap, electronegativity, hardness, softness, electrophilicity index, and dipole moment. The computed HOMO–LUMO energy gap and hardness displays pronounced odd–even oscillation behavior as a function of cluster of size. The result exhibits that Ag_5Au nanoalloy cluster is the most stable cluster, indicating maximum number of HOMO–LUMO energy gap in the range of $n = 1$–8. The high value of linear correlation coefficient between HOMO–LUMO energy gap and computed descriptors validates our analysis.

3.1 INTRODUCTION

In recent years, study of bimetallic nanoalloy clusters (Ag, Au, and Cu) has been focus of intense research due to its wide range of industrial applications in the area of nanotechnology, material science, biology, and medicine.[1–19] Among these bimetallic clusters, the compound formed between Ag and Au has received great interest nowadays because of its unique catalytic, electronic, optical, and magnetic properties. A large number of theoretical and experimental reports are available on pure gold and silver clusters. For example, Varga et al.[20] have studied diatomic molecules Cu_2, Ag_2, and Au_2 using density functional theory (DFT). Lecoultre et al.[21] have investigated optical and fluorescence spectra of Ag_n ($n = 1$–9) clusters using experimental techniques and compared the results with DFT methodology. Handschuh et al.[22] have investigated anionic Au_n^- ($n = 2$–4) clusters using photoelectron spectroscopy and two-photon ionization spectroscopy techniques. Bishea et al.[23] have investigated Au_3 cluster using resonant two-photon ionization spectroscopy technique. Zhang et al.[24] have studied neutral, cationic, and anionic Ag_n ($n = 3$–14) clusters using DFT method.

For enhancement of stability, structure, electronic, and physicochemical properties of silver and gold clusters, many studied have been performed on bimetallic Ag–Au clusters. However, reports based on Au-doped Ag nanoalloy clusters are very few. Negishi et al.[25] have investigated bimetallic anionic $Au_nAg_m^-$ ($2\leq n+m\leq 4$) clusters by using photoelectron spectroscopy technique. The authors have described the electronic properties and structure of these anionic Au–Ag clusters. The structural and electronic properties of Ag_mAu_n ($3\leq m+n\leq 5$) clusters have been investigated by Bonačić-Koutecký et al.[26] The gold clusters prefer planar geometry for larger size as compared to silver and mixed Ag–Au clusters. Lee et al.[27] have investigated pure gold and silver clusters and neutral and anionic bimetallic Au_mAg_n ($2\leq m+n\leq 7$) clusters. The result indicates that pure gold clusters prefer two-dimensional however silver clusters prefer to be in three-dimensional. Weis et al.[28] have studied $Ag_mAu_n^+$ ($m+n<6$) clusters experimentally and compared the results with DFT methodology. The most stable structure of Ag_3Au^+ cluster has Y shape structure with the gold atom located in the center, while other tetrameric clusters have rhombus structure. Zhao et al.[29] have studied structural and electronic behaviors of Au_mAg_n ($2\leq m+n\leq 8$) clusters. The gold atoms prefer to locate at outer position of Ag–Au clusters. The binding energies of clusters have growing pattern with the number of atoms. The vertical ionization potentials of Au_mAg_n ($n = 1,2,3,4$) cluster shows odd–even oscillation behavior. The geometrical structure, stabilities, highest occupied molecular orbital (HOMO)–lowest unoccupied molecular orbital (LUMO) energy gaps, hardness, and polarizabilities of Ag_2Au_n and Cu_2Au_n ($n = 1–10$) clusters have been studied by Zhao et al.[30] The result shows that Ag_2Au_4 and Cu_2Au_2 is the most stable cluster in this molecular system. The inverse correlation of the polarizabilities versus ionization potential and hardness is also observed in this analysis. Tafoughalt et al.[31,32] have investigated structural and electronic properties of Ag_nAu_m clusters up to eight atoms and $AgAu_{n-1}$ ($n = 3–13$) clusters. The clusters are having planar structure up to six number of atoms in Ag_nAu_m clusters; however, when number of Ag atoms is more than or equal to four, it favors three-dimensional structure. More number of Ag atoms in the bimetallic cluster reduces the d-orbital contribution to the highest occupied molecular orbitals of the cluster, and thus, it forms three-dimensional geometry. In the case of $AgAu_{n-1}$ cluster, the structural transition from one-dimensional to three-dimensional is observed at $n = 13$. Kuang et al.[33] have studied scalar relativistic calculation on Au_nAg ($n = 1–12$) clusters. The ground state configurations of Au_nAg clusters are having planar structure due to strong scalar relativistic effect of gold cluster. The magnetic moment of clusters show odd–even alternation behavior, indicating that clusters with odd number of atoms have zero magnetic moment, while clusters with even number of atoms have magnetic moment value as $1\mu_B$. Recently, we have also studied electronic and physicochemical properties of some bimetallic nanoalloy clusters invoking conceptual DFT (CDFT).[34-45]

It has been observed from previous works that geometrical structure, electronic and physicochemical properties of bimetallic Ag–Au nanoalloy clusters fluctuate with their size and also have different characteristics from their bulk materials.[30] In order to provide further development on Ag–Au clusters, in this report, we have systematically investigated Au-doped Ag bimetallic Ag_nAu (n = 1–8) nanoalloy clusters by using CDFT methodology. Here, we have calculated DFT-based global descriptors, namely, HOMO–LUMO energy gaps, hardness, softness, electronegativity, electrophilicity index, and dipole moment for Ag_nAu clusters up to nine atoms. To study the electrostatic interaction of the molecules, quadrupole moment of clusters has also been computed.

3.2 COMPUTATIONAL DETAILS

In recent years, DFT has emerged as one of the most successful techniques of quantum mechanics to explore the geometrical structure and physicochemical properties of bimetallic and multimetallic nanoalloy clusters. DFT methods have opened many new dimensions in the field of material science, physics, chemistry, surface science, nanotechnology, biology, and earth sciences.[46] All the modeling and structural optimization of compounds have been performed using Gaussian 03 software package[47] within DFT framework. For optimization purpose, local spin density approximation (LSDA) exchange correlation with basis set LanL2dz has been adopted. Although LSDA functionals are not so much complex, their effectiveness is proven in various applications, especially in solid-state physics,[48] where accurate transition phase transitions in solids[49] and liquid metals[50,51] are predicted, and for lattice crystals, in which 1% precision are successfully achieved.[52] The basis set LanL2dz has high accuracy for metallic clusters which has been recently reported.[10,53,54] The computation methodology used in this report is based on the molecular orbital approach, using linear combination of atomic orbitals. We have chosen Z-axis as spin polarization axis. The symmetrized fragment orbitals are combined with auxiliary core functions to ensure orthogonalization on the (frozen) core orbitals.

Invoking Koopmans' approximation,[55] we have calculated ionization energy (I) and electron affinity (A) of all the clusters. Thereafter, using I and A, the conceptual DFT-based descriptors, namely, electronegativity (χ), global hardness (η), molecular softness (S), and electrophilicity index (ω) have been computed using following ansatzes:

$$I = -\varepsilon_{HOMO} \tag{3.1}$$

$$A = -\varepsilon_{LUMO} \tag{3.2}$$

Thereafter, using I and A, the conceptual DFT based descriptors, namely, electronegativity (χ), global hardness (η), molecular softness (S), and electrophilicity index (ω) have been computed. The equations used for such calculations are as follows:

$$\chi = -\mu = \frac{I+A}{2} \tag{3.3}$$

where μ represents the chemical potential of the system.

$$\eta = \frac{I-A}{2} \tag{3.4}$$

$$S = \frac{1}{2\eta} \tag{3.5}$$

$$\omega = \frac{\mu^2}{2\eta} \tag{3.6}$$

3.3 RESULTS AND DISCUSSION

In this report, a detail computational investigation of bimetallic Ag–Au nanoalloy clusters has been done by using electronic structure theory. We have computed HOMO–LUMO energy gaps of Au-doped Ag, Ag_nAu ($n = 1$–8) clusters along with computed DFT-based global descriptors, namely, hardness, softness, electronegativity, electrophilicity index, and dipole moment, which is shown in Table 3.1. The HOMO–LUMO energy gap is a very important parameter to study the stability of bimetallic clusters. It has been already observed that clusters with low energy gap are more prone to response against any external perturbation. In other words, the HOMO–LUMO energy gap shows that an electron requires that much energy to move from occupied orbital to unoccupied orbital. The HOMO–LUMO energy gaps also display odd–even oscillation behavior as a function of cluster size, which is shown in Figure 3.1. It is distinct from Figure 3.1 that the clusters containing even number of total atoms possess higher HOMO–LUMO energy gaps as compared to the clusters having odd number of total atoms.

The cluster Ag_5Au has maximum HOMO–LUMO energy gap of 2.340 eV, whereas cluster Ag_2Au has lowest HOMO–LUMO energy gap of 0.571 eV.

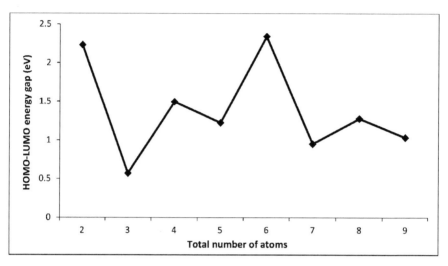

FIGURE 3.1 Odd–even oscillation behavior of HOMO–LUMO energy gap and Ag_nAu (n = 1–8) nanoalloy clusters.

The result from Table 3.1 also reveals that HOMO–LUMO energy gap of clusters maintain direct relationship with their hardness values. The hardness values also show odd–even oscillation behavior with respect to their cluster size, indicating even number of total atoms have higher values as compare to their neighbor clusters. Though there is no such available quantitative data of optical properties of aforesaid bimetallic clusters, we can assume that there must be a direct qualitative relationship between optical properties of bimetallic Ag_nAu nanoalloy clusters with their HOMO–LUMO energy gap. The assumption is based on the fact that optical properties of any materials are interrelated with drift of electrons within the systems which is influenced by the distance between valence and conduction band. There is a direct linear relationship between HOMO–LUMO energy gaps with the difference in the energy of valence-conduction band.[56] On that basis, we may conclude that optical properties of bimetallic Ag_nAu nanoalloy clusters increase with increase of their hardness values. Similarly softness data exhibits an inverse relationship toward the experimental optical properties. Although the similar kind of relationships is observed in case of evaluated electronegativity, electrophilicity index and dipole moment along with their HOMO–LUMO energy gap of clusters. The linear correlation between

HOMO–LUMO energy gaps along with their computed electrophilicity index is lucidly plotted in Figure 3.2. The regression correlation coefficient $R^2 = 0.719$ observed in Figure 3.2, which validates our computational analysis. From the obtained correlation coefficients of several DFT-based descriptors along with their HOMO–LUMO energy gaps, it can be concluded that the best linear relationship is observed in case of hardness ($R^2 = 1$) and the least one for dipole moment ($R^2 = 0.225$) of these nanoalloy clusters.

TABLE 3.1 Computed DFT-based Descriptors of Ag_nAu ($n = 1–8$) Nanoalloy Clusters.

Species	HOMO–LUMO gap (eV)	Hardness (eV)	Softness (eV)	Electro-negativity (eV)	Electrophi-licity index (eV)	Dipole moment (Debye)
AgAu	2.231	1.115	0.448	5.469	13.406	2.679
Ag_2Au	0.571	0.285	1.750	4.584	36.788	2.168
Ag_3Au	1.496	0.748	0.668	5.074	17.207	2.348
Ag_4Au	1.224	0.612	0.816	4.667	17.784	2.308
Ag_5Au	2.340	1.170	0.427	5.006	10.711	2.449
Ag_6Au	0.952	0.476	1.050	4.829	24.493	1.029
Ag_7Au	1.278	0.639	0.781	4.531	16.049	0.931
Ag_8Au	1.034	0.516	0.967	4.544	19.969	2.138

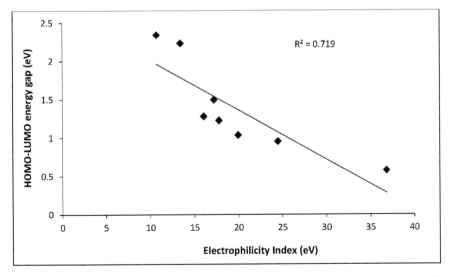

FIGURE 3.2 A linear correlation plot between electrophilicity index vs. HOMO–LUMO energy gaps (in eV).

3.4 CONCLUSIONS

We have systematically investigated electronic and optical properties of bimetallic Ag_nAu (n = 1–8) nanoalloy clusters in terms of CDFT-based descriptors, namely, HOMO–LUMO energy gap, hardness, softness, electronegativity, electrophilicity index, and dipole moment. The calculated HOMO–LUMO energy gaps of clusters have direct relationship with hardness and inverse relationship with softness values. Due to absence of any quantitative data, the optical properties of bimetallic Ag_nAu clusters have been assumed to be exactly equivalence of its HOMO–LUMO energy gap. After consideration of experimental facts, we have observed that optical properties of these clusters also have direct relationship with hardness and inverse relationship with softness values. The HOMO–LUMO energy gap and hardness displays odd–even oscillation behavior, indicating that clusters with even number of total atoms have high energy gap as compared to their neighbor clusters with odd number of total atoms. The cluster Ag_5Au with maximum value of energy gap (2.340) possesses highest stability and cluster Ag_2Au with lowest energy gap (0.571) exhibits least stability in the range of n = 1–8. The linear regression coefficient between several DFT-based descriptors along with their HOMO–LUMO energy gap is also obtained. The regression coefficient value R^2 = 0.719 is observed between electrophilicity index and HOMO–LUMO energy gap of Ag–Au clusters, which supports our computational analysis.

KEYWORDS

- **conceptual density functional theory**
- **bimetallic nanoalloy cluster**
- **AgAu**
- **HOMO–LUMO energy gap**
- **Au-doped Ag nanoalloy clusters**

REFERENCES

1. de Heer, W. A. *Rev. Mod. Phys.* **1993**, *65*, 611.
2. Brack, M. *Rev. Mod. Phys.* **1993**, *65*, 677.

3. Pal, R.; Wang, L. M.; Huang, W.; Zeng, X. C. *J. Am. Chem. Soc.* **2009**, *131*, 3396.

4. Eachus, R. S.; Marchetti, A. P.; Muenter, A. A. *Annu. Rev. Phys. Chem.* **1999**, *50*, 117.

5. Zhao, Y.; Li, Z. Y.; Yang, J. L. *Phys. Chem. Chem. Phys.* **2009**, *11*, 2329.

6. Hou, S. M.; Zhang, J. X.; Li, R.; Ning, J.; Han, R. S.; Shen, Z. Y.; Zhao, X. Y.; Xue, Z. Q.; Wu, Q. D. *Nanotechnology* **2005**, *16*, 239.

7. Scaffardi, L. B.; Pellegri, N.; de Sanctis, O.; Tocho, J. O. *Nanotechnology* **2005**, *16*, 158.

8. Fournier, R. J. *Chem. Phys.* **2001**, *115*, 2165.

9. Yuan, D. W.; Wang, Y.; Zeng, Z. *J. Chem. Phys.* **2005**, *122*, 114310.

10. Zhao, S.; Ren, Y. L.; Ren, Y. L.; Wang, J. J.; Yin, W. P. *J. Phys. Chem. A* **2010**, *114*, 4917.

11. Torres, M. B.; Fernández, E. M.; Balbás, L. C. *J. Phys. Chem. A* **2008**, *112*, 6678.

12. Hashmi, A. S. K.; Loos, A.; Littmann, A.; Braun, I.; Knight, J.; Doherty, S.; Rominger, F. *Angew. Chem.* **2009**, *351*, 576.

13. Neumaier, M.; Weigend, F.; Hamper, O.; Kappes, M. M. *J. Chem. Phys.* **2006**, *125*, 104308.

14. Autschbach, J.; Hess, B. A.; Johansson, M. P.; Neugebauer, J.; Patzschke, M.; Pyykkö, P.; Reiher, M.; Sundholm, D. *Phys. Chem. Chem. Phys.* **2004**, *6*, 11.

15. Ackerson, C. J.; Jadzinsky, P. D.; Jensen, G. J.; Kornberg, R. D. *J. Am. Chem. Soc.* **2006**, *128*, 2635.

16. Shaw, C. F. III. *Chem. Rev.* **1999**, *99*, 2589.

17. Valden, M.; Lai, X.; Goodman, D. W. *Science* **1998**, *281*, 1647.

18. Fèlix, C.; Sieber, C.; Harbich, W.; Buttet, J.; Rabin, I.; Schulze, W.; Ertl, G. *Phys. Rev. Lett.* **2001**, *86*, 2992.

19. Kim, S. H.; Medeiros-Ribeiro, G.; Ohlberg, D. A. A.; Williams, R. S.; Heath, J. R. *J. Phys. Chem. B* **1999**, *103*, 10341.

20. Varga, S.; Engel, E.; Sepp, W.-D.; Fricke, B. *Phys. Rev. A* **1999**, *59*, 4288.

21. Lecoultre, S.; Rydlo, A.; Buttet, J.; Fèlix, C.; Gilb, S.; Harbich, W. *J. Chem. Phys.* **2011**, *134*, 184504

22. Handschuh, H.; Ganteför, G.; Bechthold, P. S.; Eberhardt, W. *J. Chem. Phys.* **1994**, *100*, 7093.

23. Bishea, G. A.; Morse, M. D. *J. Chem. Phys.* **1991**, *95*, 8779.

24. Zhang, H.; Tian, D. *Comput. Mater. Sci.* **2008**, *42*, 462.

25. Negishi, Y.; Nakamura, Y.; Nakajima, A. *J. Chem. Phys.* **2001**, *115*, 3657.

26. Bonačić-Koutecký, V.; Burda, J.; Mitrić, R.; Ge, M.; Zampella, G.; Fantucci, P. *J. Chem. Phys.* **2002**, *117*, 3120.

27. Lee, H. M.; Ge, M.; Sahu, B. R.; Tarakeshwar, P.; Kim, K. S. *J. Phys. Chem. B* **2003**, *107*, 9994.

28. Weis, P.; Welz, O.; Vollmer, E.; Kappes, M. M. *J. Chem. Phys.* **2004**, *120*, 677.

29. Zhao, G. F.; Zeng, Z. *J. Chem. Phys.* **2006**, *125*, 014303.

30. Zhao, Y. R.; Kuang, X. Y.; Zheng, B. B.; Li, Y. F.; Wang, S. J. *J. Phys. Chem. A* **2011**, *115*, 569.

31. Tafoughalt, M. A.; Samah, M. *Phys. B* **2012**, *407*, 2014.

32. Tafoughalt, M. A.; Samah, M. *Comput. Theor. Chem.* **2014**, *1033*, 23.

33. Kuang, X. J.; Wang, X. Q.; Liu, G. B. *J. Alloys Compd.* **2013**, *570*, 46.

34. Ranjan, P.; Dhail, S.; Venigalla, S.; Kumar, A.; Ledwani, L.; Chakraborty, T. *Mater. Sci.-Pol.* **2015**, *33*, 719.

35. Ranjan, P.; Venigalla, S.; Kumar, A.; Chakraborty, T. *New Front. Chem.* **2014**, *23*, 111

36. Venigalla, S.; Dhail, S.; Ranjan, P.; Jain, S.; Chakraborty, T. *New Front. Chem.* **2014**, *23*, 123.

37. Ranjan, P.; Kumar, A.; Chakraborty, T. *AIP Conf. Proc.* **2016,** *1724,* 020072.
38. Ranjan, P.; Kumar, A.; Chakraborty, T. *Mat. Today Proc.* **2016,** *3,* 1563.
39. Ranjan, P.; Kumar, A.; Chakraborty, T. In *Environmental Sustainability: Concepts, Principles, Evidences and Innovations*; Mishra, G. C. Ed.; Excellent Publishing House: New Delhi, 2014; pp 239–242.
40. Ranjan, P.; Venigalla, S.; Kumar, A.; Chakraborty, T. In *Recent Methodology in Chemical Sciences: Experimental and Theoretical Approaches*; Chakraborty, T.; Ledwani, L. Eds.; Apple Academic Press and CRC Press: USA, 2015; pp 337–346.
41. Ranjan, P.; Kumar, A.; Chakraborty, T. *J. Phys. Conf. Ser.* **2016,** *759,* 012045.
42. Dhail, S.; Ranjan, P.; Chakraborty, T. In *Crystallizing Ideas—The Role of Chemistry*; Ramasami, P.; Bhowon, M. G.; Laulloo, S. J.; Wah, H. L. K.; Eds.; Springer International Publishing: Switzerland, 2016; pp 97–111.
43. Ranjan, P.; Kumar, A.; Chakraborty, T. In *Computational Chemistry Methodology in Structural Biology & Material Sciences*; Chakraborty, T.; Ranjan, P.; Pandey, A. Eds.; Apple Academic Press & CRC Press: USA, 2016. ISBN-9781771885683.
44. Ranjan, P.; Chakraborty, T.; Kumar, A. Applied Chemistry and Chemical Engineering: Experimental Technique and Methodical Developments, Vol. 4, Haghi, A. K.; Pogliani, L.; Castro, A.; Balköse, D.; Mukbaniani, O. V.; Chia, C. H. Eds.; Apple Academic Press & CRC Press: USA, 2016. ISBN-9781771885874.
45. Ranjan, P.; Kumar, A.; Chakraborty, T. *IOP Conf. Ser.: Mater. Sci. Eng.* **2016,** *149,* 012172.
46. Hafner, J.; Wolverton, C.; Ceder, G. *MRS Bull.* **2006,** *31,* 659.
47. Gaussian 03, Revision C. 02, Frisch, M. J.; Trucks, G. W.; Schlegel, H. B.; Scuseria, G. E.; Robb, M. A.; Cheeseman, J. R.; Montgomery, Jr., J. A.; Vreven, T.; Kudin, K. N.; Burant, J. C.; Millam, J. M.; Iyengar, S. S.; Tomasi, J.; Barone, V.; Mennucci, B.; Cossi, M.; Scalmani, G.; Rega, N.; Petersson, G. A.; Nakatsuji, H.; Hada, M.; Ehara, M.; Toyota, K.; Fukuda, R.; Hasegawa, J.; Ishida, M.; Nakajima, T.; Honda, Y.; Kitao, O.; Nakai, H.; Klene, M.; Li, X.; Knox, J. E.; Hratchian, H. P.; Cross, J. B.; Bakken, V.; Adamo, C.; Jaramillo, J.; Gomperts, R.; Stratmann, R. E.; Yazyev, O.; Austin, A. J.; Cammi, R.; Pomelli, C.; Ochterski, J. W.; Ayala, P. Y.; Morokuma, K.; Voth, G. A.; Salvador, P.; Dannenberg, J. J.; Zakrzewski, V. G.; Dapprich, S.; Daniels, A. D.; Strain, M. C.; Farkas, O.; Malick, D. K.; Rabuck, A. D.; Raghavachari, K.; Foresman, J. B.; Ortiz, J. V.; Cui, Q.; Baboul, A. G.; Clifford, S.; Cioslowski, J.; Stefanov, B. B.; Liu, G.; Liashenko, A.; Piskorz, P.; Komaromi, I.; Martin, R. L.; Fox, D. J.; Keith, T.; Al-Laham, M. A.; Peng, C. Y.; Nanayakkara, A.; Challacombe, M.; Gill, P. M. W.; Johnson, B.; Chen, W.; Wong, M. W.; Gonzalez, C.; Pople, J. A. Gaussian, Inc.: Wallingford CT,2004.
48. Jones, R. O.; Gunnarsson, O. *Rev. Mod. Phys.* **1989,** *61,* 689.
49. Zupan, A.; Blaha, P.; Schwarz, K.; Perdew, J. P. *Phys. Rev. B.* **1998,** *58,* 11266.
50. Theilhaber, J. *Phys. Fluids B* **1992,** *4,* 2044.
51. Stadler, R.; Gillan, M. J. *J. Phys.: Condens. Matter.* **2000,** *12,* 6053.
52. Argaman, N.; Makov, G. *Am. J. Phys.: Condens. Matter.* **2000,** *85,* 69.
53. Wang, H. Q.; Kuang, X. Y.; Li, H. F. *Phys. Chem. Chem. Phys.* **2010,** *12,* 5156.
54. Jiang, Z. Y.; Lee, K. H.; Li, S. T.; Chu, S. Y. *Phys. Rev. B* **2006,** *73,* 235423.
55. ParrR. G.; Yang, W. Eds. *Density Functional Theory of Atoms and Molecules*; Oxford University Press: Oxford,1989.
56. Xiao, H.; Kheli, J. T.; Goddard III, W. A. *J. Phys. Chem. Lett.* **2011,** *2,* 212.

CHAPTER 4

IMPROVED PERFORMANCE OF PROTON-EXCHANGE MEMBRANE FUEL CELLS (PEMFCs) USING MULTIWALLED CARBON NANOTUBES (MWCNTs) AS THE CATALYST SUPPORT

VIDYA RAJ and MANOJ KUMAR P.[*]

PSG Institute of Technology and Applied Research, Coimbatore, Tamil Nadu, India

[*]Corresponding author. E-mail: manoj@psgitech.ac.in

CONTENTS

ABSTRACT

In this work, wet chemistry route was used for the synthesis of multi-walled carbon nanotube (MWCNT)-supported Pt nanocatalyst for proton exchange membrane fuel cells. The MWCNTs were pretreated with H_2SO_4 and HNO_3 in order to obtain reactive sites for the adherence of Pt metal nanoparticles. The attachment of –OH and –COOH functional groups on MWCNTs were confirmed from the Fourier transform infrared spectrum. The synthesized 40 wt% Pt/MWCNTs electrocatalysts were characterized with high-resolution transmission electron microscope and energy dispersive spectroscopy analyses. The prepared 40 wt% Pt/MWCNT catalysts and commercial 40 wt% Pt/C catalysts were spray coated on the carbon electrodes with a Pt loading of 1 mg/cm^2. The single-cell performance with the fabricated membrane electrode assemblies was evaluated using fuel-cell test station. The improved performance was obtained in the case of prepared Pt/MWCNTs catalyst.

4.1 INTRODUCTION

In today's world, the demand for clean and sustainable energy sources has become a strong driving force in continuing economic development, and thus as well in the improvement of human living conditions. Proton exchange membrane fuel cells (PEMFCs), as clean energy-converting devices, have drawn a great deal of attention in recent years due to their high efficiency, high energy density, and low or zero emissions. Also, the solid, flexible electrolyte will not leak or crack. The only liquid in this fuel cell is water; thus, corrosion problems are minimal. PEMFCs have several important application areas, including transportation, stationary and portable power, and micropower. The two major technical gaps hindering commercialization have been identified: high cost and low reliability/durability. Fuel cell catalysts, such as platinum (Pt)-based catalysts and their associated catalyst layers, are the major factors in these challenges. Although a great deal of effort has been put into the exploration of cost-effective, active, and stable fuel cell catalysts, we have not yet had any real breakthroughs. Therefore, exploring new catalysts, improving catalyst activity and stability/durability, and reducing catalyst cost are currently the major tasks in fuel cell technology and commercialization.[1–3]

Carbon-black-supported platinum (Pt/C) particles are currently the most widely used electrocatalysts in PEMFCs. The degradation of electrocatalysts (Pt or Pt alloys) and their supports is recognized as one of the main contributors to the long-term degradation of fuel cell performance. As a result, several support materials for PEMFCs have been actively investigated, like various high surface area carbon,[4] boron-doped carbon,[5] carbon nanofibers,[6] multiwalled carbon nanotubes (MWCNTs),[7,8] CNTs directly grown carbon paper,[9,10] single-walled carbon nanotubes,[11] etc. However, MWCNTs and their composites as catalyst support material have wide-scale interest due to their unique properties such as high chemical and oxidative stability, extraordinary mechanical strength, good electronic conductivity, high surface area and relatively simple manufacturing process. Moreover, the highly inert surfaces of MWCNTs necessitate surface modification to enhance the attachment of Pt nanoparticles. The oxidation of carbon nanotubes (CNTs) with $KMnO_4$, H_2O_2, or ozone gas is well known to introduce functional groups such as hydroxyl (–OH), carboxyl (–COOH), carbonyl (–CO), and sulfate (–OSO3H) groups[11-15] on the surface of the CNTs. These groups provide nucleation anchors for the deposition of highly dispersed catalyst particles. However, the aromatic ring system of the CNTs can be disrupted by the application of extremely aggressive reagents, such as HNO_3 or H_2SO_4 or a mixture of two, and therefore, the nanotubes can be functionalized with groups such as hydroxyl (–OH), carboxyl (–COOH), and carbonyl (–C=O) that are necessary to anchor metal ions to the tube.[16] The controlled coating of metal nanoparticles onto CNTs, without aggregation of these particles would be crucial for this issue and should be researched.[17] Platinum nanoparticles can be synthesized using different approaches reported in the literature. These approaches include polyol process,[18] electrodeposition,[19] sonochemical processes,[20] sputter deposition,[21] gas reduction,[22] and solution reduction method.[9,11] All these methods have been successful in yielding Pt nanoparticles but with wide particle size ranges, essentially due to agglomeration or inefficient control on the growth of nuclei. However, colloidal process has been well established to produce uniform particles with excellent control on nuclei growth and morphology.[17,23]

In order to solve the problem of high cost of electrocatalyst, there are three alternative solutions: (1) increasing the Pt utilization, (2) Pt alloying with nonnoble transition metals, and (3) completely replacing Pt with nonplatinum catalysts. This work is mainly focused on increasing the utilization of Pt and improving the catalytic activity/performance.

Generally, there are two approaches for enhancing the utilization of platinum catalyst in PEMFCs, by decreasing the particle size of the catalyst and by achieving uniform distribution of Pt nanoparticles on the surface of support materials. So, better distribution of platinum particles and lower loading can improve the performance and reduce the cost of PEMFCs. Also with Pt, the corrosion of carbon black is even accelerated. Thus, new support materials such as MWCNTs can be used to improve the stability of catalysts.

In our approach, platinum nanoparticles of size less than 5 nm have been deposited on H_2SO_4–HNO_3-treated MWCNTs by wet-chemical method, by using chloroplatinic acid as the precursor. 40 wt% Pt/MWCNT catalysts have been synthesized and were characterized using transmission electron microscope (TEM) for analyzing the particle morphology and distribution. The single cell testing of Pt/MWCNTs-based membrane electrode assembly (MEA) and Pt/C-based MEA with catalyst loadings of 1 mg/Pt cm^2 was performed using Nafion-117 electrolyte and H_2/O_2 at 30°C.

4.2 EXPERIMENTAL

4.2.1 MWCNTs SURFACE MODIFICATION

Because of the hydrophobic properties, CNTs cannot be successfully wetted by liquids, and most metal nanoparticles do not adhere to them. So, the surface modification of MWCNTs is essential for the attachment of metal nanoparticles on it. Wet chemical oxidation is recognized as an efficient method for CNT purification and surface modification, promoting dispersion, and surface activation at the same time. Oxygen-containing functional groups (–OH, –C=O, and –COOH) can be introduced on MWCNTs through liquid-phase oxidation procedures. These oxygenated functional groups may render dispersability to the CNTs in aqueous solutions and organic solvents.[3,4]

Figure 4.1 represents the schematic representation of the functionalization of MWCNTs using H_2SO_4 and HNO_3. 1 g of MWCNTs were treated using sulfuric acid (3M) and nitric acid (3M) in a 500-mL round bottom flask equipped with a condenser, and the mixture was ultrasonicated for 20 min, then stirred for 15 min and refluxed at 120°C for 10 h. The resulting mixture was then diluted in water and washed up to neutral pH, and the sample was dried in oven at 60°C overnight into powder form.

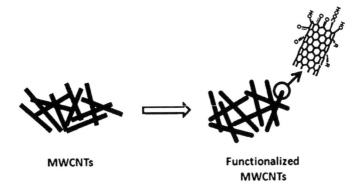

MWCNTs　　　　　　　　**Functionalized
MWCNTs**

FIGURE 4.1　Surface modification of MWCNTs.

4.2.2 DISPERSION STABILITY TEST OF FUNCTIONALIZED MWCNTS

The presence of functional groups on the CNT surface can be identified by different means. Among them, the time required for the CNTs to sediment in a polar solvent is a common technique since it is fast, cheap and provides qualitative information.

0.1 g of acid-treated MWCNTs and 0.1 g of untreated MWCNTs were immersed in 10 mL of ethanol for the comparison of the dispersion stability. Then, both the solutions were ultrasonicated for 1 h and were observed for 1 week.

4.2.3 PLATINUM NANOPARTICLES DEPOSITION

Chloroplatinic acid is used as the platinum precursor for the deposition of Pt on functionalized MWCNTs. The surface functional groups which are attached on the surface of MWCNTs roughly have a site ratio of 4 (–OH):2 (–COOH):1 (–C=O). These groups are expected to facilitate the chemical interaction between the anchoring catalyst metal ions and the modified CNT surface. A typical example of chemical interaction between anchoring catalyst metal ions with a CNT functional surface could be proposed as the following equation:[5]

$$CNT\text{-}COOH + M^+X^- \rightarrow CNT\text{-}COO^-M^+ + HX \qquad (4.1)$$

where the carboxylic group on the CNT surface exchanges a proton with a metal ion (M+).

0.1 g of functionalized MWNTs were immersed in 0.03 M of $H_2PtCl_6 \cdot 6H_2O$ solution and ultrasonicated for 30 min. After magnetic stirring for about 12 h, NaOH was added to change the pH to basic, and the platinum salt was reduced by slowly adding a solution of 0.1M $NaBH_4$. Then, the above mixture was stirred for 2 h at 60°C. When the reaction was completed, the solution was filtered with copious amount of deionized water. The recovered CNTs loaded with platinum were then dried at 80°C for about 2 h.

4.2.4   FABRICATION OF MEMBRANE ELECTRODES ASSEMBLY

4.2.4.1   HYDROPHOBIZATION OF GAS-DIFFUSION LAYERS

Porous carbon papers of 5 cm^2 active area were used as the gas-diffusion layers (GDLs). In order to get 20% weight gain of Teflon at the anode, 13.3 mL of Teflon was mixed with 25 mL of ammonia and made up to 100 mL by adding deionized water. Similarly for getting 30% weight gain of Teflon at the cathode, 20-mL Teflon was added with 25 mL of ammonia and finally made up to 100 mL. Then, carbon papers were immersed in the required solution and mechanically mixed for 3 min. Finally, Teflon-coated carbon papers were completely dried and weighed to measure the percentage weight gain of the Teflon at both the electrodes. Then, the Teflon-coated carbon papers were then sintered at 350°C for 15 min.

4.2.4.2   CATALYST INK PREPARATION AND SPRAY-COATING ON GDLS

Catalyst ink was prepared by mixing the catalyst powder (40 wt% Pt/MWCNTs or 40 wt% Pt/C catalysts) with an ionomer (5 wt% Nafion solution, DuPont) in deionized water and isopropyl alcohol. Then, the catalyst ink was ultrasonicated for 30 min. The Nafion loading in both electrodes was 30 wt%. Then, the catalyst ink was deposited on the GDLs by using microspray method to fabricate 5 cm^2 geometrically active areas. The catalyst loadings of the anode and cathode were about 1 mg of Pt/cm^2.

4.2.4.3   MEMBRANE TREATMENT

A Nafion 117 membrane (DuPont) was used as the polymer electrolyte membrane. The membrane was boiled in deionized water for 30 min and then was cleaned by immersing it in 3 wt% H_2O_2 for another 30 min at 60°C. The treated membrane was washed with deionized water for 15 min, and then, it was immersed in 0.5-M H_2SO_4 at 60°C for 30 min to ensure the full protonation of the sulfonate groups in the membrane and again rinsed in deionized water for 15 min to remove the excess acid content.

4.2.4.4   HOT-PRESSING

The catalyst-coated GDL is hot pressed together at 130°C, 0.2 kg/cm² for 3 min with the pretreated membrane using a hydraulic hot-press.

4.2.5   ASSEMBLY OF PEM FUEL CELL

A Fuel Cell Hardware Assembly consists of a pair of Graphite Blocks with a precision, machined flow-pattern, and a pair of gold-plated copper current collectors is located on the backside of the graphite flow channels; the cell leads and voltage sense leads are connected to these plates. Finally, the end plates are torque together to provide mechanical compression and connection of the fuel cell components, to seal the cell to inhibit gas leaks, and to reduce contact resistances. Gas inputs and outputs are through Swagelok fittings. A thermocouple well and two cartridge heaters are also provided. Heaters are located within the holes in the end plates and a thermocouple is used in conjunction with a temperature controller to control the cell temperature. Edge gaskets on either side of the MEA provide a gas tight seal between the flow channel and the membrane upon compression. The seal prevents gases from leaking from the cell or crossing over from one electrode to the other.

4.2.6   CHARACTERIZATION

The various functional groups attached on the surface of MWCNTs were detected using Fourier transform infrared (FTIR) spectroscopy. The surface morphology of Pt/MWCNTs was characterized using high-resolution TEM (HRTEM). Pt/MWCNTs dispersed in ethanol were applied on a lacy carbon grid for TEM characterization to examine the Pt particle size and distribution. The presence of Pt nanoparticles was confirmed with UV–Visible spectroscopy and energy dispersive spectroscopy (EDS).

4.3 RESULTS AND DISCUSSIONS

From Figure 4.2, it was observed that the transmittance peaks were obtained at 1429.49 and 3441.76 for the –COOH and –OH groups, respectively. Therefore, it can be concluded that the surface of MWCNTs was functionalized with –OH and –COOH groups when treated with H_2SO_4 and HNO_3.

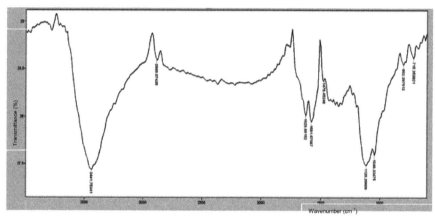

FIGURE 4.2   FTIR spectrum of H_2SO_4 and HNO_3-functionalized MWCNT.

Figure 4.3 shows that the functionalized MWCNTs have more dispersion stability compared to as-received MWCNTs. The CNTs without acid treatment (Sample 2) were sedimented completely in approximately 2 h after sonication. The CNTs treated with H_2SO_4 and HNO_3 (Sample 1) remained as a colloidal (well dispersed) solution for several weeks, without appreciable change from their dispersion state. From these results, it is clear that the presence of functional groups in the oxidized CNTs led to a reduction of van der Waals interactions among them, which promote their separation and dispersion in ethanol.

The HRTEM micrographs in Figure 4.4 show that there is a better and uniform dispersion of the Pt nanoparticles on MWCNTs compared to that of the Pt/C conventional catalyst. Pt nanoparticles were decorated on MWCNTs without much agglomeration and a narrow size distribution (2–4 nm) was obtained. From the EDS spectrum (Fig. 4.5), the confirmation of the Pt nanoparticles on the surface of MWCNTs was obtained. In Figure 4.6, the particle size distribution of Pt nanoparticles for Pt/MWCNT electrocatalyst and for conventional Pt/C electrocatalyst is given. By comparing the two histograms, it was observed that the average particle size of Pt for Pt/

MWCNT catalyst was around 1.8 nm and that for Pt/C catalyst was around 3.3 nm. Since the particle size of Pt nanoparticles is lesser in the case of Pt/ MWCNT catalyst, the surface area of the catalyst particles will be more, and hence, improved catalytic activity can be expected.

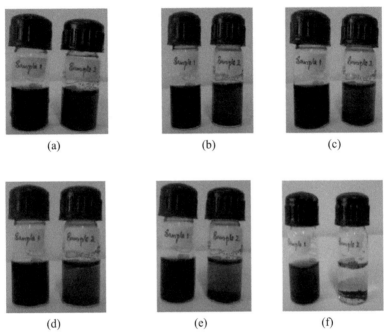

(a)   (b)   (c)

(d)   (e)   (f)

FIGURE 4.3 Comparison of dispersion unfunctionalized MWCNTs dispersed in ethanol (Sample 1 = pristine MWCNTs, Sample 2 = functionalized MWCNTs): (a) $t = 2$ h, (b) $t = 10$ h, (c) $t = 2$ days, (d) $t = 4$ days, (e) $t = 5$ days, and (f) $t = 1$ week.

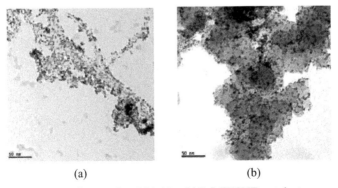

(a)   (b)

FIGURE 4.4 HRTEM micrographs of (a) 40 wt% Pt/MWCNTs catalyst prepared after the surface modification of MWCNTs and (b) 40 wt% Pt/C conventional catalyst.

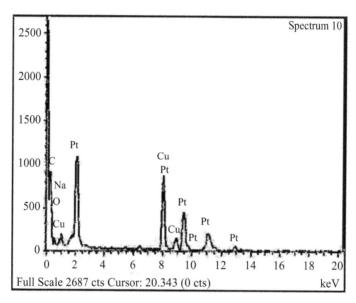

FIGURE 4.5   Energy dispersive spectroscopy of the synthesized Pt/MWCNTs electrocatalyst.

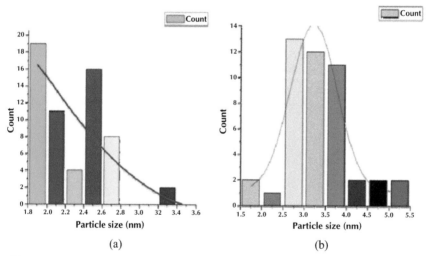

FIGURE 4.6   Particle size distribution for (a) Pt/MWCNTs electrocatalyst and (b) Pt/C conventional electrocatalyst.

In Figure 4.7, the SEM micrographs of the hydrophobized carbon paper before sintering are given. The coated Teflon solution can block the pores of the GDL. Hence, in order to prevent this, sintering of GDL is required.

(a)　　　　　　　　　(b)　　　　　　　　　(c)

FIGURE 4.7 SEM micrographs of hydrophobized carbon paper before sintering (a) 0% hydrophobized carbon paper, (b) 20% hydrophobized carbon paper, and (c) 30% hydrophobized carbon paper.

Figure 4.8a shows the comparison of the single cell performance using 40% Pt/MWCNTs catalyst and commercially available 40% Pt/C catalyst. The Pt loading was around 1 mg/cm^2 at both anode and cathode sides. They were tested under the same conditions: 30°C, ambient pressure, same GDLs. From the polarization curves, it is obvious that the Pt/MWCNTs catalyst gives higher performance than commercial Pt/C catalyst. The open-circuit voltage obtained from the Pt/MWCNT catalyst was around 0.98 V and 0.85 V was obtained from Pt/C catalyst. Also, the current density in the case of Pt/MWCNT catalyst (around 350 mA/cm^2) was found higher than that of Pt/C conventional catalyst (around 300 mA/cm^2).

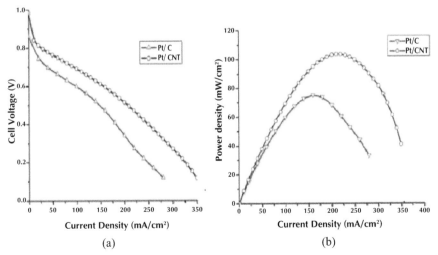

(a)　　　　　　　　　　　　　　(b)

FIGURE 4.8 (a) Fuel cell performance of commercial 40% Pt/C catalyst and 40% Pt/MWCNTs nanocatalyst and (b) power density curve of commercial 40% Pt/C catalyst and 40% Pt/MWCNTs nanocatalyst.

From Figure 4.8b, it can be seen that the peak power densities are 74 mW/cm^2 for commercial catalyst and 104 mW/cm^2 for Pt/MWCNT catalyst. As evident, the Pt nanocatalyst-based MEA performs much better compared to that of the commercial catalyst.

4.4 CONCLUSION

On the acid-treated MWCNTs, a uniform dispersion of Pt nanoparticles without much agglomeration was obtained. The dispersion stability was more in the case of functionalized MWCNTs compared to as-received MWCNTs. Thus, better Pt utilization has been achieved by uniform dispersion and improved corrosion stability is achieved by the use of MWCNTs as the catalyst support. Also, it was found that Pt/MWCNT-catalyst-based MEA showed better performance than conventional Pt/C catalyst-based MEA.

ACKNOWLEDGMENTS

The authors gratefully acknowledge the financial support from the Biomedical Department of PSG College of Technology and PSG Institute of Advanced Studies, Coimbatore, India. The transmission electron microscopy, UV–Visible spectroscopy and fuel cell testing work were performed at PSG Institute of Advanced Studies. The authors would like to thank CSIR (Council of Scientific and Industrial Research) lab, Chennai, India for the FTIR analysis.

KEYWORDS

- **carbon paper**
- **electrocatalyst**
- **gas-diffusion layer**
- **membrane electrode assembly**
- **multiwalled carbon nanotubes**
- **nafion membrane**
- **platinum nanoparticles**

REFERENCES

1. Zhang, J. *PEM Fuel Cell Electro Catalysts and Catalyst Layers: Fundamentals and Applications*; Springer-Verlag London Ltd: Guildford, Surrey, UK, 2008.
2. Chiang, Y.-C.; Ciou, J.-R. Effects of Surface Chemical States of Carbon Nanotubes Supported Pt Nanoparticles on Performance of Proton-Exchange Membrane Fuel Cells. *Int. J. Hydrogen Energy.* **2011**, *36*, 6826–6831.
3. Lee, K.; Zhang, J.; Wang, H.; Wilkinson, D. P. Progress in the Synthesis of Carbon Nanotube- and Nanofiber-Supported Pt Electrocatalysts for PEM Fuel Cell Catalysis. *J. Appl. Electrochem.* **2006**, *36*, 507–522.
4. Wang, Y.; Song, S.; Maragou, V.; Shen, P. K.; Tsiakaras, P. High Surface Tungsten Carbide Microspheres as Effective Pt Catalyst Support for Oxygen Reduction Reaction. *Appl. Catal., B: Environ.* **2009**, *89*, 223–228.
5. Acharya, C. K.; Li, W.; Liu, Z.; Kwon, G.; Turner, C. H.; Lane, A. M.; Nikles, D.; Klein, T.; Weaver, M. *J. Power Sources* **2009**, *192*, 324–329.
6. Calvillo, L.; Gangeri, M.; Perathoner, S.; Centi, G.; Moliner, R.; Lazaro, M. J. *J. Power Sources* **2009**, *192*, 144–150.
7. Wang, X.; Li, W.; Chen, Z.; Waje, M.; Yan, Y. *J. Power Sources* **2006**, *158*, 154–159.
8. Cui, S. K.; Guo, D. J. *J. Colloid Interface Sci.* **2009**, *333*, 300–303.
9. Kamavaram, V.; Veedu, V.; Kannan, A. M. *J. Power Sources* **2009**, *188*, 51–56.
10. Saminathan, K.; Kamavaram, V.; Veedu, V.; Kannan, A. M. *Int. J. Hydrogen Energy* **2009**, *34*, 3838–3844.
11. Kannan, A. M.; Veedu, V. P.; Munukutla, L.; Nejhad, M. N. G. *Electrochem. Soc.* **2007**, *10*(3), B47–B50.
12. Liu, J.; Rinzler, A. G.; Dai, H.; Hafner, J. H.; Bradley, R. K.; Boul, P. J.; Lu, A.; Iverson, T.; Shelimov, K.; Huffman, C. B.; Rodriguez-Macias, F.; Shon, Y. S.; Lee, T. R.; Colbert, D. T.; Smalley, R. E. *Science* **1998**, *280*, 1253–1256.
13. Hernadi, K.; Siska, A.; Thien-Nga, L.; Forro, L.; Kiricsi, I. *Solid State Ionics* **2001**, *141/142*, 203–209.
14. Chen, Z.; Hauge, R. H.; Smalley, R. E. *J. Nanosci. Nanotechnol.* **2006**, *6*, 1935–1938.
15. Guo, D. J.; Li, H. L. *Electroanalysis* **2005**, *17*, 869–872.
16. Rajalakshmi, N.; Ryu, H.; Srdchaijumon, M. M.; Ramaprabhu, S. Performance of Polymer Electrolyte Membrane Fuel Cells with Carbon Nanotubes as Oxygen Reduction Catalyst Support Material. *J. Power Sources* **2005**, *140*, 250–257.
17. Dobrzański, L. A.; Pawlyta, M.; Krzton, A.; Liszka, B.; Labisz, K. Synthesis and Characterization of Carbon Nanotubes Decorated with Platinum Nanoparticles. *J. Achieve. Mater. Manuf. Eng.* **2010**, *39*, 184–189.
18. Wee, J. H.; Lee, K. Y.; Kim, S. H. *J. Power Sources* **2007**, *165*, 667–677.
19. Kim, H.; Subramanian, N. P.; Popov, B. N. *J. Power Sources* **2004**, *138*, 14–24.
20. Xing, Y. *J. Phys. Chem. B* **2004**, *108*, 19255–19259.
21. Soin, N.; Roy, S. S.; Karlsson, L.; McLaughlin, J. A. Sputter Deposition of Highly Dispersed Platinum Nanoparticles on Carbon Nanotube Arrays for Fuel Cell Electrode Material. *Diamond Relat. Mater.* **2010**, *19*, 595–598.
22. Okhlopkova, L. B.; Lisitsyn, A. S.; Likholobov, V. A.; Gurrath, M.; Boehm, H. P. *Appl. Catal. A: Gen.* **2000**, *204*, 229–240.
23. Hiemenz, P. C.; Rajagopalan, R. *Principles of Colloid and Surface Chemistry*; Dekker: New York, 1997.

CHAPTER 5

GREEN SYNTHESIS AND CHARACTERIZATION OF SILVER NANOPARTICLES

VINDHYA P. S.[1,2] and D. SAJAN[1,*]

[1]*Department of Physics, Bishop Moore College, Mavelikara, Alappuzha 690110, Kerala, India*

[2]*Center for Advanced Materials Research, Department of Physics, Govt. College for Women, Trivandrum 695014, Kerala, India*

[*]*Corresponding author. E-mail: dsajand@gmail.com*

CONTENTS

ABSTRACT

In this work, colloidal silver nanoparticles were prepared by biosynthesis of *Murraya koenigii* leaf extract as reducing agent. It is proved to be a facile one step, eco-friendly procedure better than conventional physical/chemical methods. Nanoparticles were characterized by high-resolution transmission electron microscopy, X-ray diffraction (XRD), UV–Visible absorption spectroscopy, Fourier transform infra-red (FTIR), and photoluminescence spectroscopy. The absorption peak in range of 416–438 nm confirmed the reduction of Ag^+ ions. The presence of reducing agents was identified by FTIR analysis. XRD analysis shows that nanoparticles are face centered cubic structure. Transmission electron microscopy image clearly shows that the silver nanoparticles are spherical in shape. Synthesized silver nanoparticles show good antimicrobial activity against both Gram-negative organism and Gram-positive organisms.

5.1 INTRODUCTION

In recent years, noble metal nanoparticles have been extensively studied and various approaches are employed for the preparation of metal nanoparticles. Nanotechnology is the application of science to control matter at the molecular level and established recently as new interdisciplinary science.[1–4] The prefix "nano" indicates 1 billionth or 10^{-9} units. Growth of nanotechnology has opened up fundamental and applied frontiers in material science and engineering such as nano biotechnology, quantum dots, surface enhanced Raman scattering, and applied microbiology.[5–10]

Nanomaterial fabrication and their uses provide an important role in modern research by synthesis, design, and manipulation of particle. Metal nanoparticles have a high-specific surface area and surface atoms, because of their outstanding physicochemical characteristics, including optical, catalytic, electronic, magnetic, and antibacterial properties in different areas such as electronics, chemistry, energy, and medicine development.[11–13] Metal nanoparticles, particularly noble metals, have been studied mainly because of their strong optical absorption in the visible region caused by the group excitation of the free electron gas, that is, surface plasmon resonances (SPRs).[14–16] It is a collective oscillations of free electron on the metallic particle surface. Consequence of this electron oscillation the particle size and wavelength range of absorption in the spectrum can be defined. Nanoparticle size grows, the potential well dimensions increase, and its absorption is red-shifted.[17] The most effectively studied nanoparticles today are those made from noble metals, in particular silver, platinum, and gold and palladium. The silver nanoparticles have a large area of interest as they have a large number of applications: nonlinear optics, spectrally selective coating for solar energy absorption, biolabeling, and intercalation materials for electrical batteries as optical receptors, catalyst in chemical reactions, antibacterial materials, chemically stable materials, and good electrical conductors.[18]

Nanoparticle synthesis is usually carried out by physical and chemical methods including laser ablation, pyrolysis, chemical or physical vapor deposition, sol gel, lithography, and electrodeposition. But these are being expensive and toxic.[19] So recently use environmental friendly for synthesis of noble metal nanoparticles. A new branch of nanotechnology is nano biotechnology, combines biological principles with physical and chemical procedures to generate nano-sized particles with specific functions. Green nano biotechnology has increasing demand for both extracellular and intracellular microorganisms and is done by use of plant or fruit extracts and bio organisms.[20] There are several reports on the synthesis of silver nanoparticles using extracts obtained from various plant parts, for example, *Helianthus annuus* (Sunflower, *Asteraceae* or *Copositae*), *Basella alba* (Spinach, *Basellaceae*), *Oryz sativa* (Rice, *Poaceae*), *Saccharum officinarum* (sugar cane, *Poaceae*), *Sorghum bulgare* (Jowar, *Gramineae*), *Zea mays* (Corn, *Poaceae*), *Aloevera* (True or medicinal aloe, *Aspodelaceae*), *Diopyros kaki* (*Euphorbiaceae*), and *Magnolia kobus* (*Magnoliaceae*).[21]

In the present study, an environment-friendly, one-step, ultrafast, cost-efficient method for producing Ag^+ nanoparticles by green biological route, is using the extract of *Murraya koenigii* leaf. The plant extract act as reducing and capping agent for nanoparticle synthesis. The optical characterization of the synthesized nanoparticles is done by UV–Visible spectroscopy and photoluminescence spectroscopy. The functional group present in the sample was identified by Fourier transform infrared spectroscopy. The information about formation of these nanoparticles was confirmed by X-ray diffraction (XRD) spectroscopy and transmission electron microscopy. Their antimicrobial activity against human pathogenic microorganisms was investigated.

5.2 EXPERIMENTAL

5.2.1 PLANT MATERIAL AND PREPARATION OF LEAF EXTRACT

A total of 15.2632 g home-grown fresh leaves were washed thoroughly with deionized water and cut into fine pieces and were boiled into 250-mL distilled water about 15 min, and the extract is them filtered through Whatman No. 1 filter paper. The extract was stored under cool place for further experiment.

5.2.2 SYNTHESIS OF SILVER NANOPARTICLES

For the preparation, 0.0327 g of silver nitrate is weighted, and it is dissolved in 200 mL deionized water. A total of 30-mL silver nitrate solution is stirred with a magnetic stirrer and 5-mL leaf extract was added to it and again stirred. After 10 min, the color of the solution changed from colorless to brown indicating the formation of silver nanoparticles. The solution is kept at room temperature for 20 h. The concentration of the extract is varied like 10, 15, 20, 25, and 30 mL, respectively, and it is added to 30 mL of silver nitrate solution, and the same phenomena of color change occurs and the samples are shown Figure 5.2. The reduction of silver takes place and results in the formation of silver nanoparticles.

5.2.3 CHARACTERIZATION

The optical property of Ag–NPs was determined by UV–Visible and photoluminescence spectrophotometer. The chemical composition of the synthesized silver nanoparticles was studied by using Fourier transform infra-red (FTIR) spectrometer. Morphology and size of silver nanoparticles were investigated using high-resolution transmission electron microscopy. The antimicrobial activity was assessed using the agar well diffusion method. The phase variety and grain size of synthesized silver nanoparticles was determined by XRD spectroscopy. The XRD pattern measurements of drop coated film of AGNPs on glass substrate where recorded in a wide range of Bragg angle 2θ with counts. The instrument was operated at a voltage of 40 kV and a current of 30 mA with Cu$K\alpha$ radiation (1.5405 Å), the size is determined from width of X-ray peaks by Debye Scherrer's formula $t = K\lambda/B \cos\theta_B$.

5.3 RESULT AND DISCUSSION

5.3.1 BIOSYNTHESIS OF SILVER NANOPARTICLE

Figure 5.1a shows photographs of *M. koenigii* plant, and Figure 5.1b shows the reaction mixture of *M. koenigii* leaf broth and silver nitrate solution as a function of time. Visual observation of silver nanoparticles confirmed through the development of greenish aqueous suspension to brown color is due to reduction of silver ions. This is the indication of formation of silver nanoparticles.

FIGURE 5.1   (a) *Murraya koenigii* plant and (b) photograph of mixture of leaf extract and AgNo₃.

5.3.2 UV–VISIBLE SPECTRUM ANALYSIS

The optical absorption spectra of AgNPs are recorded with 200–800-nm wavelength range. Figure 5.2a shows absorption spectrum of *M. koenigii* leaf, and Figure 5.2b shows concentration of leaf extract at 5, 10, 15, 20, 25, and 30 mL shows absorption band at 416, 428, 435, 430, 432, and 437 nm, respectively. Noble metals exhibit unique optical property due to SPR. The presence of SPR in the range 400–450 nm for colloidal silver nanoparticles.[22] Broadening the peak indicates particles are polydispersed.

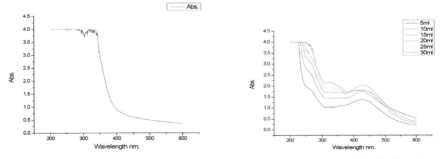

FIGURE 5.2   (a) Absorption spectra of *Murraya koenigii* leaf extract and (b) colloidal silver nanoparticle with different extract concentration.

5.3.3 PHOTOLUMINESCENCE ANALYSIS

Figure 5.3a shows photoluminescence spectrum of synthesized colloidal silver nanoparticles under an excitation wavelength 400 nm shows emission peak at 401 and 799 nm. The emission peak at 401 nm correlates with absorption maxima recorded UV–Visible spectrophotometer at 416 nm.[23]

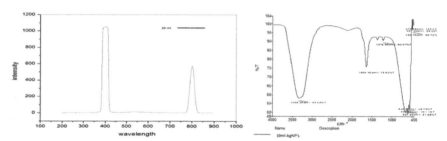

FIGURE 5.3   (a) Photoluminescence spectra and (b) FTIR spectra of silver nanoparticle.

5.3.4 FTIR ANALYSIS

Figure 5.3b shows FTIR spectra of silver nanoparticles has broad peaks at 3333.30 cm^{-1} indicate OH stretching vibrations of phenol/carboxylic group present in the extract, 1634.40 cm^{-1} indicate the presence of C=O stretching or amide bending, 551.93 cm^{-1} indicates ammine group. The other peaks at 1216.9 cm^{-1}, 482.75 cm^{-1} indicates that the formed silver nanoparticles were surrounded by proteins, terpenoids, and other secondary metabolites.[24]

5.3.5 XRD ANALYSIS

Figure 5.4 shows Bragg reflections with 2θ values of 38.31°, 44.45°, 64.59°, 77.50°, 81.74° with lattice planes indexed to (111), (200), (220), (311), (222) plane shows face-centered cubic structure (JCPDS File No. 04-0783).[25] The additional peaks in the XRD pattern represent the some impurities. The XRD pattern of silver nanoparticles synthesized using *M. koenigii* shows a characteristic peak at $2\theta = 38.5$, marked with (111) confirms the mono-crystallinity and sharpening of peak clearly indicates that the particles are in nanogreen. Table 5.1 shows calculation miller indices and particle size. From XRD analysis, the particle size of silver nanoparticle is 32.054 ± 1.151 nm.[26]

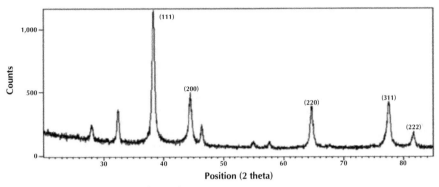

FIGURE 5.4 XRD pattern of sample.

TABLE 5.1 Calculation Miller Indices and Particle Size.

Sl no.	Position (2θ)	Position (θ)	$\sin \theta$	$\sin^2 \theta$	Ratio	$(h2 + k2 + l2)$	(hkl)	$\cos \theta_B$	B in radians	t (nm)
1	38.31	19.1555	0.3281	0.1077	1.00	3	111	0.944632	0.004049	35.83
2	44.45	22.2255	0.3783	0.1431	1.33	4	200	0.925702	0.004861	30.46
3	64.59	32.2935	0.5343	0.2854	2.65	8	220	0.845322	0.005262	30.81
4	77.50	38.75	0.6259	0.3918	3.64	11	311	0.779884	0.005934	29.62
5	81.74	40.869	0.6543	0.4282	3.98	12	222	0.756208	0.005402	33.55

5.3.6 TRANSMISSION ELECTRON MICROSCOPY ANALYSIS

The shape and size distribution of synthesized silver nanoparticle were characterized by transmission electron microscopy (TEM). Figure 5.5 shows TEM image of four silver nanoparticles and their size. From this the average size of the nanoparticle is 30.785 ± 1.268 nm, spherical in shape and nano in range.[27]

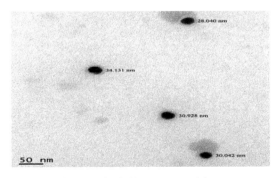

FIGURE 5.5 TEM image of synthesized silver nanoparticle.

5.3.7   ANTIMICROBIAL ACTIVITY

The antimicrobial activity was assessed using the agar well diffusion method. 10 mL of sterile nutrient broth were aseptically inoculated with test cultures and incubated at $37 \pm 0.5°C$ for 18 h. After incubation, the test bacterial cultures were spread on air-dried nutrient agar plates using a sterile cotton scab. Wells were cut using a gel puncher and about 100 mL of different solvent samples were poured on to it. The plates were incubated at $37 \pm 0.5°C$ for 12–14 h, and the zone of inhibition was measured in terms of diameter of zone as millimeters.

The synthesized nanoparticle was evaluated against *Escherichia coli, Bacillus subtilis, Enterobacterium, Klebsiella pneumonia,* and *Staphylococcus aureus* (Fig. 5.6). The diameter of inhibition zone is presented in Table 5.2. The Gram-positive bacteria show larger zones of inhibition, compared with Gram-negative bacteria which may due to the variation in cell wall composition. The zone of inhibition was found to be highest against in the case of *E. coli* and *B. subtilis* moderate against in the case of *Enterobacterium* and *K. pneumonia. S. aureus has the lowest antimicrobial activity among the five.*[28]

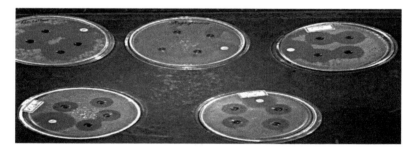

FIGURE 5.6   Antimicrobial activity of synthesized silver nanoparticles.

TABLE 5.2   Antimicrobial Activity of AgNPs Using *Murraya koenigii.*

Sl no.	Microorganisms	Zone of inhibition (mm in dia.)			
		10-mL AgNPs	15-mL AgNPs	Streptomycin (25 µg/ disc)(+ve control)	DMSO (−ve control)
1	*Escherichia coli*	28	27	38	–
2	*Enterobacterium*	23	22	29	–
3	*Klebsiella pneumoniae*	20	20	36	–
4	*Bacillus subtilis*	31	28	39	–
5	*Staphylococcus aureus*	16	13	25	–

5.4   CONCLUSION

The present study describes synthesis, characterization, and antimicrobial activity of silver nanoparticles using *M. koenigii* leaf broth. Color change is due to quantum confinement which is size dependent and affects optical property of nanoparticle. UV–Visible spectroscopy shows an absorption maximum at 438 nm. Photoluminescence study shows nanoparticle possesses fluorescent property. FTIR analysis indicates some compound from leaf extract formed coating of silver nanoparticle. TEM analysis is a good agreement with XRD study of silver nanoparticles in size 30 nm and spherical shape. Antimicrobial activity shows the synthesized nanoparticles have maximum zone of inhibition.

KEYWORDS

- **silver nanoparticles**
- **green synthesis**
- ***Murraya koenigii***
- **UV–Visible**
- **X-ray diffraction**
- **high-resolution transmission electron microscopy**
- **antimicrobial activity**

REFERENCES

1. Iravani, S. *Green Chem.* **2011**, *13*, 2638.
2. Abou El-Nour, K. M. M.; Eftaiha, A.; Al-Warthan, A.; Ammar, R. A. A. *Arab. J. Chem.* **2010**, *3*, 135–140.
3. Raveendran, P.; Fu, J.; Wallen, S. L. *J. Am. Chem. Soc.* **2003**, *125*, 13940–13941.
4. Nersisyan, H. H.; Lee, J. H.; Son, H. T.; Won, C. W.; Maeng, D. Y. *Mater. Res. Bull.* **2003**, *38*, 949–956.
5. Tran, Q. H.; Nguyen, V. Q.; Le, A.-T. *Adv. Nat. Sci.: Nanosci. Nanotechnol.* **2013**, *4*, 033001.
6. Dong, C.; Zhang, X.; Cai, H. *J. Alloys Compd.* **2014**, *583*, 267–271.
7. de Matos, R. A.; da Silva Cordeiro, T.; Samad, R. E.; Dias Vieira Jr., N.; Coronato Courrol, L. *Colloids Surf. A: Physicochem. Eng. Aspects* **2011**, *389*, 134–137.
8. Wei, D.; Sun, W.; Qian, W.; Ye, Y.; Ma, X. *Carbohydr. Res.* **2009**, *344*, 2375–2382.

9. Nima, P; Ganesan, V. *Int. J. ChemTech Res.* **2014–2015,** *7*(2), 762–768.
10. Vasireddy, R.; Paul, R.; Mitra, A. K. *Nanomater. Nanotechnol.* **2012,** *2*, Art. 8:2012.
11. Iravani, S.; Korbekandi, H.; Mirmohammadi, S. V.; Zolfaghari, B. *Res. Pharm. Sci.* **2014,** *9*(6), 385–406.
12. Anandalakshmi, K.; Venugobal, J.; Ramasamy, V. *Appl. Nanosci.* **2016,** *6*, 399–408.
13. Costlow, L.; Peter, A. *Nanomaterials and Nanostructures*; Dominant Publishers and Distributors: New Delhi, 2007.
14. Mansoori, A. Molecular based Study of Condensed Matter in Small Systems. In *Principles of Nanotechnology*; World Scientific, 2005.
15. Li, J. Ed. *Nanostructured Biomaterials*; Springer-Verlag: Berlin, Heidelberg, 2010.
16. Prasad, P. N. *Nanophotonics*; John Wiley and Sons Inc., 2004.
17. Ahmed, S.; Lkram, S. *Int. J. Pharm. Sci. Res.* **2015,** *6*(1), 14e30.
18. Ashokkumar, S.; Ravi, S.; Velmurugan, S.; *Spectrochem. Acta, A: Mol. Bimol. Spectrosc.* **2013,** *115*, 388e392.
19. Duran, N.; Marcato, P. D.; Alves, O. L.; De Souza, G. I. H.; Esposito, E. *J. Nanobiotechnol.* **2005,** *13*(3), 8.
20. Sastry, M.; Ahmad, A.; Islam, N. I.; Kumar, R. *Curr. Sci.* **2003**.
21. Kouvaris, P.; Delimitis, A.; Zaspalis, V.; Papadopoulos, D.; Tsipas, S. A.; Michailidis, N. *Mater. Lett.* **2012,** *76*, 18–20.
22. Zia, F. Ghafoor, N.; Iqbal, M.; Mehboob, S. *Appl. Nanosci.* **2016,** *6*, 1023–1029.
23. Ibrahim, H. M. M. *J. Radiat. Res. Appl. Sci.* **2015,** *8*, 265e275.
24. Krithiga, N.; Rajalakshmi, A.; Jayachitra, A. *J. Nanosci.* **2015,** *8*,928204.
25. Gavhane, A. J.; Padmanabhan, P.; Kamble, S. P.; Jangle, S. N. *Int. J. Pharm. Bio. Sci.* **2012,** *3*(3), 88–100.
26. Alagumuthu, G.; Kirubha, R. *Open J. Synth. Theory Appl.* **2012,** *1*, 13–17.
27. Song, J. Y.; Kim, B. S. *Bioprocess Biosyst. Eng.* **2009,** *32*, 79–84.
28. Elumalai, E. K.; Kayalvizhi, K.; Silvan, S. *J. Pharm. Bioallied Sci.* **2014,** *6*(4), 241–245.

PART II
Environmental Chemistry and Process

ENVIRONMENTAL CHEMISTRY, INDUSTRIAL WASTEWATER TREATMENT, AND ENVIRONMENTAL SUSTAINABILITY: A VISION FOR THE FUTURE

SUKANCHAN PALIT[*]

Department of Chemical Engineering, University of Petroleum and Energy Studies, Post-Office Bidholi via Premnagar, Dehradun 248007, India

[*]Corresponding author. E-mail: sukanchan68@gmail.com; sukanchan92@gmail.com

CONTENTS

ABSTRACT

Human civilization in today's world is moving toward visionary directions. The world of environmental engineering science is witnessing drastic and definitive changes. Global water crisis today stands in the midst of immense catastrophe. Scientific vision, scientific understanding, and scientific cognizance are all leading a long way in the true realization and true emancipation of environmental sustainability. Technology and engineering today needs to be reenvisioned and reenvisaged. The author in this treatise pointedly focuses on the immense potential and success of environmental sustainability and the wide domain of environmental chemistry in particular. Holistic sustainable development is the need of the hour globally. Technology is in the path of newer scientific regeneration. Industrial wastewater treatment today stands in the midst of immense scientific vision and deep comprehension. Zero-discharge norms and the concerns for ecological biodiversity have plunged the scientific domain in the midst of immense scientific vision. This treatise touches upon the interface between environmental chemistry and environmental sustainability with the sole target of furtherance of science and engineering. Water purification and sustainable development are two opposite sides of the visionary coin today. Wastewater treatment is the other avenues of scientific research pursuit today. The science of industrial pollution control, whether it is water, soil, and air pollution, needs to be reenshrined and revamped with the passage of science and time. The author in this treatise pointedly focuses on the success of sustainable development with rigorous water pollution control. This chapter opens up new windows of scientific innovation and scientific profundity in decades to come.

6.1 INTRODUCTION

The world of environmental engineering is today moving from one visionary frontier over another. Environmental restrictions and stringent regulations have urged the scientific domain to gear forward toward newer innovations and newer visionary future. Technological mission and scientific objectives in environmental engineering science are the forerunners toward a greater realization and a greater visionary future in the field of novel separation processes and nonconventional environmental engineering procedures. Today, science is a colossus with a definite vision of its own. Environmental engineering science in the similar vein needs to be reenvisioned and reenvisaged as science and engineering moves from one paradigmatic shift over another. The author in this treatise rigorously focuses on the wide domain of environmental sustainability and environmental chemistry with the sole aim of furtherance of science and engineering. Today, environmental engineering and environmental sustainability are the two opposite sides of the visionary coin. Technology is vastly challenged today. This treatise opens up the windows of innovation in the field of scientific endeavor in the field of environmental chemistry and industrial wastewater treatment.[18,19]

6.2   THE AIM AND OBJECTIVE OF THIS STUDY

Industrial wastewater treatment and environmental protection today stands in the midst of immense scientific vision and scientific introspection. Concerns of chemical process safety, environmental catastrophes, and the success of environmental sustainability are urging human civilization and human scientific research pursuit to gear forward toward newer innovation and inventions. Technology of environmental protection needs to be revamped with immediate effect as science treads a visionary path toward zero-discharge norms. The authors rigorously point out in this treatise a wider application of environmental chemistry and the success of industrial wastewater treatment techniques. The author also pointedly focuses on the recent scientific endeavor in the field of both environmental chemistry and sustainable development. Science of environmental protection thus opens a newer eon in the search for effective conventional and nonconventional environmental engineering techniques. The sole aim and vision of this study is to reenvision the field of environmental engineering science.[18,19]

6.3   ENVIRONMENTAL SUSTAINABILITY, THE TECHNOLOGICAL VISION, AND ENVIRONMENTAL PROTECTION

Environmental engineering and the wide scientific domain of environmental protection today stands in the midst of deep comprehension and scientific forbearance. Science and engineering today is ushering in a new era in the field of scientific forays into environmental engineering techniques and novel separation processes. Novel separation processes encompass membrane science. Global water crisis and research and development initiatives are today leading a long way in the true realization of environmental sustainability. Application of nanotechnology is another wider domain which needs to be explored with the aim of furtherance of science and environmental sustainability. The overarching goal of science and technology today is to target global needs for provision of potable water.

Environmental sustainability and environmental protection today are in the path of newer scientific vision and scientific regeneration. Environmental engineering tools such as novel separation processes and advanced oxidation processes are changing the face of scientific endeavor in our present day human civilization. Global water issues and global water challenges needs to be reenvisioned and reemphasized with the passage of scientific history and time. Today, science is a visionary research pursuit with the furtherance of progress of human civilization. Scientific and academic rigor are being challenged and reenvisaged with the progress of human civilization. Novel separation processes such as membrane science are rechallenging the worldwide issue of water crisis. Groundwater remediation and water purification are the hallmarks of today's science of water technology.[18,19]

6.4 SCIENCE, MANKIND, AND SUSTAINABILITY

Science, technology, and mankind today are moving toward a visionary future direction. Human civilization is in a state of deep distress. Environmental catastrophes and loss of ecological biodiversity are plunging human civilization toward murky depths. The science of environmental engineering needs to be revamped and rebuilt with the passage of scientific history and time. Sustainable development is the coinword of today. Technology of environmental engineering and sustainability needs to be redrawn and revisited to its utmost with the successful path of human scientific endeavor. Science today is a huge colossus with a definite vision of its own. Concerns for environmental protection, the loss of ecological biodiversity, and the wide futuristic vision of environmental science are all the technological drivers toward a greater emancipation of science and engineering. Sustainability and environmental protection are the two opposite sides of the visionary coin today. Mankind today stands in the midst of deep scientific comprehension and wide scientific vision. The scientific challenges and the scientific profundity are immense and far reaching.

Sustainable development and environmental engineering today are veritably linked by an unsevered umbilical cord. Immense scientific rigor, the futuristic vision, and the wide world of environmental protection are the forerunners and torchbearers toward a greater scientific understanding and scientific vision in the field of environmental engineering science and chemical process engineering. Environmental and energy sustainability today are in the path of newer scientific regeneration and wide scientific vision. Mankind is witnessing paradigmatic changes as respect to environmental protection and application of environmental engineering tools. Science and technology of environmental engineering are changing the scientific frontiers.

6.5 GLOBAL WATER CRISIS, RESEARCH AND DEVELOPMENT INITIATIVES, AND SUSTAINABLE DEVELOPMENT GOALS

Global water crisis today stands in the midst of immense pessimism and scientific vision. Technology of water science and environmental protection are challenged today. Global water research and development initiatives are ushering in a new eon in the field of environmental engineering techniques—conventional and nonconventional. Environmental calamities, the stringent environmental regulations, and the immense environmental concerns are all leading a long and visionary way toward the true emancipation and true realization of environmental engineering and chemical process engineering. The other side of the visionary coin of scientific progress today is sustainable development and environmental sustainability. The true challenge of science and engineering is slowly evolving. Mankind's immense scientific vision, the scientific truth, and the scientific and academic rigor are changing the face of global environmental sustainability.

Bigas et al.[1] addressed an important issue—the global water crisis—in a challenging review. The authors lucidly described water and global security, enhancement of water security through development, the human rights to alleviate the global water crisis, and the legal and ethical dimensions of a right to water. The foreword of this report was written by ex-Prime Minister of Norway Dr. Gro Harlem Brundtland, one of the leading global figures who propounded the concept of sustainable development. Today, technology has immense answers to environmental and energy sustainability. The challenge of civilization is awesome and veritably inspiring. The undeniable seriousness of the global water crisis has plunged the scientific domain toward immense concern at 1992 United Nations Conference on Environment and Development in Rio de Janeiro, at what came to be known as the Rio summit. According to the report, 20 years after the summit, the global water situation has improved but still has a long way to go because of the needs and effects of the immensely surging global population. The message which comes out of the water reports today is of immense caution and urgency with respect to how the world might prepare for and act to prevent global freshwater crisis with respect to support and quality. Technological advancements should go along with concerns for global energy and environmental sustainability. This report gleans on the urgency and risk of the water research and development initiatives and the wide vision behind such scientific forays. The magnitude of the global freshwater crisis and the risks associated with it are veritably underestimated. One billion people on earth are without reliable supplies of water, and more than 2 billion people lack proper sanitation. Here comes the need of sustainable development in a nation's progress. Water is critical to the attainment of the United Nations Millennium Development Goals whose targets are set to expire in 2015; it is already known that the world lags behind sanitation targets, which is predicted to be missed by over 1 billion people. Water security is also the foundation for food and energy security, and for overall long-term social and economic progress. Science, engineering, and technology are in a state of deep division and deep crisis as water crisis surmounts in a disastrous manner. Water encompasses health, nutrition, equity, gender equality, well-being, and economic progress, especially in the developing countries. The environmental impacts of the water crisis are exceedingly alarming. This report widely observes the success and futuristic vision of global water research and development initiatives.

United Nations University's report[2] discussed lucidly water security and the global water agenda. This is a United Nations-Water analytical brief. The salient features of this report are working definition of water security, themes for further dialogue, policy relevance of water security, and policy response options. This analytical brief serves as a starting point for dialogue on water security in the United Nations system. Technological progress, scientific vision, and the futuristic chartered course of action are the torchbearers toward a newer future dimension in global water initiatives. Global water crisis is in a state of immense catastrophe and unending distress. This analytical brief serves as a starting point for dialogue on water security and provision of water needs in the United Nations system. This brief aims to capture the wide vision of water related issues offering a holistic view for addressing water challenges under the visionary umbrella of water security. In this widely observed treatise, water security is defined as the capacity of a population to safeguard sustainable access to adequate quantities of acceptable quality water for sustainable livelihoods, human well-being, and socioeconomic development, for ensuring protection against water-borne pollution and water-related disasters. Technology and engineering science of water research today stands between wide scientific vision and deep scientific cognizance. The challenge and vision of water technology is slowly evolving in the present century. The authors deeply portray the immense success of global water research and development initiatives with the sole aim of furtherance of science. This widely envisioned definition of water security implies that water is managed sustainably throughout the water cycle and is done through an interdisciplinary focus so that it contributes toward global socioeconomic development and reinforces societal vision to environmental impacts and water-borne diseases without compromising the destruction of eco-system. The crisis and the catastrophe of water challenges throughout the world are slowly being rebuilt and revamped with more targets toward environmental sustainability. The challenge needs to be readdressed and restructured with the passage of scientific vision, scientific history, and time. This treatise investigates and targets the science of water security and trans-boundary water management and water security in conflict and disaster zones. The windows of scientific vision and scientific innovation are widely open as science moves toward a newer decade.

Rogers[3] discussed lucidly in a well-researched presentation the future of global water crisis. They stressed on the visionary fact of water issues and its meaning to the developing world in particular. Over 1 billion people throughout the world are without water. 2.4 billion people are without access to proper sanitation. In many parts of the world, access to water and power distinguishes from poor from the nonpoor. Six things which define global water crisis are as follows: (1) global climate change; (2) rapid population and economic growth; (3) increased demands for irrigation water; (4) increased demands for urban water; (5) replacing environmental flows; and lastly (6) trans-boundary conflicts. Environmental concerns, water challenges, and the success of technology will all lead a long way in the true emancipation of global water research and development initiatives today. The coinwords of today's science are scientific and technological validation. In today's world, global water crisis is a veritable curse and a bane to human civilization. Global climate change and global water shortage are the forerunners toward a greater emancipation of science today. This treatise investigates the wide vision of environmental engineering endeavor in global water challenges and global water research and development initiatives.

Sustainable development goals are faced with immense scientific challenges and deep scientific vision. Dr. Gro Harlem Brundtland's vision still needs to be readdressed and restructured with the passage of human history and time. The immense scientific rigor behind water technology and the futuristic vision of the science of water purification will definitely in future lead a long and visionary way in the true realization of sustainable development of the planet. This treatise willfully describes the immense success of global water research and development forays and the wide vision of global groundwater remediation agenda. Heavy metal groundwater remediation still remains latent and unfurled. The immense success, the definite vision, and the results of groundwater remediation techniques still need to be readdressed and reenvisioned. The proponents of the concept of "sustainability" have repeatedly over the years and in the advent of this century included the concept of alleviation of water issues in government policies. Technology of heavy metal remediation is in the verge of newer scientific regeneration and deep scientific introspection. This treatise tries to portray effectively the results and consequences in the successful application of environmental sustainability and in a similar manner the success of heavy metal groundwater remediation techniques.

6.6   RECENT SCIENTIFIC ENDEAVORS IN THE FIELD OF ENVIRONMENTAL CHEMISTRY

Environmental chemistry and engineering science today are witnessing immense scientific regeneration. Scientific vision, scientific profundity, and deep scientific introspection will lead a long way in the true emancipation of global environmental sustainability. Environmental chemistry today is plunging into the deep abyss of scientific forbearance and scientific fortitude. The author repeatedly ponders into the immense potential of science in tackling global water needs along with the furtherance of science of environmental chemistry. Chemical process engineering and nanotechnology are today two opposite sides of the visionary coin. Scientific endeavor and scientific profundity are the technology drivers of environmental chemistry and environmental engineering today. The immense challenge and the deep vision are tackled with minute details with an effective endeavor toward the march of engineering science. In the forays of science and the avenues of engineering, scientific vision is of utmost importance. Environmental chemistry is in the path of new rejuvenation and new scientific profundity. Science has today few answers to environmental catastrophes and breach of chemical process safety. Technology and engineering science needs to be revamped with the passage of scientific history and scientific vision.[18,19]

6.7   INDUSTRIAL WASTEWATER TREATMENT AND RECENT SCIENTIFIC RESEARCH PURSUITS

Industrial wastewater treatment and environmental protection are moving toward a newer visionary era today. Technology needs to be reenvisioned and reenvisaged as human civilization marches toward newer scientific destiny. Novel separation processes, chemical process engineering, and nanotechnology are revolutionizing the scientific landscape today. The challenge is moving beyond scientific imagination and scientific forbearance. Wastewater treatment, water purification, and drinking water treatment are of prime and utmost need today. Scientific research pursuit in industrial wastewater treatment and water purification needs to be reenvisioned and reemphasized with the growing concerns of environmental disasters.

The global concerns for industrial wastewater treatment and water purification are ever-growing and visionary. Global water crisis today is in a state of immense scientific distress. In such a crucial juncture of deep scientific history and time, water technology needs to be reenvisioned and reenvisaged. Shannon et al.[4] discusses with cogent foresight science and technology for water purification in the coming decades. One of the most pervasive issues afflicting people throughout the world is inadequate access to clean water and proper sanitation. Water issues are expected to grow worse in the coming decades with water scarcity occurring globally, even in regions currently considered water rich. Today, science has few answers to the ever-growing global water shortage. Widely addressing these problems, tremendous amount of scientific research pursuit needs to be readdressed and reenvisioned at the utmost.[4] Novel separation processes and membrane science is another visionary avenue of research. Advanced oxidation processes and ozonation technique in treating industrial wastewater is slowly and drastically changing the scientific frontier. The authors rigorously points forward the future of global water crisis and the acute shortage of pure drinking water. Recent advances in water treatment research offers hope in mitigating the impact of impaired waters around the world. Conventional methods of water disinfection, decontamination, and desalination can effectively address many of these issues with quality and supply. Technology of water science and environmental protection is rapidly changing. In highly industrialized world, the costs and time needed to develop state-of-the-art conventional water and wastewater treatment facilities make it difficult to address all the vexing issues. The authors touched upon the recent advances in disinfection, decontamination, reuse, and reclamation. Desalination and its wide vision is another facet of this well-observed study.[4]

6.8   THE VISIONARY WORLD OF ENVIRONMENTAL SUSTAINABILITY AND THE SCIENTIFIC RESEARCH PURSUIT

The visionary domain of sustainable development whether it is energy or environment needs to be readdressed and reenvisaged as human scientific endeavor in environmental protection reaches its zenith. Technology needs to be advanced and replete with deep scientific and academic rigor. In the progress of the human civilization and the wide thrusts in science and engineering, the need for the updated technology assumes prime importance. A scientist's vision, the roads to a nation's progress, and the world of scientific pursuit are all the torchbearers toward a greater visionary future in the field of environmental protection. Today, science and engineering stands in the midst of veritable reenvisioning. In this treatise, the author deeply ponders upon the immense success of environmental sustainability and environmental protection with the sole aim of progress of environmental engineering science and environmental separation processes. In this treatise, the author pointedly focuses on the vast and versatile world of environmental sustainability and the immense scientific rigor behind environmental engineering science.

The United Nations World Water Development Report[5] widely discussed the visionary area of water and jobs. This treatise portrays and investigates investing in water and economic growth, the global perspective on water, the wide world of economy, jobs and water, the world of water, jobs, and sustainable development, the examples of situation in Africa, Arab region, Asia, and the Pacific and Europe and North America. The authors also lucidly discusses with deep foresight capacity development needs and dialogue improvement, water efficiency and productivity, employment in water, sanitation and hygiene, scientific and technological innovation, and lastly policy responses. Technological vision and wide scientific objectives are the veritable backbones toward the emancipation of global water research and development innovations. Science, technology, and engineering science are the forerunners toward the greater innovative era of environmental protection science today. Technology and engineering science are veritably challenged today with the ever-growing concern for environmental protection. Sustainable development, human migration, conflict and natural disasters: water surpasses these, and many other issues on the global water and environmental engineering agenda. Employment is another key factor in population movements, civil unrest, and environmental sustainability. This report gives a holistic view in the wide world of sustainable development and the recent scientific research pursuit in water technology. Water and jobs are inextricably linked on various levels, whether it is from an economic, environmental, and social perspective. The challenges and vision of water technology and job creation are immense and far reaching. The science of water technology is crossing wide and vast visionary boundaries. This report estimates that well over 1 billion jobs, representing more than 40% of the world's total active workforce, are heavily water dependent. Such jobs are found in agriculture, forestry, inland fisheries, mining and resource extraction, power generation, and water supply and sanitation. The wide world scientific and technological validation is at its helm as human civilization from one scientific paradigm over another. The 2016 report of the United Nations World Water Development Report addresses a subject that has received only marginal attention, particularly at the international level; the relationship between water and jobs. Water is an important component of national and international economies and is needed to create and maintain jobs across all sectors of economy. Global perspectives on water are today far reaching and surpassing visionary frontiers. This paper widely reflects the scientific success, the vast scientific potential, and the wide academic rigor in the field of water policy, water engineering, and water technology emancipation. The author also touches upon regional perspectives in the visionary journey toward global water emancipation.

Technological advancements today are immensely diverse and far reaching. Water stands as a major component of a nation's advancement today. Provision of pure drinking water should be high on the scientific agenda of policy decisions. Environmental and energy sustainability needs to be readdressed and reenvisaged at each step of human life and scientific research pursuit today. Emancipation of the science of sustainability should be high on the cards for the advancement of human civilization today.

Rodriguez et al.[6] discussed lucidly capital, operations, and maintenance in water infrastructure. The treatise investigates the present state of water issues, financial crisis, climate crisis, food crisis and the energy shocks, and the green response. The authors also deals with investment needs and funding sources, tools for the way forward, and strong conclusions and recommendations. All nations throughout the world face a growing funding gap as they try to keep up with rehabilitation, operation, and maintenance of aging water infrastructures. New water systems need to be built to cope with growing populations, shifting consumption patterns, and a rapidly changing climate. The ever-growing concerns for environmental protection have propelled the scientific domain to gear forward for newer innovations and visionary policies. Technological and scientific advancements are gaining new heights with the passage of scientific history and human civilization.

Water infrastructure veritably stands today in the midst of deep scientific vision, introspection, and scientific cognizance. Technology needs to be revamped and reenvisioned at the utmost, and water science and water technology need to be restructured at this crucial juncture of human history and human civilization. Today, water issues are linked by an unsevered umbilical cord with the economic growth of a nation. Groundwater crisis is another vicious concern for human race today. Arsenic groundwater contamination and heavy metal contamination of drinking water are the vexing issues of science and are an enigma to the human scientific endeavor. Technology is challenged today and stands in the midst of scientific comprehension, deep introspection, and wide innovation. Mankind's immense scientific progeny, the enigma of engineering science, and the wide futuristic vision of science will all lead a long and visionary way in the true realization of water technology today.

6.9 THE WIDE WORLD OF ENVIRONMENTAL CHEMISTRY AND THE SUCCESS OF POLLUTION CONTROL

Science is marching ahead at a rapid pace this century. Industrial pollution, whether it is water, air, or soil, are baffling the progress of science and engineering today. The science of environmental protection is reaching enigmatic heights. Successive scientific generations are eager and focused on the cause of loss of biodiversity in our planet today. Industrial pollution control is the ever-growing focal point of scientific endeavor. The science of wastewater treatment, water purification, and industrial pollution control will all lead a long way in the true emancipation and true realization of sustainable development. The march of science and technology in the field of environmental chemistry is replete with deep scientific vision and scientific profundity. Environmental pollution control and the world of environmental engineering need to be reenvisioned and reenvisaged with the passage of scientific history and time.

6.10 ENVIRONMENTAL ENGINEERING TECHNIQUES AND THE WIDE SCIENTIFIC ENDEAVOR

The aim and mission of environmental engineering techniques are crossing vast and versatile frontiers today. Man's wide vision, mankind's scientific prowess, and the immense scientific and academic rigor are plunging human scientific rigor toward the murky depths of science and engineering. The scientific knots and the deep scientific profundity need to be unraveled with the march of engineering science. The overarching goal of water purification is gleaned in the following sections. The author rigorously points out the immense success of disinfection, decontamination, desalination, reuse, and reclamation.

6.10.1 DISINFECTION

The vision of disinfection is far reaching and reaching visionary scientific frontiers.[4] Global water challenges and global water research and development initiatives are changing the face of human civilization. The challenge and the vision are immense and replete with deep scientific and academic rigor. The scientific urge to excel, the empowering scientific vision, and the rigors of science are all leading a long way in the true emancipation of environmental, energy, and scientific sustainability. Disinfection, decontamination, and desalination stand as major research areas in the future march of environmental engineering science. The author pointedly focuses on the research and development initiatives and wide forays into the world of environmental protection and also the domain of water reuse and reclamation. Technology and engineering science needs to be reenvisioned and rebuilt with the visionary forays into environmental pollution control. The march of science and engineering, the immense scientific urge to excel, and the wide academic and scientific rigor will all lead a long way in the true emancipation and effective realization of industrial wastewater treatment and water purification. Desalination is of prime importance in the wake of immense global importance of water crisis.[4]

An ever-growing vision for providing safe and clean water is affordably and robustly to disinfect water from traditional and emerging pathogens, without creating more issues due to disinfection process itself. Waterborne pathogens have a disastrous effect on public health, especially in the developing countries of sub-Saharan Africa and Southeast Asia. Technological objectives and scientific motivation in disinfection need to cross visionary boundaries as science plunges forward toward profundity and vision. Waterborne infectious agents responsible for these diseases include a variety of helminthes, protozoa, fungi, bacteria, viruses, and prions. While some infectious agents have been eradicated or diminished, new ones continue to emerge and so disinfecting water has emerged as extremely effective. Viruses are of immense concern, accounting, together with prions, for nearly half of all emerging pathogens in the last two to three decades. The wide avenue of scientific vision and scientific forbearance are the pallbearers toward a new age of environmental engineering techniques such as disinfection, decontamination and water-desalination.[4]

Therefore, the effective control of waterborne pathogens in drinking water calls for the development of new disinfection strategies, including multiple-barrier approaches that provide reliable physicochemical removal (e.g., coagulation, flocculation, sedimentation, and media or membrane filtration). The use of light from visible to ultraviolet (UV) to photochemically inactive pathogens has in the recent times seen a remarkable and visionary challenge. Sequential disinfection schemes such as UV/combined chlorine and ozone/combined chlorine are being considered by many drinking water utilities as the inactivation component of their multiple-barrier treatment plants because, compared with free chlorine, both UV and ozone are extremely effective in controlling *Cryptosporidium parvum* oocytes. Water purification and water disinfection are overcoming immense hurdles in operation of water treatment plants and industrial wastewater systems. This treatise immensely points toward the water purification challenges with the sole aim of furtherance of science and effective environmental engineering techniques.[4]

Technology of water science is in the path of immense scientific regeneration and scientific rejuvenation. Frequent environmental disasters, loss of ecological biodiversity, and the concerns for environmental protection are plunging the scientific and engineering domain toward the murky depths of scientific challenges and wide scientific innovations. Both environmental engineering science and novel separation processes/environmental engineering separation processes are veritably challenged with the passage of scientific legacy and time. Human civilization today stands in the midst of immense vision and in the same time restructuring.[4]

6.10.2 DECONTAMINATION

The visionary and overarching goal for the future of decontamination is to detect and remove toxic substances from water affordably and robustly. Water pollution control and water reuse today stands in the midst of immense comprehension and scientific introspection. Groundwater contamination and drinking water crisis are challenged today and needs to be readdressed and scientifically reenvisaged with the pursuit of science and engineering.[4] Widely distributed substances, such as arsenic, heavy metals, halogenated aromatics, nitrosoamines, nitrates, phosphates, and so on, are known to cause harm to humans and the environment. Two primary problems are that the amount of suspected harmful agents is growing rapidly, and that, many of these compounds are toxic in trace quantities. Environmental engineering paradigm, the success of groundwater remediation, and the immense scientific profundity are the torchbearers toward a greater realization of environmental sustainability in future.[4]

6.10.3 DESALINATION

Desalination science and engineering are the need of the hour for many water-challenged countries of the world. The future of environmental engineering and the imminent global water crisis are changing the path of scientific research pursuit today. Desalination and drinking water treatment are the two opposite sides of the visionary coin today. Technology and engineering are retrogressive today. Science has few answers to the global water hiatus.[4] The ever-growing avenue for the future of desalination is to increase the fresh water supply via desalination or seawater and saline aquifers.[4] These sources account for 97.5% of all water on the earth, so capturing even a tiny fraction could have a lasting and a huge impact on water scarcity.[4]

Desalination of all types is often considered a capital- and energy-intensive process and typically requires the conveyance of the water to the desalination plant, pretreatment of the intake water, disposal of the concentrate (brine), and process maintenance. Science is advancing very fast in today's scientific and technological horizon. Water science and technology and global water issues are vexing issues for the furtherance of science today. A scientist's wide vision, technology's immense prowess, and the wide world of water technology are the forerunners toward a newer visionary era of environmental engineering.[4]

6.10.4 REUSE AND RECLAMATION

Water reuse and water reclamation today stand in the midst of immense scientific vision and scientific forbearance.[4] Technology of water purification today are surpassing visionary frontiers. Water research and development initiatives are gaining new scientific heights and scientific pinnacles with the passage of human history and time. Environmental catastrophes, loss of ecological biodiversity, and the wide scientific rigor behind environmental engineering science are the forerunners and torchbearers toward a greater emancipation of environmental biodiversity and environmental techniques and tools.[4]

6.11 TRADITIONAL ENVIRONMENTAL ENGINEERING TECHNIQUES

Traditional environmental engineering techniques are today surpassing visionary frontiers of science. The vision and the challenge of science are far reaching as technology needs to be reenvisioned and reenvisaged. Traditional environmental engineering techniques are activated sludge processes, flocculation, coagulation, and biochemical processes. Today, membrane science and novel separation processes also fall under that category. The endeavor of science in environmental engineering science is widely opening new vistas of scientific research pursuit in decades to come. The author deeply comprehends the wide avenues of scientific research pursuit in the field of environmental engineering.

Membrane science today is ushering in a newer eon in the field of scientific genre and deep scientific vision. A scientist's immense vision, the success of technology, and the wide academic rigor are leading a wide and visionary way in the true realization of environmental sustainability today. In this section, the author garners immense information in the successful application of membrane science in environmental protection. The Loeb–Sourirajan model in membrane science opened up a newer era in scientific research pursuit in chemical process engineering and environmental engineering. Technology of water science and environmental protection slowly initiated a revolutionary beginning in research endeavor. The crux of the domain of membrane science today is the wide world and the difficult domain of membrane fouling. The diffusion phenomenon of membrane separation processes is still today enigmatic and uncovered. Technology of membrane science is veritably complex. In this treatise, the author rigorously points out the scientific forays, the deep scientific thoughts, and the scientific fortitude behind membrane fouling.

Van der Bruggen et al.[7] discussed in lucid details in a comparative effort between distillation versus membrane filtration with the sole aim of presenting an overview of process evolutions in seawater distillation. The worldwide need for fresh water requires more and more plants for the treatment of nonconventional water sources. During the last decades, seawater has become an important source of fresh water in many arid regions. The traditional desalination processes [reverse osmosis (RO), multistage flash (MSF), multieffect distillation (MED), and electrodialysis (ED)] have resulted to reliable and established processes; current research targets on process improvements in view of a lower cost and a more environmentally benign process. Technology, science, and engineering of membrane separation processes need to be reenvisaged and reenvisioned with the passage of scientific history and scientific profundity. This paper presents an overview of recent process improvements in seawater desalination using RO, MSF, MED, and ED. Important scientific forays include the use of alternative energy sources (wind energy, solar energy, and nuclear energy) for RO and distillation processes, and the impact of different distillation processes on the environment; the implementation of hybrid processes in seawater desalination; pretreatment of desalination plants by pressure driven membrane processes compared to chemical pretreatment; new materials to prevent corrosion in distillation processes; and the prevention of fouling in RO units. Mankind, human society, and scientific endeavor are technology driven today and veritably ushers in a new dimension of future scientific thoughts. This paper widely observes cost-effectiveness of the desalination process and ensures the visionary concept of environmental sustainability. The supply of fresh water is a key element for all societies and a scientific and technological driver toward the furtherance of human civilization. Together with fresh water, groundwater is a vital need for the human society. Fresh water is needed in agriculture, as drinking water or as process water in various industries. The global concerns for provision of pure water are immense and vital. This paper reviews the important advances in seawater desalination in view of lowering the total cost and decreasing the impact on environment. Green engineering is at its worldwide best as technology today gears forward toward newer challenges. In the beginning of the paper, the author presents traditional desalination methods such as MED, MSF, RO, and other visionary techniques. Then after that, the author presents alternative energy sources for desalination. Then, the author discusses lucidly pretreatment of seawater and finally the environmental impact of desalination processes.

Van der Bruggen et al.[8] reviewed with deep and cogent insight drawbacks of application of nanofiltration (NF) and how to avoid them. The drawback of membrane science is the murky depths of membrane fouling which is still today not unraveled. Technology and science of membrane science and NF and the intricacies of separation phenomenon and the diffusion phenomenon need to be readdressed and reshaped with the wide passage of scientific history and scientific vision. In spite of all, definite and prominent environmental perspectives for NF, not only in drinking water production but also in wastewater treatment, the food industry, the chemical and pharmaceutical industry, and many other industries, there are unresolved as well as unsolved problems in membrane separation processes. In this paper, the author rigorously challenges six different avenues where intricate solutions are definitely scarce: (1) avoiding membrane fouling, (2) improving the separation between the solutes that can be achieved, (3) further treatment of concentrates, (4) chemical resistance and limited lifetime of membranes, (5) insufficient rejection of pollutants in water treatment, and (6) the immense and urgent needs for modeling and simulation techniques. Fouling is one of the major problems in any membrane separation, but for NF, it might be somewhat more complex because of the interactions leading to the fouling take place at nanoscale and therefore immensely difficult to comprehend. Foulants playing a role for NF membranes can be organic solutes, inorganic solutes, colloids or biological solutes. An extensive and thoughtful discussion on fouling is widely available in scientific literature. Fouling and adsorption can be related to the component properties, which is reflected by the correlation between the octanol–water partition coefficient (log P) and adsorption; adsorption is also related to the dipole moment and water solubility. Technology revamping is the utmost need of the hour as the challenge of science gears forward toward a newer eon. Depending on the relative size of the colloidal particles and membrane pores, colloidal fouling may occur either due to accumulation of particles on the membrane surface and the build-up of a cake or by a penetration within the membrane pores. The immense academic rigor of fouling, the intense scientific research pursuit, and the technological vision will all lead a long and visionary way in the true realization of environmental engineering science and membrane science today.

Van Geluwe et al.[9] reviewed ozone oxidation for the alleviation of membrane fouling by natural organic matter (NOM). Membrane fouling by NOM is one of the major problems that slow down the application of membrane technology in water and wastewater treatment. Ozone is able to efficiently change the physicochemical characteristics of NOM in order to reduce the membrane fouling. This paper investigates the state-of-the-art knowledge of the reaction mechanisms between NOM and molecular ozone or OH radicals, together with an in-depth study of the interactions between NOM and membranes that govern membrane fouling, including the effect of ozone oxidation in it. Scientific vision, the wide world of scientific cognizance, and the immense visionary scientific rigor are today leading a long way in the true emancipation of environmental engineering science. The authors in this paper describe the success of academic research in the field of membrane fouling and the wide world of environmental sustainability linked to it. The emerging use of ozone oxidation in water treatment offers new opportunities, because ozone is able to decompose certain membrane foulants effectively. The present treatise explores and reviews literature concerning the fouling potential of NOM in water purification and the use of ozone oxidation for the alleviation of membrane fouling by NOM.

Van der Bruggen et al.[10] widely observed with cogent insight fouling of NF and ultrafiltration membranes applied for water regeneration in the textile industry. Textile effluents usually contain high concentrations of inorganics as well as organics and are therefore difficult to treat. Membrane processes can be used for many of these wastewaters in the textile industry. Technology drivers such as membrane separation processes are changing the world of scientific forbearance and scientific justification. Textile industry is an important example in the furtherance of membrane science toward a greater visionary eon in the field of environmental engineering. Scientific vision and scientific wisdom are at its level best in the effective emancipation of environmental engineering science and membrane science today. Two facets of membrane fouling are discussed in this paper: (1) the use of NF for the treatment of exhausted dye baths, in view of water recycling and (2) the use of ultrafiltration for the removal of spin finish from wastewater resulting from rinsing of textile fibers. In the first application, fouling is assumed to be caused by adsorption of organic compounds, which has a large influence because of the high concentrations used in textile dyeing. In this article, the authors deeply investigate the NF and ultrafiltration of textile effluents and their relation to fouling mechanisms and solute transport mechanisms through the membranes. The world of scientific vision and scientific fortitude are the forerunners toward a greater visionary research trends in the field of membrane fouling.

Boussu et al.[11] deeply investigated with scientific cognizance influence of membrane and colloid characteristics on fouling of NF membranes. Colloidal fouling is still one of the major impediments for the implementation of membrane processes, for example, in the purification of surface water. Effects of fouling for several representative NF membranes during filtration of several types colloids in different circumstances (pH, ionic strength) are widely observed in this study. Four different colloidal solutions (silica–aluminum) were selected to investigate both the influence of colloid size and colloid charge. Colloidal particles are ubiquitous in nature. Colloids cover a wide size range, from a few nanometers to a few micrometers. The most colloidal fouling was observed for the hydrophobic membranes, for which a dense cake layer was formed on the entire membrane surface. Technological advancements in membrane technology are gaining immense heights as the science of membrane fouling is slowly uncovered. The challenge, the vision, and the futuristic vision in the membrane science applications are opening up new windows of innovation in environmental engineering in decades to come. In membrane water treatment processes, fouling by these colloidal particles is one of the major categories of performance deterioration. Depending on the relative size of colloidal particles and membrane pores, fouling may occur due to either accumulation of particles on the membrane surface and the build-up of a cake or by penetration within the membrane pores.

Van der Bruggen et al.[12] discussed lucidly how a microfiltration pretreatment affects the performance in NF. This is a phenomenal understanding of membrane pretreatment technology. The use of a well-chosen pretreatment system is a key element to avoid fouling in NF. This article explores the influence of a MF pretreatment by comparing the performance of three NF membranes. The wastewaters studied were bottle-rinsing water and rinsing water from the fermentation tanks, respectively. As it allows the removal of components with relatively low molecular weight, NF is a process with numerous applications in drinking water production, process water recovery in industry, and wastewater treatment. In addition to the frequently reported problems for concentrate discharge, the major limitation for implementation of NF is the vexing issue of membrane fouling.

Membrane science and industrial wastewater treatment are the two opposite sides of the visionary coin of environmental engineering science today. The technology is not new yet immature. The diffusion phenomenon of membrane separation phenomenon is still unestablished. This treatise instinctively expounds the success of recent scientific research pursuit in membrane science. The wide range of different areas of membrane science needs to be revamped with the passage of scientific endeavor.

6.12 NONCONVENTIONAL ENVIRONMENTAL ENGINEERING PROCESSES

Nonconventional environmental engineering processes include advanced oxidation processes, chemical oxidation, and integrated advanced oxidation processes. Ozonation is one of the visionary avenues of science. Nonconventional environmental engineering techniques are reshaping the face of environmental engineering scientific research pursuit. Effectivity, efficiency, and scientific vision are the hallmarks of the techniques. Today, the world is faced with the scientific enigma of groundwater contamination. Technology is challenged and veritably baffled. Science has become so retrogressive with the global concerns for environmental pollution, environmental catastrophes, and loss of ecological biodiversity. Success and potential of environmental engineering science has become immensely negative with the passage of scientific history and time. Green chemistry and green engineering are the visionary coinwords of today's scientific endeavor. Novel separation processes such as membrane science also are branches of nonconventional environmental engineering processes. Technological vision and scientific validation are the forerunners toward a greater realization of environmental treatment procedures and the holistic world of environmental protection. The scientific challenges, the scientific understanding and the scientific vision are the torchbearers toward an effective environmental protection genre today.[13]

6.13   OZONATION: THE NEXT-GENERATION SCIENTIFIC ENDEAVOR

Ozonation technology is one of the ever-growing and far-reaching endeavors of science. Scientific and technological profundities are in the process of newer scientific regeneration. Today, advanced oxidation processes and ozonation are two opposite sides of the visionary coin. Technology of advanced oxidation processes (AOPs) need to be revamped and reenvisioned at each step of scientific life and scientific forbearance. Industrial wastewater treatment and drinking water treatment today stands in the midst of scientific rejuvenation and deep scientific introspection. Ozonation technique today stands in the midst of scientific vision and versatile scientific barriers. The technological barriers are immense, and the scientific mission and vision are immense with the passage of scientific history and time. The world of scientific challenges in industrial wastewater treatment needs to be readdressed and reenvisioned with the scientific profundity and deep scientific question on efficiency of the process.

6.14   GROUNDWATER REMEDIATION AND THE VISIONARY TECHNOLOGIES FOR FUTURE

Groundwater contamination in today's human civilization stands in the midst of immense crisis and scientific introspection. Technology and engineering science are challenged as heavy metal groundwater contamination remains as a scientific enigma to the future scientific generation. Science especially environmental engineering and environmental science has no concrete answers to this monstrous catastrophe.[13] In this treatise, the author rigorously points out the scientific vision behind groundwater remediation. The immense success of environmental engineering techniques, the wide forays into chemical process engineering, and the futuristic vision of groundwater remediation techniques will all lead a long and visionary way in the true emancipation of environmental engineering and the true realization of environmental sustainability. South Asia and many parts of the developed and developing world are faced with this ever-growing crisis. Technology and science needs to be readdressed and reenvisioned with the passage of this scientific decade. Mankind's immense scientific prowess and the futuristic vision of water technology will all lead a long and visionary way in the true realization of global environmental sustainability.[13]

6.14.1   GROUNDWATER QUALITY MANAGEMENT IN SOUTH ASIA AND THE VISION FOR THE FUTURE

Science and technology of water science are in a state deep distress.[13] Groundwater quality management in South Asia particularly India and Bangladesh is challenged and needs to be revamped with great environmental engineering concern. Water science and water technology today stands in the midst of deep scientific introspection, scientific forbearance, and scientific vision. South Asia particularly the state of West Bengal in India is today in the grip of an unending and unsolved environmental crisis. Arsenic and heavy metal contamination in drinking water has taken monstrous proportions in the state of West Bengal today. This is a veritable bane and a curse to the technological and scientific advancement of the nation. Scientific research pursuit has few answers to this marauding water crisis. In such a crucial and vexing situation, effective groundwater quality management assumes immense importance in countries in South Asia affected by heavy metal groundwater contamination. Until the 1940s, the irrigation and drinking water needs and its importance were met by rivers, ponds, lakes, dugwells, and rainwater resources. However, in the middle of the 20th century, India was faced with tremendous circumstances which are providing of food to its ever-increasing population and the effective prevention of water-borne diseases. This immense problem resulted in immense challenges for the country's water policy. The Government of India's next great challenge was to address the issue of prevention of diseases such as cholera, dysentery, and diarrhea, which resulted in many deaths between 1940 and 1950.[13,14]

6.14.1.1 GROUNDWATER: THE ONLY HOPE AND THE IMMENSE NEED

To overcome these problems greatly impending economic growth and overall development, the national government and the national policy of Indian nation, in association with international agencies, began transitioning to the proper utilization of the country's immense groundwater resources. This resulted in a rapid and a gigantic transformation.[13,14] Tubewells were dug in unimaginable proportion in the state of West Bengal, India, and the neighboring country Bangladesh. Inadequate sanitation, however, contributed to untold and immense misery after India's independence. In other social sphere, industrial contaminants and industrial effluents brought serious concern for the environmental engineer and the science of environmental protection as a whole. In 1969, the National Rural Drinking Water Supply Programme was launched in India with the collaboration with UNICEF. National Water Policy was immense reenvisioned and reemphasized in 1987 and 2002. Technology overcame the deep failures in water policy in India and many countries in South Asia due to arsenic and other heavy metal groundwater poisoning. As increased groundwater use enhanced the agricultural and health benefits, quality concerns destroyed the very concern of groundwater quality management in the country due to immense microbial contamination. Along with this concern, the greater concern arose in 1976 with the surfacing of the scientific realization of arsenic groundwater contamination. It was taken seriously from the 1990s.[13,14]

6.15 REMEDIATION TECHNOLOGIES FOR HEAVY-METAL-CONTAMINATED GROUNDWATER AND THE WIDE VISION OF GROUNDWATER REMEDIATION

Remediation technologies for heavy-metal-contaminated groundwater are wide and visionary. The contamination of groundwater by heavy metal, originating either from natural soil sources or from anthropogenic sources, is a matter of immense concern to the public health and human civilization. Remediation of contaminated groundwater and the provision of clean drinking water stands as a high priority since billions of people throughout the world depend upon this basic need. Extremely complex soil chemistry and aquifer characteristics has propelled the scientific vision and the intense scientific endeavor.[13,14] Keeping the sustainability issues in mind and the environmental ethics, the technologies encompassing natural chemistry, bioremediation, and biosorption are highly recommended to be adopted in appropriate cases. Science of environmental protection needs to be reenvisioned at each forays in human life. In the environment, the heavy metals are generally more persistent and available than organic compounds such as pesticides or petroleum products. Technology of water science is vast and varied. These heavy metals can become mobile in soils depending on soil pH and their speciation. So, a fraction of the total mass can leach to aquifer or can become bioavailable to living organism.[13,14] Heavy metal poisoning can result from drinking-water contamination (e.g., Pb pipes, industrial, and consumer wastes), intake via the food chain, or high ambient air concentrations near emission sources. Over the last few years, many remediation technologies were applied throughout the world to deal with contaminated soil and aquifers. Many treatises on this domain are available along with sustainability issues and topics on environmental ethics on remediation. Heavy metals occur in the earth's crust and may get solubilized in groundwater through natural processes or by change in soil pH. Moreover, groundwater can get severely contaminated with heavy metals from landfill leachate, sewage, leachate from mine tailings, deep well disposal of liquid wastes, seepage from industrial waste lagoons, or from industrial spills and leakages. A wide variety of reactions in soil environment, for example, acid/base, precipitation/dissolution, oxidation/reduction, sorption, or ion exchange processes can influence the speciation and mobility of metal contaminants. The technologies for treatment of heavy-metal-contaminated groundwater includes chemical treatment technologies, in situ treatment by using reductants, reduction by dithionite, reduction by gaseous hydrogen sulfide, reduction by using iron-based technologies, soil washing, in situ soil flushing, in situ chelate flushing, in situ chemical fixation, biological, biochemical, and biosorptive treatment technologies.[13,14]

6.16 APPLICATIONS OF NANOTECHNOLOGY IN WASTEWATER TREATMENT: A WIDE VISION FOR THE FUTURE

Nanotechnology is the next-generation visionary technology. Research forays into the domain of nanoscience and nanotechnology have surpassed wide and vast visionary frontiers. In this treatise, the author pointedly focuses on the research thrusts in the area of nanotechnology and industrial wastewater treatment. Technological vision, vast and versatile scientific objectives, and the futuristic vision of nanotechnology will lead a long and visionary way in the true realization of environmental sustainability and wastewater treatment. Nanotechnology is ushering in a new wave of futuristic thoughts and revolutionary emancipation. Nanoscience and nanotechnology are today avenues of immense scientific forbearance and scientific understanding.

6.17   MEMBRANE SCIENCE AND SCIENTIFIC VISION

Membrane science and its wide scientific vision need to be revamped with each future step of environmental engineering. Technology is so advanced today. The backbone of scientific and technological validation today rests upon scientific vision and deep scientific cognizance. Water science and technology is of utmost importance today with the growing concerns for environmental protection and ecological balance. Membrane science is of immense importance and novel separation processes are gaining immense heights with the passage of scientific history and technological validation. Global water crisis and groundwater contamination are in a state of immense crisis and deep scientific introspection. Technological and scientific validation needs to be reenvisioned and readdressed with the growing concerns for environmental protection. Mankind's wide scientific prowess, technological advancements, and the wide applications of environmental engineering tools are all leading a long and visionary way in the true emancipation of environmental science today. Novel separation processes and its vast and versatile scientific vision are ushering in a new era in the field of chemical process engineering. The immense success, the vast enigma of science, and the deep futuristic vision will all lead an effective way in truly addressing the success of application of environmental sustainability in our present day human civilization. Technology of groundwater remediation is another wide visionary area which needs to be attacked with utmost importance.[16,17]

The development of the Loeb–Sourirajan synthetic membrane in 1960 provided a valuable separation tool to the chemical engineering separation processes.[15] It faced tremendous resistance in its earliest days. The situation is totally different today: membranes are more robust, modules and equipment are better designed and science has a better explanation to fouling phenomenon and how to reduce its effects. Most importantly, costs have drastically come down, partly because of maturing of technology and partly because of intense competition from an increasing number of membrane suppliers and original equipment manufacturers.[15] Developments in NF, gas separations, pervaporation, and bipolar membrane ED have widened the application scenario.[15]

6.18 CLASSIFICATIONS OF MEMBRANE SCIENCE

Membrane science is today in the path of scientific rejuvenation and newer scientific regeneration. Global water crisis and global water challenges are reshaping the scientific mind-set of the present scientific generation. Vision needs to be readdressed at the utmost as science and technology reshapes itself with the passage of scientific history and scientific forbearance. Water technology and its effectivity is the vital and veritable need of the hour. Membrane science falls under novel separation processes. This is a conventional area of scientific endeavor. Nonconventional environmental engineering tools such as advanced oxidation processes are changing the scientific panorama. Human civilization is today moving toward a newer era of scientific fortitude and scientific cognizance (Table 6.1).

TABLE 6.1 Characteristics of Membrane Processes.

Process	Driving force	Retentate	Permeate
Osmosis[1]	Chemical potential	Solutes/water	Water
Dialysis	Concentration difference	Large molecules/water	Small molecules/water
Microfiltration	Pressure	Suspended particles/water	Dissolved solutes/water
Ultrafiltration	Pressure	Large molecules/water	Small molecules/water
Nanofiltration	Pressure	Small molecules/divalent salts/dissociated acids/water	Monovalent ions/undissociated acids/water
Reverse osmosis	Pressure	All solutes/water	Water
Electrodialysis	Voltage/current	Nonionic solutes/water	Ionized solutes/water
Pervaporation[1]	Pressure	Nonvolatile molecules/water	Volatile small molecules/water

6.19 NOVEL SEPARATION PROCESSES AND THE VISION FOR THE FUTURE

Novel separation processes are changing the future of environmental engineering. Scientific history, scientific vision, and technological motivation will all lead a long and visionary way in the true realization of novel separation processes and future sustainable development whether it is energy or environment. The success of scientific research pursuit lies in the hands of deep scientific validation and introspection. The scientific prowess, the futuristic vision of novel separation processes, and the success of scientific validation are the forerunners toward a greater visionary era in the field of global water research and development initiatives. Novel separation processes and advanced oxidation processes are already gearing forward toward a newer eon in the field of environmental engineering applications and its wide scientific profundity and vision. Mankind's immense scientific prowess and scientific justification and the progress of academic rigor in the field of membrane science are aptly opening up new avenues of scientific innovation and instinct in decades to come. Water science and technology is veritably linked with membrane science today by an umbilical cord. The author in this treatise widely observes and pointedly focuses on the immense scientific potential and the scientific regeneration in the field of global water research and development initiatives. Technology is today gaining immense heights reaching beyond scientific imagination. Another facet of scientific research pursuit is that membrane science and water purification are today linked by an unsevered umbilical cord. Mankind's immense scientific prowess, the wide scientific rigor in water science and technology, and the vast and versatile scientific vision are the global forerunners of environmental protection. Novel separation processes are the next-generation technology and the futuristic wonders of environmental engineering science.[16,17]

6.20 FUTURE TRENDS IN RESEARCH IN ENVIRONMENTAL CHEMISTRY

Research endeavor and research forays in environmental chemistry are crossing vast and versatile visionary frontiers. Technology is challenged today and the future of environmental protection stands in a state of deep distress. Scientific progeny, scientific profundity, and the success of science and engineering are gearing scientific research pursuit toward a newer visionary eon. Water science and technology is another area of scientific genre which needs to be explored at this crucial juncture of scientific regeneration. Environmental chemistry is that branch of scientific endeavor which is surpassing visionary frontiers and needs investigation with every step of scientific research pursuit. Science is expanding leaps and bounds this century. The success of green chemistry and green technology are changing the face of human scientific endeavor and ushering in a new age of science.

6.21 FUTURISTIC VISION AND THE WORLD OF CHALLENGES IN INDUSTRIAL WASTEWATER TREATMENT

Environmental chemistry and green chemistry are the vastly challenging areas of science and technology today. Environmental restrictions, the concerns of ecological biodiversity, and the world of challenges in status of global ecology are veritably changing the face of scientific endeavor in environmental protection and industrial pollution control. Green chemistry and green engineering are challenging areas of science today which needs to be reenvisioned and reenvisaged with the passage of scientific history and scientific rejuvenation. The futuristic vision and futuristic challenges should be toward more innovations and more diversified knowledge in the field of industrial pollution control, industrial wastewater treatment, and water purification. Technology is so much retrogressive and baneful today. This area of technological validation and scientific motivation needs to be explored with the cause of furtherance of science and engineering. Today, science is a huge colossus with a definite vision and strong willpower of its own. In such a crucial juncture, industrial wastewater treatment and environmental protection assumes veritable importance. Global water crisis, groundwater contamination, and heavy metal groundwater attenuation are leading a long and visionary way in the true emancipation of environmental engineering science and sustainable development today. The world of challenges in industrial wastewater treatment is vast, versatile, and far reaching. The answers to environmental sustainability are widely observed with the true challenging scientific genre in this treatise. Technological challenges in environmental and energy sustainability are boundless and need intense investigation with the utmost need of furtherance of science and engineering. Futuristic vision and the deep challenges in industrial wastewater treatment and environmental protection should be toward more scientific innovations and the holistic sustainable development.

6.22 FUTURE FLOW OF THOUGHTS AND VISION OF ENVIRONMENTAL SCIENCE

Environmental science and engineering are witnessing immense challenges today. Environmental protection, industrial pollution control, and drinking water treatment are the utmost need of the hour. Technological pinnacles, scientific forbearance, and deep scientific sagacity are the widely observed facets of science and technology today. Environmental regulations, the rigors of environmental engineering science, and the wide scientific and academic rigor in the field of science and technology are all today leading a long way in the path toward environmental sustainability. Environmental engineering science is moving toward a newer eon with the passage of scientific history and scientific rejuvenation. Environmental science and technology needs to be reenvisioned and reenvisaged as human civilization plunges toward a newer scientific genre and a newer scientific reflection.

6.23   CONCLUSION

Environmental engineering science and sustainable development are the two avenues of scientific endeavor which need to be reenvisioned and reenvisaged with the passage of scientific history, scientific profundity, and scientific cognizance. Environmental regulations, stringent environmental restrictions, and the global concern for environmental protection are gearing toward a long and visionary way in the true challenge and true emancipation of environmental engineering science and environmental sustainability today. The global status of environment stands in the midst of crisis and deep distress. Arsenic groundwater contamination and heavy metal attenuation in groundwater are challenging the scientific frontiers. In such a vexing situation, science and engineering rigors need to be widely observed and deeply contemplated. Global water crisis is witnessing new future trends and new future directions of research endeavor. Heavy metal groundwater contamination has a negative effect on the progress of human civilization. It is an extremely difficult situation in South Asia and many developing and developed nations of the world. Mankind's scientific candor, the challenge, and vision of water resources engineering and the success of environmental engineering separation processes are the forerunners of a greater visionary era in the domain of environmental engineering science. Technology is beleaguered and belittled as civilization faces immense global water challenges, and there are significant breaches as regards sustainable development. The science of environmental chemistry and green chemistry are in a state of immense scientific introspection and vast challenges. Sustainable development goals are in a state of immense scientific enigma. The global water concerns are veritably changing the scientific landscape of application of environmental engineering tools. The technologies need to be readdressed and reenvisioned as human civilization plunges into the deep and murky depths of scientific and technological validation. The future of environmental engineering science is wide, bright, and far reaching as technology and science overcomes vast and versatile visionary boundaries.

ACKNOWLEDGMENTS

The author acknowledges the contribution of Chancellor, Vice-Chancellor, Faculty, and students of University of Petroleum and Energy Studies, Dehradun, India without whom this writing project would not have been complete. The contribution of Shri Subimal Palit, the author's late father, and an eminent textile engineer from India is widely acknowledged.

KEYWORDS

- **environment**
- **vision**
- **sustainability**
- **wastewater**
- **drinking water**

REFERENCES

1. Bigas, H. *The Global Water Crisis: Addressing an Urgent Security Issue*; Papers for the InterAction Council, 2011–2012.
2. Water Security and the Global Water Agenda. *A United Nations—Water Analytical Brief*; United Nations University, 2013.
3. Rogers, P. *Is There a Global Water Crisis?* Tufts University Graduate Program in Water Issues, December 3, 2004.
4. Shannon, M. A.; Bohn, P. W.; Elimelech, M.; Georgiadis, J. A.; Marinas, B. J. *Science and Technology for Water Purification in the Coming Decades*; Nature Publishing Group, 2008; pp 301–310.
5. Water and Jobs. *The United Nations World Water Development Report*; Published by UNESCO: Paris, France, 2016.
6. Rodriguez, D. J.; van der Berg, C.; McMahon, A. S. *Investing in Water Infrastructure: Capital, Operations and Maintenance*; Water Partnership Program, The World Bank, 2012.
7. Van der Bruggen. B.; Vandecasteele. C. Distillation vs. Membrane Filtration: Overview of Process Evolutions in Seawater Distillation. *Desalination* **2002**, *143*, 207–218.

8. Van der Bruggen, B.; Manttari, M.; Nystrom, M. Drawbacks of Applying Nanofiltration and How to Avoid Them. *Sep. Purif. Technol.* **2008**, *63*, 251–263.

9. Van Geluwe, S.; Braeken, L.; Van der Bruggen, B. Ozone Oxidation for the Alleviation of Membrane Fouling by Natural Organic Matter: A Review. *Water Res.* **2011**, *45*, 3551–3570.

10. Van der Bruggen, B.; Cornelis, G.; Vandecasteele. C.; Devreese, I. Fouling of Nanofiltration and Ultrafiltration Membranes Applied for Wastewater Regeneration in the Textile Industry. *Desalination* **2005**, *175*, 111–119.

11. Boussu, K.; Belpaire, A.; Volodin, A.; Haesendonck, C.; Van der Meeren, P.; Vandecasteele, C.; Vander Bruggen, B. Influence of Membrane and Colloid Characteristics on Fouling of Nanofiltration Membranes. *J. Membr. Sci.* **2007**, *289*, 220–230.

12. Van der Bruggen, B.; Segers, D.; Vandecasteele, C.; Braeken, L.; Volodin, A.; Haesendonck, C. How a Microfiltration Pretreatment Affects the Performance In Nanofiltration. *Sep. Sci. Technol.* **2004**, *39*(7), 1443–1459.

13. Hashim, M. A.; Mukhopadhyay, S.; Sahu, J. N.; Sengupta, B. Remediation Technologies for Heavy Metal Contaminated Groundwater. *J. Environ. Manage.* **2011**, *92*, 2355–2388.

14. Chakraborti, D.; Das, B.; Murril, M. T. Examining India's Groundwater Quality Management. *Environ. Sci. Technol.* **2011**, *45*, 27–33.

15. Cheryan, M. *Ultrafiltration and Microfiltration Handbook*; Technomic Publishing Company, Inc: USA, 1998.

16. Palit, S. Filtration: Frontiers of the Engineering and Science of Nanofiltration—A Far-reaching Review. In *CRC Concise Encyclopedia of Nanotechnology*; Ubaldo Ortiz-Mendez, Kharissova, O. V., Kharisov. B. I., Eds.; Taylor and Francis: USA, 2016, pp 205–214.

17. Palit, S. Advanced Oxidation Processes, Nanofiltration, and Application of Bubble Column Reactor. In *Nanomaterials for Environmental Protection*; Kharisov, B. I.; Kharissova, O. V.; Rasika Dias, H.V., Eds.; Wiley: USA, 2015, pp 207–215.

18. www.google.com

19. www.wikipedia.com

CHAPTER 7

HEAVY METAL AND ARSENIC REMEDIATION TECHNOLOGIES FOR CONTAMINATED GROUNDWATER: A SCIENTIFIC PERSPECTIVE AND A VISION FOR THE FUTURE

SUKANCHAN PALIT*

Department of Chemical Engineering, University of Petroleum and Energy Studies, Energy Acres, Post-Office Bidholi via Premnagar, Dehradun 248007, India

Corresponding author. E-mail: sukanchan68@gmail.com; sukanchan92@gmail.com

CONTENTS

ABSTRACT

Environmental engineering science and the domain of environmental protection today are undergoing drastic changes. Environmental regulations and stringent restrictions are veritably urging the scientific domain to gear forwards toward newer challenges and newer innovations. Human civilization is challenged and the vision of science is targeted toward newer scientific rejuvenation and scientific regeneration. Global water crisis today stands in the midst of deep scientific introspection and immense challenges. This treatise investigates the wide domain of heavy metal and arsenic remediation technologies with the sole aim of furtherance of science and technology. Scientific challenges are ushering in a new eon in the field of global water research and development initiatives. Human mankind's immense scientific prowess, the wide futuristic vision, and the challenges of engineering science are opening up new windows of scientific innovation in years to come. True realization and true emancipation of environmental engineering tools are the utmost need of the hour. The author deeply portrays the sources, chemical property, and speciation of heavy metals in groundwater. The treatment technologies delineated are chemical treatment technologies, in situ treatment by using reductants, reduction by dithionite, reduction by using iron-based technologies, removal of chromium by ferrous salts, soil washing, in situ soil flushing, in situ chelate flushing, in situ chemical fixation, and the vast domain of biological, biochemical, and biosorptive treatment technologies (Hashim et al. *J. Environ. Manage.* 2011, *92*, 2355). The vast and versatile academic and scientific rigor in heavy metal remediation technologies are being challenged and the scientific domain is taking vast strides in true realization of zero-discharge norms and successful environmental sustainability. This treatise unravels the murky depths of environmental engineering science with a wide vision toward true emancipation of science and technology.

7.1 INTRODUCTION

Global water crisis and water shortage issues are challenging the wide visionary scientific landscape of human civilization today. Water science and water technologies are the forerunners toward a greater emancipation of environmental engineering science today. Arsenic and heavy metal groundwater contamination today stands in the midst of immense scientific introspection and scientific profundity. Technology and engineering science of environmental protection will go a long and visionary way in the true realization of environmental and energy sustainability today. In this well-observed and well-researched treatise, the author rigorously points out the immense success, the wide potential, and the futuristic vision of heavy metal remediation technologies with the immediate need of realization of environmental sustainability. Engineering science is veritably challenged today. Environmental and energy sustainability are the visionary coinwords of today. Today sustainable development also indicates provision of clean drinking water and emancipation and realization of zero-discharge norms and successful industrial wastewater treatment. In this paper, the author pointedly focuses on the success of heavy metal remediation technologies in provision of clean drinking water to the common people. Technology and science of heavy metal remediation technologies are not new yet immature. The scientific and academic rigors are immense and groundbreaking. Water issues are the backbones of a progressive nation. South Asia particularly India and Bangladesh are in the deadly jaws of arsenic drinking water crisis. This treatise aims toward the veritable success of heavy metal and arsenic remediation technologies with the sole concern of provision of clean drinking water and also the furtherance of scientific rigor in remediation science.[18,19]

7.2   THE AIM AND OBJECTIVE OF THE STUDY

Human civilization today stands in the midst of immense scientific introspection and wide scientific vision. Global water issues are paralyzing a nation's vibrant economy. South Asia particularly India and Bangladesh are in the midst of a serious environmental disaster and a deep crisis with the evergrowing concern of arsenic and heavy metal drinking water poisoning. Science and technology has few answers to the widely evolving crisis. Technology of heavy metal remediation is the backbone of this study. The author pointedly focuses on the immense potential, the wide success, and the futuristic vision of heavy metal groundwater remediation. Science is a huge colossus with a definite vision of its own. The challenge and the vision of environmental engineering science today are crossing wide and vast scientific frontiers. The prime aim and objective of this study is to delineate lucidly the immense global crisis of arsenic and heavy metal contamination in South Asia and many developing and developed countries throughout the world. Technological vision, scientific profundity, and the deep scientific objectives will go a long and visionary way in the true emancipation and the true realization of environmental sustainability in today's world. This treatise also truly and widely observes the success of remediation technologies whether it is chemical or biological, with a sole aim toward furtherance of science, engineering and particularly environmental engineering. The aim and objective of the entire scientific domain as well as our entire scientific research pursuit should be toward greater realization of techniques to target environmental protection and holistic sustainable development. In such a crucial juncture of global environmental crisis, technological profundity and vision are of immense importance in the futuristic success of human civilization. A scientist's immense instinctive prowess and human mankind's scientific struggle will lead a long way in the true realization of energy and environmental sustainability.[18,19]

7.3 HUMAN MANKIND, SUSTAINABILITY, AND THE VISION FOR THE FUTURE

Global sustainability issues are transforming the scientific landscape and the wide scientific horizon. Global water research and development initiatives are veritably part and parcel of sustainable development and the wide scientific vision of environmental sustainability. The challenge of human civilization is today targeted toward successful energy and environmental sustainability. Human mankind's immense scientific prowess, the vast technological vision, and the futuristic vision of environmental engineering science are veritably changing the path of human scientific endeavor. Arsenic and heavy metal groundwater contamination in South Asia and many developing and developed nations are today of immense concern to the global scientific domain. Environmental disasters such as Tsunami are reshaping the cause of environmental sustainability. The vision for the future for drinking water treatment and industrial wastewater treatment is widely challenged. The vision of the foremost proponent of sustainable development, Dr Gro Harlem Brundtland, the former Prime Minister of Norway needs to readdress at this remarkable and vexing juncture of human history. Technology of water issues are deeply challenged as human life passes through one of the most difficult terrains of scientific history and time. On the other hand, energy sustainability is the only answer to the widespread concern of depletion of fossil fuel resources such as petroleum and coal. Civil society and governments throughout the world are reshaping the global economy with the sole aim toward furtherance of scientific applications, energy and environmental sustainability. The immense scientific and academic rigor behind environmental sustainability needs to be rigorously addressed as human life moves ahead in the path of life and struggle. Technology is immensely baffled with the rising concern for depletion of fossil fuel resources and the loss of ecological bio-diversity. The progress of human civilization today has a definite umbilical cord with sustainable development. In this treatise, the author rigorously points forward the global concerns for heavy metal groundwater contamination and the marauding and vexing issue of arsenic drinking water poisoning in South Asia in particular. Environmental engineering tools are the only definite answers to the successful realization of environmental sustainability.[18,19]

7.4   SCIENTIFIC VISION AND THE GLOBAL WATER CRISIS

Scientific vision and scientific cognizance are the visionary coinwords of today's scientific research pursuit in global water research and development initiatives. The cause of provision of clean drinking water to the common mass assumes immense importance as human civilization crosses one visionary frontier over another. Water technology should be targeted toward heavy metal groundwater remediation today. Water is a vital component to the progress of human mankind. In this paper, the author pointedly focuses on the immense potential and the success of heavy metal remediation technologies with the sole aim toward the furtherance of science and engineering. In South Asia, particularly Bangladesh and India, environmental disasters are causing immense havoc to the human society. Technology is extremely baffled as science and engineering moves forward toward a newer scientific regeneration and scientific rejuvenation. The author in this treatise rigorously points out the global water challenges and global scientific forays into the visionary arena of groundwater remediation.[18,19]

7.5   HEAVY METAL GROUNDWATER CONTAMINATION AND THE FUTURE OF ENVIRONMENTAL ENGINEERING

Future of environmental engineering is at a state of immense disaster and scientific comprehension. Mankind's wide scientific cognizance, the vast technological vision of environmental engineering, and the grave environmental concerns will go a long way in the true realization and true emancipation of environmental engineering science today. Science, technology, and human scientific vision are today in a state of immense distress as well as scientific introspection. Technology of heavy metal groundwater remediation needs to be reenvisioned and readdressed with the passage of human scientific history and time. Environmental engineering is today linked by an unsevered umbilical cord with membrane science. Membrane science and its separation phenomenon are veritably changing the scientific landscape and wide scientific frontiers. Today, the success of human civilization is dependent upon the provision of basic human needs such as water and power. In such a crucial and thoughtful situation, energy and environmental sustainability assumes vital importance. Today, human civilization needs to address holistic sustainable development at the earliest. Environmental disasters, environmental sustainability, and loss of environmental biodiversity are the grave concerns and the utmost need of the hour. Environmental engineering tools such as desalination, groundwater remediation, novel separation processes, and advanced oxidation processes (AOPs) will all lead a long and visionary way in the true emancipation of science and engineering today.

7.6 ARSENIC GROUNDWATER CONTAMINATION AND SUBSEQUENT REMEDIATION IN SOUTH ASIA

Arsenic as well as heavy metal groundwater contamination is in its worst state in South Asian countries particularly India and Bangladesh. Medical emergency and global drinking water concerns needs to be addressed at the utmost. Until the 1940s, the irrigation and drinking water needs of India were met by rivers, ponds, lakes, dugwells, and rainwater sources. The global concern for water is immense and far-reaching. Technology has few answers to this worldwide crisis. Scientific imagination and scientific forbearance has no answers to this monstrous crisis. In the middle of the 20th century, India continually faced two exceedingly daunting challenges: providing food for an ever-increasing population and decreasing the burden of highly prevalent water-borne diseases. Technology of water science is highly advanced today. Yet the marauding scenario of arsenic groundwater contamination in South Asia has no answers to the technological advancements in the global water research and development initiatives. Chemical process engineering, environmental engineering science, and novel separation processes are all leading a long way in the true realization and visionary emancipation of environmental protection today. The world of challenges and the scientific vision in water research and development initiatives need to be readdressed and reenvisioned with the passage of this century, scientific history, and time. Success of human scientific endeavor, the challenge and objective of technology, and the wide futuristic path of science are all the torchbearers toward a greater realization of environmental technology. South Asia particularly West Bengal state of India and Bangladesh are in the throes of unimaginable environmental disaster of massive proportion. Research endeavor needs to be scientifically streamlined with the passage of this visionary century.

7.7 ARSENIC CONTAMINATED GROUNDWATER REMEDIATION TECHNOLOGIES

Heavy metal and arsenic remediation technologies are the visionary scientific endeavor of today. The hurdles of scientific vision and scientific forbearance need to be addressed and reenvisaged with the progress of science and technology. Remediation technologies today are innovative and global water challenges are ushering in a new era. Scientific vision and scientific profundity needs to be reenvisioned and revisited with each step of scientific history and time. Heavy metal and arsenic remediation technologies are the utmost need of the hour.

Water Stewardship Information Series[20] lucidly described Arsenic in groundwater in North America. In water, arsenic has no smell or taste and can be detected through a chemical test. The ambient concentration of arsenic in surface and ground waters in Canada is very low, usually ranging from 0.001 to 0.002 mg/L.[20] Concentrations in groundwater are often higher than those measured in surface waters. The Canadian drinking water guideline for arsenic sets a maximum acceptable concentration of 0.001 mg/L.[20] Localized high concentrations of arsenic have been found in well water from several regions in British Columbia, almost associated with arsenic-containing bedrock formations. This example veritably shows that arsenic is not limited to South Asia but an evergrowing inferno in the Western world and developed countries. The challenge of arsenic groundwater remediation in South Asia and the developed world is immense as science plunges into the deep abyss of scientific vision and scientific forbearance.[20]

Mukherjee et al.[1] discussed with deep comprehension and cogent insight arsenic groundwater contamination with a global perspective with emphasis on the Asian scenario. Technology of arsenic groundwater remediation is highly advanced today with the evergrowing environmental engineering science concern.[1] This paper presents an overview of the current scenario of arsenic concentration in countries across the world with a vital emphasis on Asia. South Asia is under the veritable grip of world's largest environmental disaster.[1] Along with the present situation in severely affected countries in Asia, such as Bangladesh, India, and China, recent instances from Pakistan, Myanmar, Afghanistan, Cambodia, and so on are presented in deep details.

There have been few review works covering the arsenic-contamination scenario around the world.[1] The vision of science is today challenged as scientific innovations in heavy metal and arsenic remediation reached a new vision and a new pinnacle. In this treatise, the authors rigorously points out to the immense scientific ingenuity in both the awareness and the innovations in heavy metal groundwater remediation. With the discovery of newer sites in recent past, the arsenic contamination scenario around the world, especially in Asian countries: Bangladesh, West Bengal, India, and sites in China. The situation in West Bengal is veritably disastrous. Between 2000 and 2005, arsenic-related groundwater problems have emerged in different Asian countries including new sites in China, Mongolia, Nepal, Cambodia, Myanmar, Afghanistan, DPR Korea, and Pakistan. There are widespread reports of arsenic contamination from Kurdistan province of Western Iran and Vietnam where millions of people have a serious risk of chronic arsenic poisoning. The scenario is grave and is evergrowing in alarming proportions.[1]

Shallow groundwater with high arsenic concentrations from naturally occurring sources is the primary source of drinking water for millions of people in Bangladesh. It has definitely resulted in a major public health disaster with as many as 70 million people possibly at risk.[1] The International Atomic Energy Agency is supporting international efforts and the Government of Bangladesh to find alternative, safe, and widely sustainable sources of drinking water. Technology and science of arsenic remediation is slowly opening up new scientific ventures and opening up new windows of innovation and scientific vision in years to come.[1]

The arsenic poisoning in groundwater is world's largest public health disaster. Efforts to reduce the number of people, especially children, dying from diarrhea and other water borne diseases from contaminated surface water has resulted in a huge increase in the number of tube wells sunk in the last decade. There are now about 0.9 million public and 1.6 million private wells and more than 90% of the population use groundwater for drinking. Until the discovery of arsenic in groundwater in 1993, well water was safe. But now the disaster disastrously multiplied:[1]

- Arsenic contamination of groundwater has affected 59 of the 64 districts in Bangladesh where arsenic levels have been found to be above nationally accepted limit.[1]
- It is widely reported that above 21 million people are currently exposed to arsenic contamination and approximately 70 million people may be at a disastrous risk.[1]

It may take many years for the effects of drinking arsenic-contaminated water to show and the true extent of the problem is till now not revealed. The science of arsenic poisoning is still not fully discovered. Technological prowess and motivation are in a state of immense disaster as arsenic poisoning devastates the scientific fabric and scientific profundity. Nevertheless, evidence of chronic arsenic toxicity is accumulating and includes melanosis (abnormal black-brown pigmentation of skin), hyperkeratosis (thickening) of palm and sole, gangrene, and skin cancer. Progress of human civilization is at stake as the developing countries like Bangladesh are in a state of immense environmental catastrophe. Malnutrition and hepatitis B, both of which are prevalent in Bangladesh, accentuate the problem of arsenic poisoning. Scientific vision, scientific forbearance, and deep scientific cognizance are in a state of immense distress as human civilization is faced with an immense unsolved environmental disaster. Of over-riding importance of public health is the need to find adequate and sustainable sources of arsenic-free drinking water.[1]

The source of the arsenic in groundwater in Bangladesh is natural and arises because the water flows through arsenic-rich sediments. The exact mechanism by which arsenic is transferred into the water is not yet fully understood. The engineering science of groundwater remediation is still at its latent stage. It may be deeply thought that the increase of groundwater use in recent years may have increased the level of arsenic poisoning. The science of arsenic poisoning in Bangladesh is still today unraveled. It is a veritable imperative to find alternative, safe sources of water as quickly as possible.[1]

Safiuddin et al.[2] discussed with deep and cogent insight groundwater arsenic contamination in Bangladesh, its causes, its effects, and remediation. The absolutely serious arsenic contamination of groundwater in Bangladesh has come out recently as the biggest natural catastrophe in our planet. The people in 59 out of 64 districts comprising 126,134 km^2 of Bangladesh are suffering due to the arsenic contamination in drinking water.[2] Seventy five million people are at dire risk and 24 million are potentially exposed to arsenic poisoning. The situation is extremely grave in Bangladesh. Science and technology has no answers to this marauding calamity. Most of the recognized stages of arsenic poisoning are identified in Bangladesh, and the risk of arsenic poisoning is increasing day by day. The extreme severity of arsenic poisoning is severely demanding extensive research and development effort in this field. The present study is a critical overview of groundwater arsenic contamination in Bangladesh. Arsenic contamination of groundwater in Bangladesh was first detected in 1993. Further investigations were carried out in the following years. The institutions that contributed in these investigations are School of Environmental Studies, Jadavpur University in Kolkata, Bangladesh Atomic Energy Commission, Dhaka Community Hospital, Department of Public Health Engineering (DPHE), and National Institute of Preventive and Social Medicine. DPHE collected and analyzed 31,651 well water samples with the assistance of WHO, UNICEF, and DFID.[2] The laboratory reports confirmed that the groundwater in Bangladesh is severely contaminated by arsenic. Scientific vision and wide scientific imagination are today belittled due to this monstrous arsenic groundwater poisoning in South Asia.[2] The challenge is immense and overarching. The millions of shallow and deep wells that had been sunk in various parts of the country are dispensing their own brand of special poison that is arsenic. Recent studies in Bangladesh indicate that the groundwater is severely contaminated with arsenic above the maximum permissible limit of drinking water. In 1996, altogether 400 measurements were conducted in Bangladesh. Arsenic concentrations in about half of the measurements were above the maximum permissible level of 0.05 mg/L in Bangladesh. In 1998, British Geological Survey collected 2022 water samples from 41 arsenic affected districts.[2] Laboratory tests revealed that 35% of these water samples were found to have arsenic concentrations above 0.05 mg/L. Technological profundity and deep scientific objectives are the hallmarks toward an effective groundwater heavy metal remediation in today's scientific research pursuit. Human mankind's greatest environmental disaster today stands in the midst of immense scientific introspection and effective scientific vision. Fundamental issues in heavy metal remediation need to be dealt with surgical precision as human civilization ushers in a new era of science and engineering.[2]

Safiuddin et al.[2] clearly explained to the global scientific community the unending and unimaginable disaster in West Bengal, India, and Bangladesh. The success of scientific endeavor is at a disastrous stake as science of heavy metal remediation faces immense scientific barriers. Technological vision and scientific objectives needs to be addressed and reenvisioned as Bangladesh faces one challenge over another as a developing country and a Third World country. The authors rigorously points out the intricacies of science of arsenic poisoning and its remediation in minute details. The recent statistics on arsenic contamination indicate that 59 out of 64 districts of Bangladesh have been affected by arsenic contamination. Approximately, arsenic has contaminated the ground water in 85% of the total area of Bangladesh and about 75 million people are at a dire risk. The citizens of Bangladesh are encompassed by this disastrous peril. It is estimated that the magnitude of arsenic problem in Bangladesh surpasses the aggregate problem of all the 20 countries of the world where arsenic groundwater contamination has been reported.[2] The alarm bells are veritably ringing in the scientific scenario of Bangladesh. Technology today is totally baffled as science of remediation is gearing up for new challenges in South Asia and the developed world.[2]

Science of heavy metal remediation is ushering in a new era of scientific vision and deep scientific understanding. Human mankind's immense technological prowess needs to be streamlined as global water issues faces tremendous challenges. The vision of science, the wide futuristic vision of environmental engineering science, and the immense academic rigor of remediation will all lead a long and visionary way in the true emancipation of environmental sustainability and holistic sustainable development today. Today environmental sustainability needs to be of prime importance as human civilization passes through one crisis over another.

7.8   RECENT SCIENTIFIC ENDEAVOR IN THE FIELD OF HEAVY METAL GROUNDWATER REMEDIATION TECHNOLOGIES

Scientific endeavor and scientific vision in the field of global water challenges are crossing wide and vast scientific frontiers. Today, the challenge of human civilization lies in the hands of scientists and engineers. Technology of water science is vastly advanced today and in the similar vein global water issues are witnessing drastic and dramatic challenges. History of human civilization is veritably ushering in a new eon in the field of water research and development initiatives.

Hashim et al.[3] discussed with cogent hindsight in a review remediation technologies for heavy metal contaminated groundwater. The poisoning and contamination of groundwater by heavy metal, originating either from natural soil sources or from anthropogenic sources is a matter of vital and veritable concern to the public health and hygiene. Remediation of contaminated groundwater is of highest priority since billions of people throughout the world use it for drinking water purpose. In this paper, the author rigorously points out 35 approaches for groundwater remediation and classified under three large categories, that is, chemical, biochemical/biological/biosorption, and physicochemical treatment processes. Technology of global water research and development initiatives is widely challenged today and needs to be restructured with the passage of scientific history. The world of challenges needs to be revamped and restructured as human civilization moves toward a newer age of scientific emancipation.[3] In this paper, the author pointedly focuses on the different chemical and biological remediation technologies with the sole aim toward the furtherance of science. Selection of a relevant technology for contamination remediation at a particular site is one of the most challenging jobs due to extremely complex soil chemistry and aquifer characteristics and no cardinal rule can be suggested regarding this issue. The science of groundwater remediation is replete with scientific hurdles. Yet the scientific truth and the scientific forbearance will lead a long and visionary way in the true realization of the science of groundwater remediation and environmental engineering. In the past decade, iron-based technologies, microbial remediation, biological sulfate reduction, and various adsorbents played primary and vital roles. Keeping environmental sustainability in mind, the technologies encompassing natural chemistry, bioremediation, and biosorption are recommended to be adopted in appropriate situations. The success of environmental engineering science and environmental sustainability is at stake.[3] Heavy metal is a general collective terminology, which applies to the group of metals and metalloids with atomic density greater than 4000 kg/m^3 or five times more than water and they are natural components of the earth's crust.[3] In this paper, the authors deeply ponder on the potential and success of application of environmental engineering tools in groundwater remediation.[3] Although some of these metals act as essential micronutrients for living beings, at higher concentrations they can lead to severe poisoning.[3] The most toxic forms of these metals in their ionic species are the most stable oxidation states, for example, Cd^{2+}, Pb^{2-}, Hg^{2+}, Ag^+, and As^{3+} in which, they veritably react with the body's biomolecules to form extremely stable biotoxic compounds which are extremely difficult to dissociate. In the surrounding environment, the heavy metals are generally more persistent than organic components such as pesticides or petroleum by-products. They can become mobile in soils depending on soil pH and their speciation. So a fraction of the total mass can leach to aquifer or can be widely available biologically to living organisms. Technology of remediation is opening up new avenues and new vision in years to come.[3]

According to the authors,[3] heavy metal poisoning can result from drinking water contamination (e.g., Pb pipes, industrial, and consumer wastes), intake via the food chain or high ambient air concentrations near emission sources.[3]

Over the past few decades, many remediation technologies were applied all over the world to deal with contaminated soil and aquifer. Many documents and reviews deal with the deep scientific understanding and deep cognizance of groundwater remediation. The papers dealt with organic and inorganic pollutants.[3]

Heavy metals occur in the earth's crust and may get solubilized in groundwater through natural processes or by change in soil pH. Moreover, groundwater can get contaminated with heavy metals from landfill leachate, sewage, leachate from mine tailings, deep-well disposal of liquid wastes, seepage from industrial waste lagoons, or from industrial spills and leaks. A variety of reactions in soil environment, for example, acid/base, precipitation/dissolution, oxidation/reduction, sorption, or ion exchange processes can influence the speciation and mobility of wide variety of metal contaminants.[3]

Technology and science are surpassing wide and vast visionary frontiers today. Heavy metal and arsenic groundwater remediation are the utmost need of the hour. The situation is extremely grave in Bangladesh and India. Human mankind's immense scientific and academic rigor at veritably challenged with the passage of scientific profundity and time. In this treatise, the author deeply observes the vast technological and scientific challenges behind heavy metal and arsenic groundwater remediation. Science has highly developed in this century. Hashim et al.[3] repeatedly address this burning as well as a vexing issue. The authors of this paper[3] with deep and cogent insight bring forward the wide vision of heavy metal remediation technologies. Technology needs to be widely streamlined and innovations need to be challenged as human civilization moves forward.

Zvinowanda et al.[4] discussed with lucid insight the effectivity of a novel adsorbent for heavy metal remediation in aqueous environment. The objective of this study was to investigate the possibility of using maize tassel as an alternative adsorbent for the removal of chromium(VI) and cadmium(II) ions from aqueous solutions.[4] The effect of pH, solution temperature, contact time, initial metal ion concentration, and adsorbent dose on the adsorption of chromium(VI) and cadmium(II) by tassel was investigated with batch methods.[4] Science of adsorption is surpassing wide and vast scientific frontiers.[4] Lots of scientific endeavor has been recorded in scientific literature throughout the world. This treatise is a remarkable example of the science of mass transfer and separation phenomenon. A heavy metal is a metal of relatively high specific gravity greater than 5 or high atomic weight, one that is poisonous like lead, mercury, chromium, and cadmium.[4] Chromium and cadmium have been selected in this study because of their inherent toxicity to most living organisms. Scientific candor, scientific ingenuity, and scientific truth are at its pinnacle as human mankind steps into a newer visionary eon in the field of environmental engineering science and chemical process engineering. These mentioned metals are major pollutants in South Africa's water resources due to their use in many manufacturing industries such as metal plating, mining operations and tanneries. Technology and engineering science are undergoing immense transformation in this decade as human scientific research pursuit overcomes scientific hurdles and barriers.[4] Cr(III) is essential trace element needed for glucose metabolism in human, plants and animals.[4] It is relatively innocuous and immobile when compared to Cr(VI) compounds.[4] Cr(VI) moves exceedingly rapid through soils and aquatic environments and is a strong oxidizing agent very much capable of being absorbed through the skin. The maximum concentration limit for chromium(VI) for discharge into inland surface waters is 0.1 mg/L and in potable water is 0.05 mg/L. The wide and major sources of Cd(II) and Cr(VI) compounds are ferrochrome processing and metal plating industries. Because of the high positive charge of Cr(VI), it is hydrolyzed to form oxo-anions, CrO_4^{2-}, $HCrO_4^{-}$, and $Cr_2O_3^{2-}$ depending on the pH of the media. Acute exposure to Cr(VI) causes nausea, diarrhea, liver and kidney damage, dermatitis, internal hemorrhage, and respiratory problems.[4]

Chemical process engineering and environmental engineering science are today in the path of immense scientific regeneration and vision. Groundwater remediation is a veritably immense chemical process engineering issue. Scientific ingenuity in the field of arsenic groundwater remediation is witnessing a new beginning. The author of this well-observed and well-written treatise rigorously points out the present global water issues and the global water research and development forays. Scientific validation and technological ingenuity are befitting to the progress of human civilization and immense scientific and academic rigor in the field of global groundwater remediation.

Khan et al.[5] discussed with cogent insight role of plants, mycorrhizae, and phytochelators in heavy metal contaminated land remediation. Phytoremediation is a site remediation strategy, which employs plants to remove nonvolatile and immiscible soil contaminants.[5] This extremely sustainable and cost-effective process is emerging as a viable alternative to traditional contaminated land and soil remediation methods. This paper provides a deep insight in the area of phytoaccumulation, most of which are implemented in Europe and USA.[5] The world of scientific challenges in phytoremediation is slowly witnessing one drastic change over another. Contaminated soil can be remediated by chemical, physical, and biological tools. The available techniques may be grouped into two categories: (1) ex situ techniques which require removal of the contaminated soil for treatment on- or off-site and (2) in situ methods, which remediate without excavation of the contaminated soil. This paper widely focuses on the bioremediation of heavy metal contaminated soils using in situ tools. This treatise rigorously focuses on the wide challenging domain of bioremediation of heavy metals from the main group of inorganic components. Science and engineering of environmental protection is undergoing scientific metamorphosis as biological and chemical treatments of contaminated soils ushers in a newer vision and newer eon. This research work also targets phytoremediation.[5] In recent years, phytoremediation/phytoextraction, that is, the use of plants to cleanup soils contaminated with nonvolatile hydrocarbons and immobile organics is showing immense and feasible promises as a new method for in situ cleanup of large volumes of low to moderately contaminated subsurface layer.[5]

Davis et al.[6] discussed lucidly in a well-observed review the biochemistry of heavy metal biosorption by brown algae. The passive removal of toxic heavy metals such as Cd^{2+}, Cu^{2+}, Zn^{2+}, Pb^{2+}, Cr^{3+}, and Hg^{2+} by inexpensive biomaterials, termed biosorption, requires that the substrate displays high metal uptake and selectivity, as well as suitable mechanical properties for applied remediation strategies.[6] Scientific validation and sound technological objectives are gaining immense heights as bioremediation and biological treatment of contaminated soils ushers in a visionary era in environmental engineering science. This area needs to be immensely developed and remediation science needs to be readdressed and reenvisioned. A detailed deliberation of the macromolecular conformation of the alginate biopolymer is offered in order to explain the heavy metal selectivity displayed by brown algae. Scientific truth, scientific progeny, and scientific profundity are at its level best as remediation science enters into a newer era.[6]

Rajendran et al.[7] discussed in a well-researched review microbes in heavy metal remediation. The challenge and the vision of science need to be restructured as environmental engineering science undergoes vast changes. Bioremediation is not a new area of scientific endeavor yet needs deep articulation. Heavy metal contamination due to natural and anthropogenic sources is a worldwide concern. Release of heavy metal without proper treatment poses a significant threat to public health and hygiene. It is due to deep persistence, biomagnifications, and accumulation in food chain. Nonbiodegradability and immense sludge production are the two main drawbacks of heavy metal treatment. Microbial metal bioremediation is a major step toward environmental protection and in a similar vein toward the furtherance of science. The need for bioremediation is immense and ever-growing. Greater awareness of the ecological effects of toxic metals and their biomagnifications through food chain as well as highly publicized environmental disasters such as mercury pollution in Minamata, Japan has veritably prompted a demand for decontamination of heavy metals in the aquatic system.[7] The challenge, the vision, and the objectives of bioremediation science are undergoing immense metamorphosis as human scientific research pursuit evolves into a new era of scientific regeneration. Technological challenges are enormous as well as the area of scientific validation. Also other avenue of scientific research in environmental engineering is the immense health and safety issues.[7] Metal concentration has been linked to birth defects, cancer, skin lesions, retardation leading to disabilities, liver and kidney damage, and a host of other health problems. The Centre for Disease Control and the Agency for Toxic Substances and Disease Registry estimate in a well-observed survey that 15–20% of US children have lead levels greater than 15 mg/dl in blood, which is immensely toxic. The mechanisms by which metal ions bind to the cell surface include electrostatic interactions, van der Waals forces, covalent bonding, redox interactions, and extracellular precipitation, or combination of these processes. The science of environmental protection needs to be revamped and reenvisioned as human civilization moves from one paradigm toward another.[7]

7.9 RECENT SCIENTIFIC RESEARCH PURSUIT IN THE DOMAIN OF ARSENIC GROUNDWATER REMEDIATION

Scientific research pursuit in the domain of arsenic groundwater remediation is gaining new and visionary heights as human mankind witnessing immense scientific struggles and scientific barriers. Technology needs to be reenvisioned as human civilization passes through a new visionary era. Scientific and academic rigor, deep futuristic vision, and wide scientific introspection will all lead a long and visionary way in the true realization of global water challenges today.

Arsenic contamination in groundwater in the Ganga–Brahmaputra fluvial plains in India and the Padma–Meghna fluvial plains in Bangladesh and its consequences to human health have been reported as one of the world's biggest natural groundwater calamities of human mankind. The situation is extremely grave and science has no answers to it. Today, scientific objectives and technological vision are in a state of immense disaster. Scientific determination, scientific truth, and deep scientific understanding need to be widely addressed as human civilization moves from one scientific paradigm over another.

7.10 SCIENCE, HUMAN SOCIETY, AND GLOBAL WATER CHALLENGES

In this century, technology is gearing forward toward newer challenges. The challenge and vision lies in the hands of scientists and engineers. Today global water crisis has devastated the scientific fabric and scientific generation of today. The world of immense challenges needs to be reenvisioned and restructured with the passage of human scientific research pursuit. Human scientific vision and scientific ingenuity needs to be readdressed as science and human society faces the challenge of the century—the global arsenic crisis. Doctors, scientists, and environmental engineers are widely observing this unimaginable scientific as well as environmental crisis in West Bengal State in India and Bangladesh. Vision of science, mankind's scientific ingenuity, and the evergrowing concerns for environmental protection will all lead a long and visionary way in the true realization of environmental sustainability today. Sustainable development today encompasses infrastructural development, provision of basic human needs and education. It now focuses on provision of clean drinking water also. Success of civilization's progress depends on energy and environmental sustainability today.

7.11 CONVENTIONAL ENVIRONMENTAL ENGINEERING TECHNIQUES

Science and technology of environmental engineering are moving at a rapid pace. Scientific ingenuity is at a state of immense distress as global water issues stands as an immense enigma to scientific progress. Technology diversification and scientific objectives needs to be reemphasized and reenvisaged as human civilization treads forward toward newer scientific destiny. In this section, the author deeply ponders upon the success of conventional environmental engineering tools in industrial wastewater treatment and water purification. Membrane separation science and novel separation processes are a major conventional environmental engineering technique today. Scientific challenges and scientific validation are the hallmarks of a wide and visionary future in membrane separation processes.

Cheryan[8] discussed lucidly the success of ultrafiltration and nanofiltration in environmental protection and other varied applications. The author discusses membrane chemistry, structure, and function, membrane properties, performance and engineering models, fouling and cleaning, process design, and wide variety of applications. The visionary development of the Sourirajan–Loeb synthetic membrane in 1960 provided a valuable separation tool to the process industries. The vision and the challenge immensely faced tremendous scientific barriers and hurdles. After that discovery, development occurred at a rapid pace. The situation is different today: membranes are more robust, modules and equipment are better designed, and fouling phenomenon is better understood. This reference work is of highest value and is visionary.

In this treatise and a major reference work, the author rigorously delineates the different remediation techniques of heavy metal groundwater contamination and also presents before the reader the immense success of traditional and nontraditional environmental engineering techniques,[8] thus the critical overview on membrane science. Membrane separation phenomenon and its diffusion mass transfer are today unexplored.[8] Technology and science of membrane separation phenomenon are veritably stunted. Cheryan[8] with deep scientific insight unfolds the success of separation phenomenon in ultrafiltration and nanofiltration. Technologically, the techniques are latent yet crossing visionary frontiers. Applications of membrane separation processes are vast and versatile. The challenge and vision are immense and groundbreaking. The Loeb–Sourirajan model of membrane science revolutionized the entire scientific fabric of environmental engineering science. Technological splendor and scientific profundity veritably changed the scientific domain as membrane science gained unimaginable heights. The science of novel separation processes is still undeveloped.[8]

Filtration is defined as the separation of two or more components from a fluid stream based basically on size differences.[8] In conventional usage, membrane separation processes refers to the separation of solid immiscible particles from liquid or gaseous streams. Technology of membrane science is veritably crossing wide and vast visionary frontiers.[8] Membrane separation processes are classified into: reverse osmosis, ultrafiltration, nanofiltration, microfiltration, pervaporation, and electrodialysis. Desalination science is another wide avenue of scientific endeavor.[8] Today, in many countries of the world, desalination and global water research and development initiatives are veritably linked by an unsevered umbilical cord.[8] Countries in developed and developing world are in the throes of immense water shortage and also encompassed by lack of clean drinking water. Global water crisis today stands in the midst of deep scientific introspection and scientific sagacity. In conventional terms, filtration usually refers to the separation of solid immiscible particles from liquid or gaseous streams. Membrane filtration extends this application further to include the separation of dissolved solutes in liquid streams and for effective separation of gas mixtures.[8]

The basic role of a membrane is to act as a selective barrier.[8] It should permit passage of certain components and retain certain other components of a mixture. By scientific intuition, either the permeating stream or the retained phase should be enriched in one or more components. What demarcates the more common pressure—driven membrane processes: microfiltration, ultrafiltration, nanofiltration, and reverse osmosis—is the application of hydraulic pressure to speed up the transport process.[8] However, the nature of the membrane itself controls which components permeate and which are retained. In an ideal terminology, reverse osmosis retails all components other than the solvent (e.g., water) itself, while ultrafiltration retains only macromolecules or particles larger than about 10–200 Å units (about 0.001–0.002 μm).[8] Microfiltration, on the other hand, is designed to retain particles in the "micron" range, that is, suspended particles in the range of 0.10 μm to about 5 μm (particles larger than 5–10 μm are better separated using conventional cake filtration methods).[8]

The barriers of membrane separation processes are membrane fouling. Scientific challenges and scientific truth are in the process of new rejuvenation. The immense scientific success and scientific profundity of membrane separation phenomenon are veritably opening new doors of innovation in decades to come. Technological vision and wide scientific objectives are the forerunners toward a greater visionary future toward novel separation processes. Vision of science, the technological forays in environmental engineering science, and the wide futuristic vision will all lead a long and visionary way in the true emancipation of global water challenges today. The concept of fouling and its removal today stands in the midst of vision and forbearance. In this treatise, the author widely pronounces the success of today's scientific vision in realization of global water research and development initiatives. Success of technology and scientific validation are witnessing immense scientific challenges and barriers. In this treatise, the author rigorously points out the wide and vast scientific vision in the furtherance of membrane science.

Van der Bruggen et al.[9] deeply discussed with cogent foresight drawbacks of applying nanofiltration and how to avoid them. In spite of all the evergrowing perspectives for nanofiltration, not only in drinking water production but also in industrial wastewater treatment, the food industry, the chemical and pharmaceutical industry, and many other process industries, there are still some unresolved issues in successful membrane separation processes. In this paper, the author pointedly focuses on six definite challenges for nanofiltration which are (1) avoiding membrane fouling, (2) improving the separation of solutes, (3) further treatment of concentrates, (4) chemical resistance and limited lifetime of membranes, (5) insufficient rejection of pollutants, and (6) the immediate need of the modeling and simulation tools.[9] Science of membrane fouling is expanding and ever-growing. Research and development initiatives are surpassing veritable scientific frontiers. In this paper, the author, with deep scientific imagination and scientific fortitude discusses the intricacies of membrane fouling phenomenon. This area of scientific endeavor is veritably unexplored. Scientific vision, scientific truth, and wide scientific forbearance will lead a long and visionary way in the true realization of successful membrane separation processes. Fouling is a major impediment to membrane separation process. Scientific domain throughout the world are pursuing veritably in the research of membrane fouling. The scientific and academic rigors in membrane fouling are today ushering in a new era of scientific innovation. In this treatise, the author rigorously puts forward to the readers the success of the present day research pursuit in the membrane with the sole aim toward furtherance of environmental engineering science and chemical process engineering.[9]

Mohammadi et al.[10] discussed with lucid details membrane fouling. Fouling of ultrafiltration membranes in milk industries is mostly due to precipitation of microorganisms, proteins, fats, and minerals on the membrane surfaces.[10] Thus, chemical cleaning of membranes is absolutely essential. In this paper, the author widely observes results obtained from investigations on a polysulfon ultrafiltration membrane fouled by precipitation of milk components. In this well-researched paper, the effect of different cleaning agents on the recovery of fouled membranes has been studied. Membrane fouling present important limitations on the technology available in the scientific domain. Fouling is defined as existence and growth of microorganisms and irreversible collection of materials on the membrane surface. Today, technological advancements are ushering in a new era of science and engineering. Fouling is one such area of scientific endeavor in the field of environmental engineering and membrane science. Fouling results in flux decline. To overcome this problem, cleaning process needs to be addressed and envisioned. Technological validation and the vision of science today are the torchbearers toward a greater visionary era in environmental protection. The authors of this treatise deeply discussed fouling and cleaning phenomenon in details.[10]

7.12 NONTRADITIONAL ENVIRONMENTAL ENGINEERING TECHNIQUES AND THE VISION FOR THE FUTURE

Nontraditional environmental engineering techniques are today replete with immense scientific vision and scientific forbearance. Global water shortage, environmental crisis, and the stringent environmental regulations have urged the entire scientific domain to delve deep into the murky depths of scientific cognizance and profundity. The challenge needs to be reenvisioned at each step of scientific path. Environmental sustainability today encompasses also the provision of clean drinking water as well as the success of global water research and development initiatives. Human mankind today stands in the midst of immense scientific comprehension and introspection. Nontraditional environmental engineering tools encompass advanced oxidation techniques. In this treatise, the author describes three major research works in the field of AOPs.

Gilmour et al.[11] delineated with lucid details application perspectives in water treatment using AOPs. AOPs using hydroxyl radicals and other oxidative radical species are being studied widely for removal of organic compounds from industrial wastewater. However, large-scale applications are not being established till today.[11] The vision of science needs to be readdressed with each step of scientific history and time. This research focuses on the evaluation of the upstream processing and downstream post treatment analysis of selected AOPs. In the first stage of research, the performance of a proprietary catalyst (VN-TiO$_2$) was compared with the industry standard P25 TiO$_2$, for the use in a pilot-scale immobilized photocatalytic reactor.[11]

Stasinakis[12] discussed and described with deep insight use of selected AOPs for wastewater treatment. AOPs are widely used for removal of recalcitrant compounds in wastewater. The wide vision of AOP science is today ushering a new era in environmental engineering science.[12] The aim of this study was to review the use of titanium dioxide/UV light process, hydrogen peroxide/UV light process and Fenton's reactions in industrial wastewater treatment. Scientific vision, the wide technological vision, and the futuristic vision of environmental engineering science are the forerunners of the larger targets of environmental sustainability today. The main reactions and the operating parameters affecting these processes are reported while recent research trends presented.[12]

Al-Kdasi et al.[13] delineated with deep insight treatment of textile wastewater by AOPs. The use of conventional textile wastewater treatment processes becomes drastically and rigorously challenged to environmental scientists with increasing stress of effluent and water quality by environmental authorities. The science of AOPs is unfolding toward a newer era of scientific vision and scientific understanding. Technology is so advanced today with the sole visionary aim toward furtherance of engineering science. Conventional treatments such as biological techniques are fruitless as 53% of 87 colors are totally nonbiodegradable. The author of this treatise pointedly focuses on the particular case of dye degradation which is a scientifically vexing issue. AOPs have been found to be extremely effective with the passage of scientific history and vision.[13]

7.13 ENVIRONMENTAL SUSTAINABILITY AND RECENT SCIENTIFIC ENDEAVOR

Environmental sustainability today stands in the midst of immense scientific introspection and vision. The challenge of environmental engineering science is evergrowing with the progress of science and technology. Sustainable development whether it is energy or environment is the prime need of the hour in attaining self-sufficient economy of a developed as well as developing nation. Scientific endeavor in sustainability needs to be streamlined in order to attain the wide emancipation of environmental engineering science. Heavy metal and arsenic groundwater contamination is a bane toward human scientific endeavor as well as scientific rigor. The vision of Dr Gro Harlem Brundtland, former Prime Minister of Norway, needs to be readdressed and revamped as Sustainable Development Goals gains new scientific pinnacles. In this treatise, the author widely observes the success of global water challenges as regards heavy metal groundwater remediation.

7.14 STATUS OF GLOBAL WATER RESEARCH AND DEVELOPMENT INITIATIVES

Global water research and development initiatives today stand in the midst of deep scientific vision and profundity. Environmental catastrophes and environmental regulations are urging the scientific domain to gear for newer challenges. The challenge and the vision of water purification are wide and bright. Scientific innovations need to be reenvisioned and streamlined with the passage of scientific history and time. Arsenic groundwater poisoning has urged the human scientific research pursuit to gear forward toward new avenues of scientific vision and endurance. Science and engineering of environmental protection are veritably crossing wide and vast visionary boundaries. Environmental engineering science today is in the path of newer scientific regeneration. Shannon et al.[14] discussed with immense scientific forbearance the future of science and engineering of water purification. Problems with water will aggravate in near future with water scarcity growing rapidly over the scientific horizon. Water shortage will devastate the global scenario whether it is developing or developed nation. Even in water-rich countries, the issue of global water crisis looms large over the scientific horizon. Shannon et al.[14] present to the reader a broad view of the global water research and development status. The authors touched upon disinfection, decontamination, reuse, and reclamation of industrial wastewater and polluted water bodies. The science of desalination needs to be restructured as human civilization treads toward a newer scientific era. Environmental engineering science and the wide world of water purification will surely open up new areas of scientific imagination in decades to come.[14,17–19]

7.15   TECHNOLOGICAL AND SCIENTIFIC VALIDATION OF ENVIRONMENTAL ENGINEERING TOOLS

Environmental engineering tools are ushering in a new era in the field of research pursuit in environmental protection. Technological and scientific validations are the coinwords of this century. Conventional as well as nonconventional environmental engineering techniques are surpassing wide scientific vision and vast scientific frontiers. Today global water shortage is linked by an umbilical cord with effective environmental engineering tools. Provision of clean drinking water is today a parameter toward the economic growth of a nation. In this treatise, the author pointedly focuses on the success and immense potential behind remediation technologies, conventional, and nonconventional environmental engineering techniques.[15,16]

7.16   FUTURISTIC RESEARCH AND DEVELOPMENT TRENDS IN WATER SCIENCE

Water science and technology is moving toward a newer visionary eon as environmental engineering science crosses vast and versatile scientific boundaries. Technology revamping and scientific validation are the utmost need of the hour. Water science and green engineering are the coinwords of this century as global water crisis faces newer challenges and newer vision. Desalination and disinfection science are the visionary coinwords of the future of environmental engineering. Technology revamping and global water research and development initiatives are the other sides of the visionary coin of global water issues. Green technology and green engineering are today replete with scientific vision and scientific fortitude. This treatise widely observes the scientific world of environmental sustainability and the hurdles and barriers in the implementation of successful sustainability whether it is energy or environment. Human mankind's immense scientific prowess, the scientific greatness of environmental engineering, and the futuristic vision of remediation technologies will all lead a long way in the true realization of environmental sustainability today.

7.17 SUMMARY AND CONCLUSION

Remediation technologies are changing the face of human scientific vision and scientific ingenuity. Technology needs to be revamped and is the utmost need of the hour. Scientific and academic rigor of arsenic groundwater remediation is of utmost importance with the progress of human civilization. Scientific truth, scientific ingenuity, and scientific vision are the utmost parameters in a nation's progress. Provision of clean drinking water is of prime importance in the success of human civilization and human scientific and academic advancement.

ACKNOWLEDGMENT

The author acknowledges the contribution of Shri Subimal Palit, the author's late father and an eminent textile engineer who taught the author the rudiments of chemical engineering. His contributions to the author's career are immeasurable.

KEYWORDS

- vision
- arsenic
- heavy metal
- water
- remediation
- contamination

REFERENCES

1. Mukherjee, A.; Sengupta, M. K.; Hossain, M. A.; Ahamed, S.; Das, B.; Nayak, B.; Lodh, D.; Rahman, M. M.; Chakraborti, D. Arsenic Contamination in Groundwater: A Global Perspective with Emphasis on the Asian Scenario. *J. Health Popul. Nutr.* **2006**, *24*(2), 142–163.
2. Safiuddin, M.; Karim, M. M. In *Groundwater Arsenic Contamination in Bangladesh: Causes, Effects and Remediation*, Proceedings of the 1st IEB International Conference

and 7th Annual Paper Meet; Institution of Engineers, Bangladesh: Chittagong, Bangladesh, 2001.

3. Hashim, M. A.; Mukhopadhayay, S.; Sahu, J. N.; Sengupta, B.; Remediation Technologies for Heavy Metal Contaminated Groundwater. *J. Environ. Manage.* **2011**, *92*, 2355–2388.

4. Zvinowanda, C. M.; Okonkwo, J. O.; Shabalala, P. N.; Agyei, N. M. A Novel Adsorbent for Heavy Metal Remediation in Aqueous Environments. *Int. J. Environ. Sci. Technol.* **2009**, *6*(3), 425–434.

5. Khan, A. G.; Kuek, C.; Chaudhry, T. M.; Khoo, C. S.; Hayes, W. J. Role of Plants, Mycorrhizae and Phytochelators in Heavy Metal Contaminated Land Remediation. *Chemosphere* **2000**, *41*, 197–207.

6. Davis, T. A.; Volesky, B.; Mucci, A. A Review of the Biochemistry of Heavy Metal Biosorption by Brown Algae. *Water Res.* **2003**, *37*, 4311–4330.

7. Rajendran, P.; Muthukrishnan, J.; Gunasekaran, P. Microbes in Heavy Metal Remediation. *Indian J. Exp. Biol.* **2003**, *41*, 935–944.

8. Cheryan, M. *Ultrafiltration and Microfiltration Handbook*; Technomic Publishing Company Inc.,: USA, 1998.

9. Van der Bruggen, B.; Manttari, M.; Nystrom, M. Drawbacks of Applying Nanofiltration and How to Avoid Them: A Review. *Sep. Purif. Technol.* **2008**, *63*, 251–263.

10. Mohammadi, T.; Madaeni, S. S.; Moghadam, M. K. Investigation of Membrane Fouling. *Desalination* **2002**, *153*, 155–160.

11. Gilmour, C. R. Water Treatment Using Advanced Oxidation Processes: Application Perspectives. Master of Engineering Thesis, The University of Western Ontario, 2012.

12. Stasinakis, A. S. Use of Selected Advanced Oxidation Processes (AOPs) for Wastewater Treatment—A Mini-Review. *Global NEST J.* **2008**, *10*(3), 376–385.

13. Al-Kdasi, A.; Idris, A.; Saed, K.; Guan, C. T. Treatment of Textile Wastewater by Advanced Oxidation Processes—A Review. *Global Nest: Int. J.* **2004**, *6*(4), 222–230.

14. Shannon, M. A.; Bohn, P. W.; Elimelech, M.; Georgiadis, J. A.; Marinas, B. J.; Science and Technology for Water Purification in the Coming Decades. *Nat. Publ. Group.* **2008**, 301–310.

15. Palit, S. Nanofiltration and Ultrafiltration—The Next Generation Environmental Engineering Tool and A Vision for The Future. *Int. J. Chem. Tech. Res.* **2016**, *9*(5), 848–856.

16. Palit, S. Filtration: Frontiers of the Engineering and Science of Nanofiltration—A Far-Reaching Review. *CRC Concise Encyclopedia of Nanotechnology*; Ubaldo, O.-M., Kharissova, O. V., Kharisov, B. I., Eds.; Taylor and Francis: USA, 2016, pp 205–214.

17. Palit, S. Advanced Oxidation Processes, Nanofiltration, and Application of Bubble Column Reactor. *Nanomaterials for Environmental Protection*; Kharisov, B. I., Kharissova, O. V., Rasika Dias, H. V., Eds., Wiley: USA, 2015, pp 207–215.

18. www.google.com

19. www.wikipedia.com

20. *Water Stewardship Information Series, Arsenic in Groundwater*; The Ministry of Environment, Province of British Columbia: Canada, 2007.

CHAPTER 8

INFLUENCE OF SOIL WASHING ON IRON MOBILIZATION AND OTHER METALS/METALLOIDS AND THEIR IMPACTS ON BIODEGRADABILITY ENHANCEMENT DURING AN AQUEOUS ADVANCED ELECTROCHEMICAL TREATMENT

EMMANUEL MOUSSET[1,2,*], NIHAL OTURAN[1],
and MEHMET A. OTURAN[1]

[1]Université Paris-Est, Laboratoire Géomatériaux et Environnement (LGE), EA 4508, UPEM, 5 bd Descartes, 77454 Marne-la-Vallée Cedex 2, France

[2]Laboratoire Réactions et Génie des Procédés, UMR CNRS 7274, Université de Lorraine, 1 rue Grandville BP 20451, 54001 Nancy Cedex, France

*Corresponding author. E-mail: emmanuel.mousset@univ-lorraine.fr

CONTENTS

ABSTRACT

The influence of extracting agents on metals/metalloids solubilization efficiency is carried out, and their influence on an electrochemical advanced oxidation treatment, namely, electro-Fenton, is investigated. Two washing agents have been compared, a nonionic surfactant (e.g., Tween 80) and a cyclodextrin (hydroxypropyl-beta-cyclodextrin (HPCD)). After a detailed laboratory investigation, the biodegradability could reach 33% in both kinds of soil washing solutions, a threshold value that is sufficient to consider a biological posttreatment. Tween 80 solution lead to higher ecotoxicity and lower biodegradability improvement which could be explained by the presence of more organic pollutants, soil organic matter, and heavy metals such as Pb due to its higher extraction power. It could also be attributed to the higher ability of Tween 80 molecules to generate more toxic by-products than HPCD.

8.1   INTRODUCTION

The remediation of contaminated sites is a common concern and represents a challenge for the next years since the number of polluted sites increases together with human activities. In the last two decades, the number of potentially contaminated sites increased seven times in most of the developed countries.[1] The European Environment Agency estimated that over 3,000,000 sites are potentially contaminated in Europe in 2006 and around 250,000 contaminated sites among them may need urgent remediation.[2] The main causes of sites contamination are anthropogenic activities and most of these sites are located close to or in urban areas.[3] Hydrophobic organic compounds (HOCs) such as aliphatic hydrocarbons and polycyclic aromatic hydrocarbons (PAHs) are typical pollutants found in soil. Their nonpolar and hydrophobic properties with a high octanol/water partition coefficient (Log K_{ow}) make them persistent in the environment.[4] Moreover, their high carbon partition coefficient (Log K_{oc}) makes them strongly bound to soil, which remain the main sink.[5]

Such toxic and persistent compounds cause potential risk for environment and human health. The toxicity of PAHs can be explained by intercalation of the PAH aromatic ring system into the DNA duplex.[6] This formation of DNA adducts is a key event in mutagenicity and carcinogenicity by PAHs.[7] Sixteen of them are listed as priority substances by the Environmental Protection Agency of United States (USEPA) such as benzo(*a*) pyrene. It makes the soil quality regulations more and more stringent all around the world.[8]

In this context, different techniques have been proposed to deals with HOCs-contaminated soils such as physical processes, thermal treatments, biological processes, and physicochemical treatments.[9,10] The physical methods like containment and landfilling do not remove the pollutant from the soil but only avoid the expansion of the pollution, while the "pump-and-treat" technique with water process is not efficient enough since PAHs pollutants are hydrophobic and strongly sorbed into soil.[9,10] Thermal treatments are usually more expensive and energy consuming. In addition, it strongly alters the physical and chemical nature of the treated soil. Biological treatments are generally slow and not efficient enough with xenobiotics compounds like heavy PAHs.[9,10] Physicochemical treatments like solidification/stabilization do not treat the soil but only restrain the pollution diffusion.[9,10] Besides, soil washing (SW), and soil flushing (SF) recently appeared to be competitive techniques especially in terms of cost and time of remediation.[9,11,12] Moreover, they are more environment-friendly than the thermal treatments, assuming that the enhancing agents used are biodegradable. Surfactants are the most conventional extracting agent being employed in SW/SF studies.[11,13] Nonionic surfactant have shown good efficiency, especially because of their low adsorption onto soil, their low critical micelle concentration, and their high HOC extraction efficiency.[13] Among these surfactants, Tween 80 usually outperforms its congeners such as Tween 20, Tergitol NP10, Brij 35, Tyloxapol, Igepal CA-720, and Triton X-100.[14,15] More recently, cyclodextrins, such as hydroxypropyl-beta-cyclodextrin (HPCD) as the most cost-effective one, have emerged as new extracting agent due to their toroidal shape and physicochemical properties.[11]

Since the enhanced SW/SF techniques only allow the extraction of pollutants without eliminating it, a post-treatment is required.[11,12] Advanced oxidation processes (AOPs) have gained interest in the last decades in order to treat such organic recalcitrant contaminants by hydroxyl radicals ($^{\bullet}$OH) oxidation.[16] This radical is a very strong oxidizing agent (2.8 V vs. standard hydrogen electrode (SHE)) and has the advantage to be nonselective and react very quickly with carbon–carbon double bonds with rate constants ranging between 10^8 and 10^{10} L mol^{-1} s^{-1},[17] for most of HOCs.[12] Recently, electrochemical AOPs (EAOPs) have emerged because they continuously produce $^{\bullet}$OH in situ with electron as a clean reagent.[17–20] Among them, electro-Fenton (EF) is a promising environment-friendly process based on the Fenton reaction that generates $^{\bullet}$OH (eq 8.1):[21–23]

$$H_2O_2 + Fe^{2+} \rightarrow Fe^{3+} + {}^{\cdot}OH + HO^- \qquad (8.1)$$

The Fenton's reagent is produced in situ at the cathode via oxygen reduction into H_2O_2 (eq 8.2) and ferrous ion (Fe^{2+}) regeneration from reduction of ferric ion (Fe^{3+}) (eq 8.3) produced by Fenton's reaction:[24–26]

$$O_2 + 2H^+ + 2e^- \rightarrow H_2O_2 \qquad (8.2)$$

$$Fe^{3+} + e^- \rightarrow Fe^{2+} \qquad (8.3)$$

Thanks to these improvements, higher degradation and mineralization rates and yield (>99%) of HOCs can be reached.[24,27–32] In addition, there is no sludge production as compared to the conventional chemical Fenton process, since Fe^{2+} is only needed at a catalytic amount—if not originally present in the wastewater[33,34]—thanks to its fast regeneration (eq 8.3). Recently, the combination between EAOPs and SW/SF technologies have been reported,[33–44] but no authors have carried out the catalytic benefit that could come from the metals/metalloids extracted from the soil during the implementation of EF process. Moreover, the combination of EAOPs and a post-biological treatment arouse interest as it would completely mineralize recalcitrant HOCs while reducing energy cost.[37,39,45,46] Therefore, in an integrated SW/SF-EF-biological process, the impact of those metals/metalloids extracted during the SW/SF step would have an impact on the biodegradability of solutions.

In this context, this study intends to bring new knowledge according to the two following objectives: (1) assess the amount of iron extracted by Tween 80 and HPCD as compared to ultrapure water using sequential extractions technique and (2) evaluate the potential impact of metals/metalloids present in SW solutions on the biodegradability and ecotoxicity of solutions treated by EAOPs.

8.2 MATERIALS AND METHODS

8.2.1 CHEMICALS

Sodium sulfate (Na_2SO_4), 2-(p-toluidino)naphthalene-6-sulfonic acid sodium (TNS), Tween 80 (polyoxyethylene(20)sorbitan monooleate) (molar weight (MW) = 1310 g mol^{-1}), acetonitrile (CH_3CN), and acetic acid (CH_3COOH) were purchased from Aldrich. HPCD was provided by Xi'an

Taima Biological Engineering Company (MW = 1250 g mol^{-1}). Ammonium acetate (CH$_3$CO$_2$NH$_4$) was supplied by Acros. Analytical reagents like *n*-hexane, acetone ((CH$_3$)$_2$O), hydroxylammonium chloride (NH$_2$OH·HCl), and sodium hydroxide (NaOH) were provided by VWR International company. *N*-Allylthiourea was provided by Alfa Aesar. Sodium phosphate dibasic (Na$_2$HPO$_4$), ammonium chloride (NH$_4$Cl), heptahydrated magnesium sulfate (MgSO$_4$·7H$_2$O), dehydrated calcium chloride (CaCl$_2$·2H$_2$O), potassium peroxodisulfate (K$_2$S$_2$O$_8$), phenanthroline 1,10, and iron (Fe) standard solution were purchased from Merck. Potassium chloride (KCl) (>99.0%, Fluka) was also used. HNO$_3$ (70%) and fluorhydric acid (HF) (48%) from Fisher Scientific, nitric acid (HNO$_3$) (65%) from Fluka and chlorhydric acid (HCl) (32%) from Riedel-de-Haën were employed. Mohr's salt was provided by Acros. All the reagents were of analytical grade. In all experiments, ultrapure water from a Millipore Simplicity 185 (resistivity > 18 MΩ cm) system was used.

8.2.2   SOIL PREPARATION AND CHARACTERISTICS

The polluted soil was sampled from a PAHs and hydrocarbons contaminated site. Before its utilization, the soil was sieved under 2 mm and homogenized by a sample divider (Retsch). The soil physicochemical characteristics obtained from an external certified laboratory (ALcontrol Laboratories) are described in Table 8.1.

The total amount in soil of these PAHs was determined by Soxhlet extraction (Behr, Labor-Technik). Two grams of dried soil were mixed with 5 g of anhydrous sodium sulfate to prevent trace of humidity. A mixture of *n*-hexane/acetone (70/70 mL) was then added. Extractions were performed for 16 h (4–5 cycle/h) in triplicate. The calculated amount of PAHs was compared to the values obtained with the ALcontrol Laboratories data. The highest content measured was considered for each PAH. The final concentrations of selected pollutants are given in Table 8.1. The total concentration of the 16 PAHs listed by USEPA was 1090 mg kg^{-1} Dry Weight (DW).

Total petroleum hydrocarbons (C10–C40) contents were present at a level of 850 mg kg^{-1} DW, while the total amount of benzene, toluene, ethylbenzene, and xylene (BTEX) was around 0.56 mg kg^{-1} DW, which was much below the regulations.

TABLE 8.1   Soil Characteristics.

Particle size distribution (%)	Clay (<2 μm)	19.7
	Fine silt (2–20 μm)	23.3
	Coarse silt (20–50 μm)	7.5
	Fine sand (50–200 μm)	12.3
	Coarse sand (0.2–2 mm)	37.1
Organic matter (OM) (%)		4.71
TOC (mg kg^{-1} DW)		26,000
Total nitrogen (mg N kg^{-1} DW)		1350
CEC (soil pH) (meq kg^{-1})		203
Saturation of clay–humic complex (%)	Ca^{2+}	68.4
	K^+	5.2
	Mg^{2+}	22.9
	Na^+	3.5
	H^+	0.0
	Total	100
Metals (mg kg^{-1} DW)	Aluminum	13,000
	Arsenic	18
	Cadmium	<0.4[*]
	Chrome	21
	Copper	41
	Mercury	2.4
	Lead	87
	Nickel	18
	Iron	9550
	Zinc	160
	Silver	<5[*]
Organic pollutants (mg kg^{-1} DW)	Total PAHs (16 listed by USEPA)	1090
	Total petroleum hydrocarbons (C10–C40)	850
	BTEX	0.56

[*]Below detection limit (DL).

8.2.3   SOIL WASHING (SW) EXPERIMENTS

SW experiments were performed in a 500-mL glass bottle at a soil/liquid ratio equal to 10% (40 g DW/400 mL), which is the most conventional ratio value used by authors.[11] Solutions of Tween 80 or HPCD (7.5 ± 0.2 g L^{-1}) were used at the same initial mass concentration, equivalent to 5.7 ± 0.15 mmol L^{-1} and 6.0 ± 0.16 mmol L^{-1}, respectively. The mixtures were rotated in a Rotoshake RS12 (Gerhardt, Germany) at 10 rpm for 24 h. Then, the particles settled for 12 h, and the supernatants were filtered with a 0.7-μm glass microfiber filter. SW experiments with ultrapure water were also performed under the same conditions as the controls. The supernatants were then characterized by monitoring total organic carbon (TOC), HPCD, Tween 80, pH, conductivity, iron, and other metals/metalloids followed by electrochemical treatments.

8.2.4 ELECTROOXIDATION TREATMENT

EF experiments of SW solutions were performed at room temperature ($22 \pm 1°C$), in a 0.40-L undivided glass electrochemical reactor at current controlled conditions. The cathode was a 150-cm^2 carbon-felt piece (from Carbone-Lorraine, France). Boron-doped diamond (BDD) plate (5 cm × 4 cm) anode was used, since this electrode was determined to be the best option compared to Pt and dimensionally stable anode (Ti/IrO_2/RuO_2) anodes in our previous study.[39] The anode was centered in the cell and surrounded by the cathode covering the inner wall of the cell. The electrochemical cell was monitored by a power supply HAMEG 7042-5 and applied current density was set to 6.67 mA cm^{-2}, taking into account the cathode surface area as the working electrode. An inert electrolyte (0.150 mol L^{-1} Na_2SO_4) was added to the medium since the conductivity of solutions was too low.[34] Prior to each experiment, the solutions containing HPCD were saturated in O_2 by supplying compressed air (10 min at 0.25 L min^{-1}). Since too much foam is formed during bubbling system, the solutions containing Tween 80 were not saturated with O_2. However, solutions were stirred continuously and vigorously by a magnetic stirrer to compensate O_2 depletion, as mentioned in our previous study.[40] A heat exchanger system is provided to keep the solution at constant room temperature by using fresh water. The pH of initial solution was not adjusted to pH 3 as usual. No iron was added since it was assumed that iron was already present in SW solutions.[33,34] The results about initial iron content in solutions are presented in Section 8.3.1.1.

The schematic representation of the integrated process (SW + EF treatment) is shown in Figure 8.1, considering the study of a possible biological posttreatment assessed by bioassays.

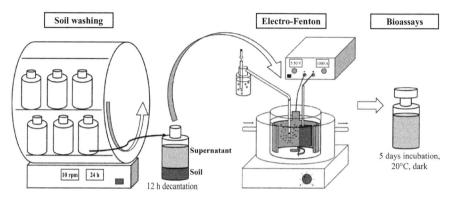

FIGURE 8.1 Schematic representation of the integrated process: SW combined to EF treatment considering possibility of biological posttreatment by bioassays studies.

8.2.5 SEQUENTIAL EXTRACTIONS

A three-stage sequential extraction speeding up with focused ultrasound method was employed[47] to study iron fractionation[48] in the contaminated soil by combining both protocols. The applied operating conditions are described in Table 8.2. The amount of soil used was 0.25 g DW. Each soil after SW with Tween 80 (7.5 g L^{-1}), HPCD (7.5 g L^{-1}), and ultrapure water was studied. These three kinds of soil were previously dried at 105°C during 48 h before sequential extraction. The ultrasound system was a Bandelin UW70 probe with a Bandelin Sonopuls GM70 equipment providing a sonication power of 20 W.

TABLE 8.2 Operating Conditions Used for Ultrasound Accelerated Sequential Extraction Methods.

Stage	Reagents			Ultra-sound time (min)
	Compound	Concentration	Volume (mL)	
Stage 1: acid soluble fraction (e.g., carbonates)	CH$_3$COOH	0.11 M	10	7
Stage 2: reducible fraction (e.g., Fe–Mn oxides)	NH$_2$OH·HCl (pH 1.5)	0.5 M	10	7
Stage 3: oxidizable fraction (e.g., OM)	H$_2$O$_2$ + CH$_3$COONH$_4$ (pH 2)	30% w/v + 1 M	5 + 2.5	2 + 6

The solution pH in stages 2 and 3 were adjusted with HNO_3 (70%). Between each stage, the samples were centrifuged (3000 g) during 15 min at 20°C with Jouan-KR22i equipment. The supernatants were then filtered (0.45 μm) and diluted in volumetric flask with the respective reagents used for the concerned stage. These liquid samples were kept for further atomic absorbance spectrometric (AAS) measurements. The soil samples were then rinsed with ultrapure water by centrifuging (3000 g, 15 min, 20°C) a second time.

The total concentration of iron in soil was performed with 0.25 g of dried contaminated soil. The mineralization was done in a Multiwave 3000 (Anton Paar) at 1400 W during 30 min with a mixture of HNO_3 (65%), HCl (32%), and HF (48%) with the respective following ratio: 5 mL/2 mL/1 mL. The F^- ions were then complexed with boric acid (0.7 mol L^{-1}) with a ratio of 6 mL of H_3BO_3 per mL of HF. These samples were then mineralized at 1400 W during 20 min. The samples were then diluted in 50-mL volumetric flask and filtered (0.2 μm) before their quantification by AAS. The total iron content in soil was reported in Table 8.1.

The AAS analyses of iron were performed with a Varian SpectrAA 220 with Flame provided by air/acetylene gas and a hollow cathode SpectrAA lamp as a radiation source for iron element. An external calibration was done with standard solutions of iron. Each sample was then analyzed in triplicate with a standard deviation less than 5%.

8.2.6 BIOASSAYS

8.2.6.1 ECOTOXICITY ASSAYS

Acute ecotoxicity assays were performed by using Microtox® standard method (ISO 11348-3) with marine bacteria *Vibrio fischeri* from LUMIStock LCK-487 (Hach Lange). A Berthold Autolumat Plus LB 953 equipment was used. 22% of NaCl was added in each sample to ensure an osmotic protection for bacteria. Before each ecotoxicity measurement, all the samples were adjusted with NaOH to circum-neutral pH and samples from EF experiment were filtered with RC filters (0.2 μm) to remove iron precipitates.[49] In each batch test, the inhibition percentage of a blank (sample without the compound studied) was also measured and used for percentage of inhibition calculation based on 15 min of exposure.

8.2.6.2 BIODEGRADABILITY TESTS

The biodegradability was given by the ratio between biochemical oxygen demand after 5 days (BOD_5) and the chemical oxygen demand (COD). BOD_5 was determined with a respirometric method (OECD 301F, ISO 9408) by manometric measurement with the OxiTop® IS 6 system (WTW). The system measured the difference of pressure due to the consumption of oxygen by aerobic microorganisms (eq 8.4):

$$\text{BOD} = \frac{M(O_2)}{RT_m}\left(\frac{V_{tot} - V_{sample}}{V_{sample}} + á\,\frac{T_m}{T_0}\right)\Delta p(O_2) \qquad (8.4)$$

where $M(O_2)$ is the MW of O_2 (32,000 mg mol^{-1}), R is the gas constant (83.144 L mbar (mol K)$^{-1}$), T_0 is the reference temperature (273.15 K), T_m is the measuring temperature, V_{tot} is the bottle volume (nominal volume in mL), V_{sample} is the sample volume in mL, α is the Bunsen absorption coefficient (0.03103) and, $\Delta p(O_2)$ is the difference of the oxygen partial pressure (mbar).

The CO_2 released in the meantime by microorganisms was trapped in a rubber sleeve in which NaOH pellets were added. An inoculum is added in each sample solution just before starting the experiment. It consists of bacteria extracted from uncontaminated soil by adding KCl at 9 g L^{-1} (30 mL with 3 g of dried soil) and by using an IKA-MS1 minishaker (1800 rpm during 1 min). In order to promote the bacterial growth, nutrients were added. It consists of an aqueous solution containing a phosphate buffer solution (pH 7.2) and a saline solution prepared according to Rodier et al.[50] This solution was then saturated in oxygen. All the samples were adjusted to circum-neutral pH. N-Allylthiourea (10 mg L^{-1}) was added to prevent nitrification. The samples were then incubated at 20°C (±0.1) during 5 days in dark conditions. In order to consider the organic matter (OM) extracted from soil and the endogenous respiration, the BOD_5 measured in each blank was deduced from the BOD_5 of the samples. The BOD_5 of blanks was insignificant and caused no interferences.

COD measurements were achieved by a photometric method using a Spectroquant® NOVA 60 (Merck) equipment. Two millimeters of diluted samples were added in each COD cell test (Merck). The test tubes were then heated at 148°C for 2 h with a Spectroquant® TR 420 (Merck).

Since the H_2O_2 was produced in situ during EF experiment and the radicals formed during EF treatments have a limited lifetime, these oxidants cause no interferences during the BOD_5 or COD measurements.

8.2.7 ANALYTICAL PROCEDURES

8.2.7.1 TOTAL DISSOLVED IRON CONCENTRATION

Total dissolved iron was measured by molecular absorption spectrometry with phenanthroline 1,10, according to Rodier et al.[50] 50 mL of samples were acidified at pH 1 (HCl) and 5 mL of potassium peroxodisulfate (40 g L^{-1}) is added. The samples were then boiled during 40 min and let cool down at room temperature. Ammonium acetate was added in order to have a solution at pH around 4.5. Then, 2 mL of 1,10-phenanthroline (0.5%) was added and kept in dark conditions during 15 min. The absorbance measurements were performed with a spectrophotometer UV–Visible Lambda 10 at 510 nm. A blank without iron was prepared by following the same protocol and was deduced from the absorbance value of the samples. An external calibration curve was done with Mohr's salt (1 mM).

8.2.7.2 pH AND CONDUCTIVITY OF SOLUTIONS

The pH of solutions was measured with a CyberScan pH 1500 pH-meter from Eutech Instruments. Before each use, the pH-meter was calibrated with standard buffer solutions at 6.87 and 4.1. All the samples and buffer solutions were at room temperature ($22 \pm 1°C$) before each measurement. The standard deviations of replicates were always less than 0.15. These values were too low to be readable on graphs.

Conductivity measurements were done with a MeterLab CDM 210 from Radiometer analytical SA. The conductivity values were adjusted according to the temperature of solutions.

8.2.7.3 QUANTIFICATION OF EXTRACTING AGENTS

The HPCD and Tween 80 concentrations were determined by a fluorimetric essays based on enhancement of the fluorescence intensity of TNS, when they are complexed with the cyclodextrin[44] and Tween 80.[38] This method allows quantifying HPCD and slightly modified HPCD (hydroxylated) in the same time, since the nonpolar HPCD cavity brings about a TNS fluorescence intensity enhancement until the CD cavity is cleaved by the degradation technique. A Kontron SFM 25 spectrofluorimeter was set out at 318 nm for excitation and 428 nm for emission for both HPCD and Tween 80 quantification. Each sample is diluted in TNS at 3×10^{-6} M and 5×10^{-5} M for HPCD and Tween 80, respectively. The fluorescence intensity of soil OM (SOM) and organic pollutant are not significant in this range of concentration.[38] Since TNS is photosensitive, TNS and the diluted samples were therefore stored in dark conditions.

8.2.7.4   TOC ANALYSIS

The TOC values were determined by thermal catalytic oxidation using a Shimadzu V_{CSH} TOC analyzer. The temperature was set at 680°C (±1°C) and Pt was used as catalyst. Calibrations were performed by using the potassium hydrogen phthalate solutions (50 mgC L^{-1}) as standard. All samples were acidified to a pH value of 2 with H_3PO_4 (25%) to remove inorganic carbon contents. The samples were then analyzed by nonpurgeable organic carbon method. The injection volumes were 50 µL. All samples values are given with a coefficient of variance below to 2%.

8.3   RESULTS AND DISCUSSION

8.3.1   IMPACT OF EXTRACTING AGENTS ON THE MOBILIZATION/SOLUBILIZATION OF METALS/METALLOIDS

8.3.1.1   IMPACT OF EXTRACTING AGENTS ON THE MOBILIZATION/SOLUBILIZATION OF IRON NEEDED FOR EF PROCESS

The total dissolved iron concentrations were 0.020 ± 0.013 mmol L^{-1} and 0.060 ± 0.013 mmol L^{-1} for SW solutions with HPCD and Tween 80, respectively. It corresponds to an average of 0.011 ± 0.007 mg kg^{-1} DW and 0.033 ± 0.007 mg kg^{-1} DW of extracted iron from soil with HPCD and Tween 80, respectively. These values are very low compared to the initial iron content in soil that is around 9550 mg kg^{-1} DW. However, both amounts are sufficient to perform an EF treatment, since a concentration of 0.05 mmol L^{-1} of iron(II) was found to be optimal in order to degrade and mineralize HPCD SW solutions polluted by phenanthrene in a former study.[40]

It can also be noticed that Tween 80 allows a relatively higher extraction of iron than HPCD. One of the hypotheses could be iron fractionation in soil. To validate the assumption sequential extractions of iron in soils coming from three kinds of SW experiments according to the extracting agent used (HPCD (7.5 g L^{-1}), Tween 80 (7.5 g L^{-1}), and ultrapure water) were performed. These results are presented in Figure 8.2. In acid soluble fraction (first stage), the following rank, in terms of concentrations of iron in soil, was observed: Tween 80 > HPCD > ultrapure water. Regarding the reducible fraction (second stage), the following rank is obtained: ultrapure water > HPCD > Tween 80, and about the oxidizable fraction (third stage): ultrapure water > Tween 80 ≈ HPCD. These trends could be explained by the higher solubilization capacity of Tween 80 toward SOM. Indeed, the iron oxides (reducible fraction) that can be present in SOM[51] are mobilized by Tween 80 and can then be considered as soluble iron when first stage of sequential extraction is performed. This could also confirm the slightly higher amount of iron in SW solution with Tween 80.

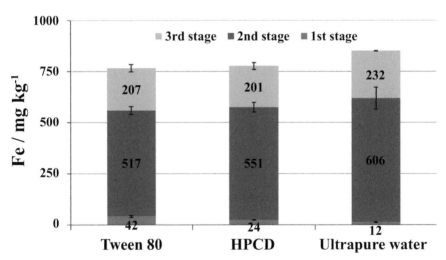

FIGURE 8.2   Sequential extraction of iron in soil after SW with different solutions: Tween 80 (7.5 g L^{-1}), HPCD (7.5 g L^{-1}), ultrapure water. First stage: acid soluble fraction, second stage: reducible fraction and, third stage: oxidizable fraction.

Besides, the averages of total iron concentration in soil after SW with Tween 80, HPCD, and ultrapure water were 766, 776, and 850 mg kg^{-1} DW, respectively. The higher iron concentration after ultrapure water washing as compared to the extracting agents' SW experiments highlights the influence of such agents on mobilization/solubilization of inorganic compounds. Besides, the very low difference between Tween 80 and HPCD solutions confirms the low difference of extracted iron in both solutions.

8.3.1.2   IMPACT OF EXTRACTING AGENTS ON THE SOLUBILIZATION OF OTHER METALS/METALLOIDS

Knowing that iron could be extracted from soil with washing agent such as HPCD and Tween 80, it was interesting to investigate the possibility to extract other metals and metalloids from the soil. Therefore, the analysis of 10 elements (Al, As, Co, Cr, Cu, Hg, Mg, Ni, Pb, and Zn) was conducted in SW solutions and their concentrations are displayed in Figure 8.3. Few elements were below the detection limit such as As, Cr, Hg, and Ni, whatever the extracting agent used. Those elements had a concentration in soil not higher than around 20 mg kg^{-1} in soil, which could explain that they were not detectable in SW solutions. Contrastingly, other elements such as Al, Cu, Mg, Pb, and Zn were quantifiable in SW solutions; their concentrations in the initial soil were higher, that is, 13,000 mg kg^{-1} for Al, 41 mg kg^{-1} for Cu and 87 mg kg^{-1} for Pb, while Mg was one of the major constituent in saturation in the clay–humic complex. Al concentration was initially very high in the soil due to the high content of clay (20%), but its concentration in SW solutions was quite low (17 and 61 μg L^{-1} with SW-HPCD and SW-Tween 80, respectively), especially as compared to Mg and Zn extraction power (0.5–5.2 mg L^{-1}). It could be attributed to the stronger binding of Al that is covalently bonded within each mineral clay layer.[52] Furthermore, the amount of extracted metals was higher using Tween 80 washing agent than with HPCD, except for Zn element. It corroborates the higher iron content in SW-Tween 80 than in SW-HPCD solutions (Section 8.3.1.1). The lower solubilization power of Tween 80 toward Zn could not be explained and further experiments would be required to better figure out the mechanism of Zn extraction during washing with nonionic surfactants or beta-cyclodextrins. Interestingly, Pb was only quantified in Tween 80 SW solutions. This heavy metal (HM) is known to be toxic even at such low concentration[53] and could affect the toxicity/biodegradability of SW solutions as discussed in Section 8.3.3.

Though, the conventional EF process is carried out with one of the forms of the Fe^{3+}/Fe^{2+} redox couple ($E^0 = 0.77$ V/SHE), any appropriate homogeneous metallic redox couple ($M^{(n+1)+}/M^{n+}$) could be used according to Fenton-like reaction (eq 8.5) in order to form ·OH:[54]

$$M^{n+} + H_2O_2 \rightarrow M^{(n+1)+} + HO^- + \cdot OH \qquad (8.5)$$

The efficacy of such metals is influenced by the standard potential of $M^{(n+1)+}/M^{n+}$ redox couple used and by scavenging catalyst effect.[54] Among the elements detected in SW solutions, copper could be a candidate to enhance the ˙OH production in bulk solution during an EF treatment.[54] However, it has been previously demonstrated that Cu was not sufficient in promoting the ˙OH formation.[54] According to the cathodic potential optimally use in EF treatment ($-0.5/0.6$ V/SHE),[55] higher standard reduction potential of Cu(II) ($E^0(Cu^{2+}/Cu^+) = 0.34$ V/SHE) and Cu(I) ($Cu^+/Cu = 0.52$ V/SHE) ions could justify the cathodic deposition of copper by favoring their easy reduction.[54] Therefore, none of the extracted metal elements (except iron) would participate to the promotion of ˙OH formation during the electrooxidation treatment that is performed in Section 8.3.2.

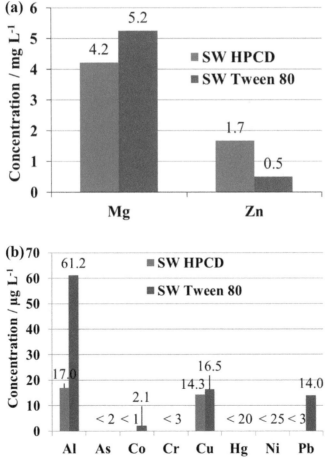

FIGURE 8.3   Influence of SW on metals/metalloids extraction by HPCD (7.5 g L[1]) or Tween 80 (7.5 g L[−1]). Absence of bar means that the concentration value is below the detection limit.

8.3.2 OXIDATION OF SW SOLUTIONS

8.3.2.1 DEGRADATION/MINERALIZATION EFFICIENCY OF SW SOLUTIONS

The electrochemical degradation and mineralization of real HPCD or Tween 80 SW solutions coming from a historically PAHs-contaminated soil are depicted in Figure 8.4. As above mentioned (Section 8.3.1), the presence of dissolved iron in solutions justifies that the electrochemical technology can be ascribed to an EF process by involving the ·OH production through

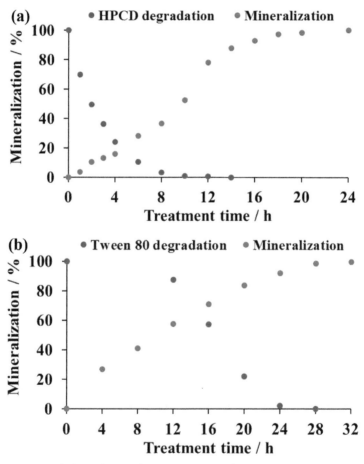

FIGURE 8.4 Evolution of extracting agent degradation and mineralization during the electrochemical treatment of SW HPCD (7.5 g L⁻¹) and SW Tween 80 (7.5 g L¹) solutions. *Operating conditions:* carbon felt cathode, BDD anode, current density: 6.67 mA cm⁻², $V =$ 400 mL.

the Fenton reaction (eq 8.1). Though initial SW solutions contain similar washing agent mass concentration (7.5 ± 0.2 g L^{-1})—equivalent to 5.7 ± 0.15 mmol L^{-1} and 6.0 ± 0.16 mmol L^{-1}, for HPCD and Tween 80, respectively—the degradation rate of HPCD (0.42 h^{-1}) was 4.2 times quicker than the Tween 80 decay (0.10 h^{-1}). Since Tween 80 allowed extracting much more organic pollutants and SOM than HPCD,[34] it could increase the $^{\cdot}OH$ oxidation competition between them, making the degradation slower. Interestingly, the mineralization rates were very similar during the first 10 h of treatment with SW-HPCD and SW-Tween 80 solutions and equaled 0.084 ± 0.3 h^{-1}. The mineralization of SOM and organic pollutants along with Tween 80 in SW-Tween 80 solutions could explain that the mineralization rate is still similar to SW-HPCD solution although the washing agent decay rate constant are not the same. Moreover, a mineralization yield higher than 99% could be observed after 20 h of treatment of SW-HPCD solution while it was after 24 h of treatment with SW-Tween 80 solutions. It highlights the slower mineralization at the end of treatment with SW-Tween 80 solution, which could be explained by the formation of more recalcitrant short-chain organic molecules toward $^{\cdot}OH$ radicals.[33,56] Still, quasicomplete mineralization could be reached with both kinds of SW solutions due to the strong oxidizing media promoting by (1) homogeneous $^{\cdot}OH$ formed through Fenton reaction in bulk solution (eq 8.1) and (2) heterogeneous $^{\cdot}OH$ generated at the surface of BDD anode (eqs 8.6, 8.7, and 8.8), due to its high O_2 evolution overpotential (2.3 V/SHE):[17,57]

$$BDD + H_2O \rightarrow BDD(^{\cdot}OH) + H^+ + e^- \qquad (8.6)$$

$$BDD(^{\cdot}OH) + \text{organic compound} \rightarrow BDD + \text{oxidation products} \qquad (8.7)$$

$$\text{Oxidation products} + {^{\cdot}OH}/BDD(^{\cdot}OH) \rightarrow \rightarrow \rightarrow CO_2 + H_2O \qquad (8.8)$$

8.3.2.2 EVOLUTION OF PH DURING OXIDATIVE TREATMENT

The pH of solution is determinant in processes involving Fenton reaction, due to the pH-dependency of iron ion species. A pH higher than 4 promotes the precipitation of ferric hydroxide ($Fe(OH)_3$), while a pH below 2 the protonation of H_2O_2 lead to the formation of $H_3O_2^+$ that is less reactive towards Fe^{2+} and make decrease the rate of Fenton's reaction.[17] The evolution of pH during the EF treatment of real SW-HPCD and SW-Tween 80 solutions was monitored and is displayed in Figure 8.5. The pH of SW solutions was not adjusted to 3 in order to avoid the use of acid reagent that increases the operating cost. Interestingly, a quick drop of pH is noticed after less than 2 h of electrolysis, from around the initial value of 8 to a plateau around 3. This is an interesting feature since this pH is optimal to operate EF treatment.[17] The pH decrease is due to the formation of carboxylic acids that can be formed very quickly, especially from the opening of aromatic rings during the degradation of pollutants. The presence of carboxylic acids and aromatics molecules in OM—much more present in Tween 80 solutions (due to its higher extraction capacity)—can also contribute to the acidification of solutions as previously stated.[34]

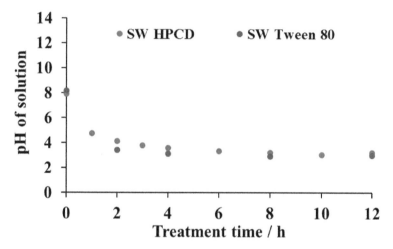

FIGURE 8.5 Evolution of solution pH during the electrochemical treatment of SW HPCD (7.5 g L^{-1}) and SW Tween 80 (7.5 g L^{-1}) solutions. *Operating conditions:* carbon felt cathode, BDD anode, current density: 6.67 mA cm^{-2}, V = 400 mL.

8.3.3   IMPACT OF OXIDATIVE TREATMENT ON ECOTOXICITY AND BIODEGRADABILITY EVOLUTION

The environmental impact such as ecotoxicity and biodegradability of effluent is a topic of major concerns since the release of water after EF treatment need to follow the regulations in order to avoid damaging the environment. Additionally, it is a key point if a postbiological treatment is considered in order to reduce the global electric energy consumption.[46,58]

In this context, two kinds of bioassays have been performed with the real SW solutions during the electrolysis treatment: (1) acute ecotoxicity tests by evaluating the inhibition of *V. fischeri* bacteria with bioluminescence device

(Microtox® tests) and (2) biodegradability assays represented by the BOD_5/ COD ratio.

8.3.3.1   INFLUENCE OF EXTRACTING AGENT ON ECOTOXICITY/ BIODEGRADABILITY EVOLUTION

The influence of extracting agent (HPCD and Tween 80) on the ecotoxicity and biodegradability evolution during oxidation of SW solutions has been assessed and the results are represented in Figure 8.6. The ecotoxicity and biodegradability of solutions seems inversely correlated whatever the kinds

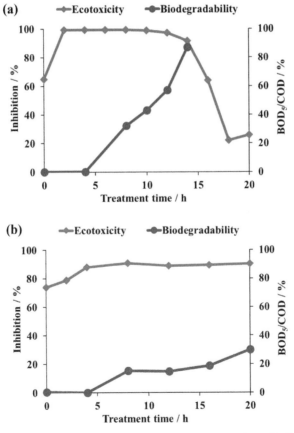

FIGURE 8.6   Influence of extracting agent ((a) HPCD (7.5 g L^{-1}) or (b) Tween 80 (7.5 g L^{-1})) on *Vibrio fischeri* inhibition and biodegradability (BOD_5/COD) evolution during EF treatment of SW solutions. *Operating conditions:* carbon felt cathode, BDD anode, current density: 6.67 mA cm^{-2}, V = 400 mL.

of SW solutions since the bacteria inhibition is very high at the beginning of the treatment along with the BOD_5/COD ratio that is very low. This lag phase could be attributed to the formation of hydroxylated by-products that are formed at the beginning of the EF treatment. These kind of compounds are known to be often toxic even more than the initial molecules.[34,39] Then, in the case of HPCD solutions, the toxicity starts decreasing, while the biodegradability increases even more quickly. At this time, all the xenobiotic compounds are degraded into smaller linear aliphatic compounds that are much easier to biodegrade.[39,56] In contrast, the toxicity of Tween 80 solutions was not decreasing and remained stable all along the treatment, even after quasicomplete mineralization. Biodegradability results confirm this trend since it barely increases in the meanwhile. Still, the biodegradability could reach 33% after 8.5 and 20 h with HPCD and Tween 80 solutions, respectively, that is a threshold value suitable to consider a biological posttreatment.[50]

Several reasons could explain the much lower biodegradability of Tween 80: (1) the higher solubilization power of Tween 80 that extracted more toxic and recalcitrant pollutants,[33] (2) the lower ability of cyclodextrins to generate toxic intermediates,[40] and/or (3) the presence of Pb in initial SW-Tween 80 solutions (Section 8.3.1). In order to better figure out the explanation, further experiments have been performed by comparing the evolution of ecotoxicity and biodegradability during the EF treatment of synthetic and real SW solutions in the following section.

8.3.3.2   IMPACT OF SOILS' METALS/METALLOIDS ON BIODEGRADABILITY EVOLUTION

The influence of metals/metalloids from real SW solutions on the biodegradability evolution during an EF treatment was indirectly carried out by comparing the biodegradability evolution of synthetic SW solutions with real SW solutions (Fig. 8.7). The comparison between synthetic (containing PAH pollutant) and real (containing mainly PAHs pollutants and SOM) SW-HPCD solutions (Fig. 8.7a) showed that the biodegradability evolution was similar until 8 h of EF treatment. It was slower after 8 h of treatment with real solutions, and it reached 90% of biodegradability against 100% of biodegradability after 14 h with real and synthetic solutions, respectively. It could be attributed to the presence of SOM that could lead to less biodegradable intermediates in real samples. Considering Tween 80 solutions (Fig. 8.7b), the real solutions containing mainly PAHs pollutants, SOM, and

HMs (such as Pb as mentioned in Section 8.3.1) lead to lower biodegrad-ability evolution than in synthetic solution (containing PAH pollutant). It could be accredited to the presence of SOM as noticed with HPCD solu-tions and/or the presence of HMs as well. Moreover, the biodegradability evolution of synthetic SW-Tween 80 solution was still quite low (44%) even after 20 h of EF treatment. All these statements would make Tween 80 as another responsible constituent in the low biodegradability evolution due to the formation of more toxic intermediates than HPCD. This was further studied in Section 8.3.3.3.

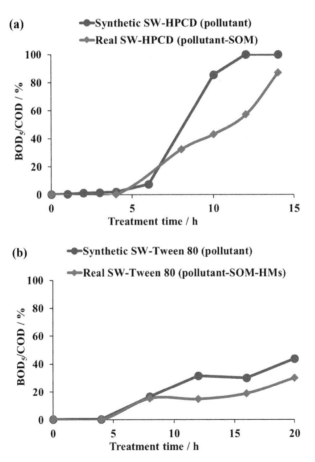

FIGURE 8.7 Influence of metals/metalloids on biodegradability (BOD_5/COD) evolution during EF treatment of (a) HPCD (9 g L^{-1} (synthetic)* and 7.5 g L^{-1} (real)) and (b) Tween 80 (9 g L^{-1} (synthetic)* and 7.5 g L^{-1} (real)) SW solutions. *Operating conditions:* carbon felt cathode, BDD anode, current density: 6.67 mA cm^{-2}, V = 400 mL.*Reprinted from Mousset et al.[39,59]

8.3.3.3 EFFECT OF SURFACTANT CONCENTRATION ON BIODEGRADABILITY EVOLUTION

In order to better understand the influence of Tween 80 on the biodegradability evolution during an EF treatment, different concentrations of Tween 80 have been varied from 0.8 to 12 g L^{-1} in synthetic solutions (Fig. 8.8). Interestingly, a 12-g-L^{-1} concentration of Tween 80 gave higher biodegradability yields (47%) against 27% and 20% with Tween 80 at 0.8 g L^{-1} and 6 g L^{-1}, respectively. It is noticed that a higher concentration of Tween 80 was not synonym of a lower biodegradability evolution and was even higher after a certain duration of treatment (10 h). It further highlighted that the Tween 80 at a concentration of 7.5 g L^{-1} employed in real SW-Tween 80 solutions as shown in Section 8.3.3.2 could also participate in the low biodegradability improvement by favoring the formation of less biodegradable by-products.

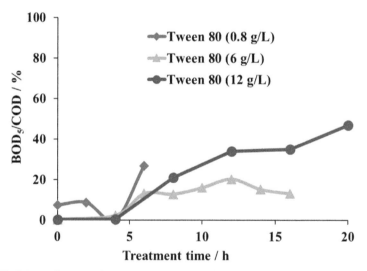

FIGURE 8.8 Influence of Tween 80 concentration on biodegradability (BOD$_5$/COD) evolution during EF treatment. *Operating conditions:* carbon felt cathode, BDD anode, current density: 6.67 mA cm^{-2}, $V = 400$ mL.

8.4 CONCLUSIONS

The quantity of metals/metalloids extracted by extracting agent-enhanced SW technique was monitored and their influence on the subsequent EF treatment efficiency was evaluated. Dissolved iron was present in both kinds of washing agent (e.g., HPCD and Tween 80) which was enough to involve production of ·OH through Fenton's reaction in EF process. The nonionic surfactant could solubilize more metals/metalloid than HPCD in SW solutions. After investigation, the impact of real SW solutions on EF treatment efficiency could be explained according to the two following reasons: (1) the higher extraction power of Tween 80 that extracted more recalcitrant pollutants, SOM, and HMs such as Pb as compared to HPCD and (2) the higher ability of Tween 80 to generate toxic intermediates than HPCD.

Further studies would be required to separate the different component of SW solutions (organic pollutants, SOM, HMs, extracting agent) and/or to selectively treat the organic pollutants while recycling the extracting agent and valorizing the metals. It therefore opens up possibilities to treat and valorize mixed pollution in contaminated soil.

KEYWORDS

- soil preparation
- soil washing experiments
- electrooxidation treatment
- electro-Fenton
- sequential extractions
- bioassays

REFERENCES

1. Swartjes, F. A. *Dealing with Contaminated Sites: From Theory Towards Practical Applications*; Swartjes, F. A., Ed.; Springer: Dordrecht, Netherlands, 2011.
2. European Environment Agency. *Overview of Activities Causing Soil Contamination in Europe*; Copenhagen, 2007.
3. Srogi, K. Monitoring of Environmental Exposure to Polycyclic Aromatic Hydrocarbons: A Review. *Environ. Chem. Lett.* **2007**, *5*(4), 169–195.
4. Manoli, E.; Samara, C. Polycyclic Aromatic Hydrocarbons in Natural Waters: Sources, Occurrence and Analysis. *TrAC Trends Anal. Chem.* **1999**, *18*(6), 417–428.
5. INERIS (Institut National de l'environnement industriel et des Risques). *Toxicological and Environmental Data of Chemical Substances—PAH (Fiches de Données*

Toxicologiques et Environnementales Des Substances Chimiques—HAP); Verneuil-en-Halatte, France, 2005.

6. Cai, Y.; Zheng, H.; Ding, S.; Kropachev, K.; Schwaid, A. G.; Tang, Y.; Mu, H.; Wang, S.; Geacintov, N. E.; Zhang, Y.; et al. Free Energy Profiles of Base Flipping in Intercalative Polycyclic Aromatic Hydrocarbon-Damaged DNA Duplexes: Energetic and Structural Relationships to Nucleotide Excision Repair Susceptibility. *Chem. Res. Toxicol.* **2013**, *26*(7), 1115–1125.

7. WHO (World Health Organization). *Guidelines for Indoor Air Quality: Selected Pollutants*, 2010.

8. Venny; Gan, S.; Ng, H. K. Modified Fenton Oxidation of Polycyclic Aromatic Hydrocarbon (PAH)-Contaminated Soils and the Potential of Bioremediation as Post-Treatment. *Sci. Total Environ.* **2012**, *419*, 240–249.

9. Colombano, S.; Saada, A.; Guerin, V.; Bataillard, P.; Bellenfant, G.; Beranger, S.; Hube, D.; Blanc, C.; Zornig, C.; Girardeau, I. *Quelles Techniques Pour Quels Traitements—Analyse Coûts-Bénéfices*, 2010.

10. Cadiere, F.; Dueso, N.; Margot, D.; Marion, R.; Colombano, S.; De La Hougue, C.; Laffaire, D.; Brun, J.; Bourdin, C.; Valtech, G. R. S. *Taux D'utilisation et Coûts Des Différentes Techniques et Filières de Traitement Des Sols et Des Eaux Souterraines Pollués En France*, 2011.

11. Mousset, E.; Oturan, M. A.; Van Hullebusch, E. D.; Guibaud, G.; Esposito, G. Soil Washing/Flushing Treatments of Organic Pollutants Enhanced by Cyclodextrins and Integrated Treatments: State of the Art. *Crit. Rev. Environ. Sci. Technol.* **2014**, *44*(7), 705–795.

12. Trellu, C.; Mousset, E.; Pechaud, Y.; Huguenot, D.; Hullebusch, E. D. Van; Esposito, G.; Oturan, M. A. Removal of Hydrophobic Organic Pollutants from Soil Washing/Flushing Solutions : A Critical Review. *J. Hazard. Mater.* **2016**, *306*, 149–174.

13. Paria, S. Surfactant-Enhanced Remediation of Organic Contaminated Soil and Water. *Adv. Colloid Interface Sci.* **2008**, *138*(1), 24–58.

14. Alcántara, M. T.; Gómez, J.; Pazos, M.; Sanromán, M. A. Combined Treatment of PAHs Contaminated Soils Using the Sequence Extraction with Surfactant-Electrochemical Degradation. *Chemosphere* **2008**, *70*(8), 1438–1444.

15. Dhenain, A.; Mercier, G.; Blais, J.-F.; Bergeron, M. PAH Removal from Black Sludge from Aluminium Industry by Flotation Using Non-Ionic Surfactants. *Environ. Technol.* **2006**, *27*(9), 1019–1030.

16. Oturan, M. A.; Aaron, J.-J. Advanced Oxidation Processes in Water/Wastewater Treatment: Principles and Applications. A Review. *Crit. Rev. Environ. Sci. Technol.* **2014**, *44*(23), 2577–2641.

17. Brillas, E.; Sirés, I.; Oturan, M. A. Electro-Fenton Process and Related Electrochemical Technologies Based on Fenton's Reaction Chemistry. *Chem. Rev.* **2009**, *109*(12), 6570–6631.

18. Martinez-Huitle, C. A.; Rodrigo, M. A.; Sires, I.; Scialdone, O. Single and Coupled Electrochemical Processes and Reactors for the Abatement of Organic Water Pollutants : A Critical Review. *Chem. Rev.* **2015**, *115*(24), 13362–13407.

19. Sirés, I.; Brillas, E.; Oturan, M. A.; Rodrigo, M. A.; Panizza, M. Electrochemical Advanced Oxidation Processes: Today and Tomorrow. A Review. *Environ. Sci. Pollut. Res. Int.* **2014**, *21*(14), 8336–8367.

20. Moreira, F. C.; Boaventura, R. A. R.; Brillas, E.; Vilar, V. J. P. Electrochemical Advanced Oxidation Processes: A Review on Their Application to Synthetic and Real Wastewaters. *Appl. Catal. B: Environ.* **2017**, *202*, 217–261.

21. Rodrigo, M. A.; Oturan, N.; Oturan, M. A. Electrochemically Assisted Remediation of Pesticides in Soils and Water: A Review. *Chem. Rev.* **2014**, *114*(17), 8720–8745.

22. Nidheesh, P. V.; Gandhimathi, R. Trends in Electro-Fenton Process for Water and Wastewater Treatment: An Overview. *Desalination.* **2012**, *299*, 1–15.

23. Oturan, M. A. Ecologically Effective Water Treatment Technique Using Electrochemically Generated Hydroxyl Radicals for In Situ Destruction of Organic Pollutants: Application to Herbicide 2,4-D. *J. Appl. Electrochem.* **2000**, *30*(4), 475–482.

24. Mousset, E.; Ko, Z. T.; Syafiq, M.; Wang, Z.; Lefebvre, O. Electrocatalytic Activity Enhancement of a Graphene Ink-Coated Carbon Cloth Cathode for Oxidative Treatment. *Electrochim. Acta.* **2016**, *222*, 1628–1641.

25. Mousset, E.; Wang, Z.; Lefebvre, O. Electro-Fenton for Control and Removal of Micropollutants—Process Optimization and Energy Efficiency. *Water Sci. Technol.* **2016**, *74*(2), 1–7.

26. Sirés, I.; Garrido, J. A.; Rodríguez, R. M.; Brillas, E.; Oturan, N.; Oturan, M. A. Catalytic Behavior of the Fe^{3+}/Fe^{2+} System in the Electro-Fenton Degradation of the Antimicrobial Chlorophene. *Appl. Catal. B: Environ.* **2007**, *72*(3–4), 382–394.

27. Oturan, N.; Brillas, E.; Oturan, M. A. Unprecedented Total Mineralization of Atrazine and Cyanuric Acid by Anodic Oxidation and Electro-Fenton with a Boron-Doped Diamond Anode. *Environ. Chem. Lett.* **2012**, *10*(2), 165–170.

28. Sopaj, F.; Oturan, N.; Pinson, J.; Podvorica, F.; Oturan, M. A. Effect of the Anode Materials on the Efficiency of the Electro-Fenton Process for the Mineralization of the Antibiotic Sulfamethazine. *Appl. Catal. B: Environ.* **2016**, *199*, 331–341.

29. Trellu, C.; Péchaud, Y.; Oturan, N.; Mousset, E.; Huguenot, D.; van Hullebusch, E. D.; Esposito, G.; Oturan, M. A. Comparative Study on the Removal of Humic Acids from Drinking Water by Anodic Oxidation and Electro-Fenton Processes: Mineralization Efficiency and Modelling. *Appl. Catal. B: Environ.* **2016**, *194*, 32–41.

30. Olvera-Vargas, H.; Oturan, N.; Oturan, M. A.; Brillas, E. Electro-Fenton and Solar Photoelectro-Fenton Treatments of the Pharmaceutical Ranitidine in Pre-Pilot Flow Plant Scale. *Sep. Purif. Technol.* **2015**, *146*, 127–135.

31. Ma, L.; Zhou, M.; Ren, G.; Yang, W.; Liang, L. A Highly Energy-Efficient Flowthrough Electro-Fenton Process for Organic Pollutants Degradation. *Electrochim. Acta.* **2016**, *200*, 222–230.

32. García-Rodríguez, O.; Bañuelos, J. A.; El-Ghenymy, A.; Godínez, L. A.; Brillas, E.; Rodríguez-Valadez, F. J. Use of a Carbon Felt-Iron Oxide Air-Diffusion Cathode for the Mineralization of Malachite Green Dye by Heterogeneous Electro-Fenton and UVA Photoelectro-Fenton Processes. *J. Electroanal. Chem.* **2016**, *767*, 40–48.

33. Huguenot, D.; Mousset, E.; van Hullebusch, E. D.; Oturan, M. A. Combination of Surfactant Enhanced Soil Washing and Electro-Fenton Process for the Treatment of Soils Contaminated by Petroleum Hydrocarbons. *J. Environ. Manage.* **2015**, *153*, 40–47.

34. Mousset, E.; Huguenot, D.; Van Hullebusch, E. D.; Oturan, N.; Guibaud, G.; Esposito, G.; Oturan, M. A. Impact of Electrochemical Treatment of Soil Washing Solution on PAH Degradation Efficiency and Soil Respirometry. *Environ. Pollut.* **2016**, *211*, 354–362.

35. Rosales, E.; Pazos, M.; Longo, M. A.; Sanroman, M. A. Influence of Operational Parameters on Electro-Fenton Degradation of Organic Pollutants from Soil. *J. Environ. Sci. Health. A: Tox. Hazard. Subst. Environ. Eng.* **2009**, *44*(11), 1104–1110.

36. Murati, M.; Oturan, N.; van Hullebusch, E. D.; Oturan, M. A. Electro-Fenton Treatment of TNT in Aqueous Media in Presence of Cyclodextrin. Application to Ex-Situ Treatment of Contaminated Soil Abstract. *J. Adv. Oxid. Technol.* **2009**, *12*(1), 29–36.

37. Trellu, C.; Ganzenko, O.; Papirio, S.; Pechaud, Y.; Oturan, N.; Huguenot, D.; van Hullebusch, E. D.; Esposito, G.; Oturan, M. A. Combination of Anodic Oxidation and Biological Treatment for the Removal of Phenanthrene and Tween 80 from Soil Washing Solution. *Chem. Eng. J.* **2016**, *306*, 588–596.

38. Mousset, E.; Oturan, N.; van Hullebusch, E. D.; Guibaud, G.; Esposito, G.; Oturan, M. A. A New Micelle-Based Method to Quantify the Tween 80® Surfactant for Soil Remediation. *Agron. Sustain. Dev.* **2013**, *33*(4), 839–846.

39. Mousset, E.; Oturan, N.; van Hullebusch, E. D.; Guibaud, G.; Esposito, G.; Oturan, M. A. Treatment of Synthetic Soil Washing Solutions Containing Phenanthrene and Cyclodextrin by Electro-Oxidation. Influence of Anode Materials on Toxicity Removal and Biodegradability Enhancement. *Appl. Catal. B: Environ.* **2014**, *160–161*, 666–675.

40. Mousset, E.; Oturan, N.; van Hullebusch, E. D.; Guibaud, G.; Esposito, G.; Oturan, M. A. Influence of Solubilizing Agents (Cyclodextrin or Surfactant) on Phenanthrene Degradation by Electro-Fenton Process—Study of Soil Washing Recycling Possibilities and Environmental Impact. *Water Res.* **2014**, *48*(1), 306–316.

41. López-Vizcaíno, R.; Sáez, C.; Cañizares, P.; Rodrigo, M. A. The Use of a Combined Process of Surfactant-Aided Soil Washing and Coagulation for PAH-Contaminated Soils Treatment. *Sep. Purif. Technol.* **2012**, *88*, 46–51.

42. dos Santos, E. V.; Saez, C.; Martinez-Huitle, C. A.; Canizares, P.; Rodrigo, M. A. Combined Soil Washing and CDEO for the Removal of Atrazine from Soils. *J. Hazard. Mater.* **2015**, *300*, 129–134.

43. dos Santos, E. V.; Sáez, C.; Martínez-Huitle, C. A.; Cañizares, P.; Rodrigo, M. A. The Role of Particle Size on the Conductive Diamond Electrochemical Oxidation of Soil-Washing Effluent Polluted with Atrazine. *Electrochem. Commun.* **2015**, *55*, 26–29.

44. Hanna, K.; Chiron, S.; Oturan, M. A. Coupling Enhanced Water Solubilization with Cyclodextrin to Indirect Electrochemical Treatment for Pentachlorophenol Contaminated Soil Remediation. *Water Res.* **2005**, *39*(12), 2763–2773.

45. Ganzenko, O.; Huguenot, D.; van Hullebusch, E. D.; Esposito, G.; Oturan, M. A. Electrochemical Advanced Oxidation and Biological Processes for Wastewater Treatment: A Review of the Combined Approaches. *Environ. Sci. Pollut. Res. Int.* **2014**, *21*(14), 8493–8524.

46. Olvera-Vargas, H.; Oturan, N.; Buisson, D.; Oturan, M. A. A Coupled Bio-EF Process for Mineralization of the Pharmaceuticals Furosemide and Ranitidine: Feasibility Assessment. *Chemosphere* **2016**, *155*, 606–613.

47. Pérez-Cid, B.; Lavilla, I.; Bendicho, C. Speeding up of a Three-Stage Sequential Extraction Method for Metal Speciation Using Focused Ultrasound. *Anal. Chim. Acta.* **1998**, *360*, 35–41.

48. Mossop, K. F.; Davidson, C. M. Comparison of Original and Modified BCR Sequential Extraction Procedures for the Fractionation of Copper, Iron, Lead, Manganese and Zinc in Soils and Sediments. *Anal. Chim. Acta.* **2003**, *478*(1), 111–118.

49. Dirany, A.; Efremova Aaron, S.; Oturan, N.; Sirés, I.; Oturan, M. A.; Aaron, J. Study of the Toxicity of Sulfamethoxazole and Its Degradation Products in Water by a Bioluminescence Method During Application of the Electro-Fenton Treatment. *Anal. Bioanal. Chem.* **2011**, *400*(2), 353–360.

50. Rodier, J.; Legube, B.; Merlet, N. *Analyse de L'eau (Water Analysis)*, 9th ed.; Dunod, Paris (in French), 2009.
51. Gu, B.; Mehlhorn, T.; Liang, L.; McCarthy, J. Competitive Adsorption, Displacement, and Transport of Organic Matter on Iron Oxide: I. Competitive Adsorption. *Geochim. Cosmochim. Acta.* **1996**, *60*(11), 1943–1950.
52. Sposito, G. *The Chemistry of Soils*, 2nd ed.; Oxford University Press: New York, NY, 2008.
53. Flora, G.; Gupta, D.; Tiwari, A. Toxicity of Lead : A Review with Recent Updates. *Interdiscip. Toxicol.* **2012**, *5*(2), 47–58.
54. Pimentel, M.; Oturan, N.; Dezotti, M.; Oturan, M. A. Phenol Degradation by Advanced Electrochemical Oxidation Process Electro-Fenton Using a Carbon Felt Cathode. *Appl. Catal. B: Environ.* **2008**, *83*(1–2), 140–149.
55. Mousset, E.; Wang, Z.; Hammaker, J.; Lefebvre, O. Physico-Chemical Properties of Pristine Graphene and Its Performance as Electrode Material for Electro-Fenton Treatment of Wastewater. *Electrochim. Acta.* **2016**, *214*, 217–230.
56. Oturan, M. A.; Pimentel, M.; Oturan, N.; Sirés, I. Reaction Sequence for the Mineralization of the Short-Chain Carboxylic Acids Usually Formed upon Cleavage of Aromatics During Electrochemical Fenton Treatment. *Electrochim. Acta.* **2008**, *54*(2), 173–182.
57. Panizza, M.; Cerisola, G. Direct and Mediated Anodic Oxidation of Organic Pollutants. *Chem. Rev.* **2009**, *109*(12), 6541–6569.
58. Trellu, C.; Ganzenko, O.; Papirio, S.; Pechaud, Y.; Oturan, N.; Huguenot, D.; van Hullebusch, E. D.; Esposito, G.; Oturan, M. A. Combination of Anodic Oxidation and Biological Treatment for the Removal of Phenanthrene and Tween 80 from Soil Washing Solution. *Chem. Eng. J.* **2016**, *306*, 588–596.
59. Mousset, E. *Integrated Processes for Removal of Persistent Organic Pollutants: Soil Washing and Electrochemical Advanced Oxidation Processes Combined to a Possible Biological Post-Treatment*; University of Paris-Est—University of Cassino and The Southern Lazio—UNESCO-IHE for Water Education, 2013.

PART III
Physical Chemistry from Different Angles

CHAPTER 9

AN EXPERIMENTAL STUDY ON DRILLING OF GLASS-FABRIC-REINFORCED EPOXY COMPOSITES FILLED WITH MICROFILLERS

G. RAVICHANDRAN[1,*], K. RAJU[2], Y. S. VARADARAJAN[3], and B. SURESHA[3]

[1]Faculty of Engineering, Christ (Deemed to be University), Bangalore 560074, India

[2]St. Joseph Engineering College, Mangalore 575028, India

[3]The National Institute of Engineering, Mysore 570008, India

[*]Corresponding author. E-mail: ravig_s@rediffmail.com

CONTENTS

ABSTRACT

The objective of this study is to evaluate the performance of high speed steel (HSS) and carbide drills by measuring the various drilling parameters such as thrust, torque, delamination, specific cutting pressure, and surface roughness. Two microfillers, namely, alumina and graphite were used. The amount of fillers addition was 5 wt% in glass-fabric-reinforced epoxy composite. All composites were fabricated using hand lay-up technique followed by compression molding. Drilling experiments were conducted with HSS and carbide drills under various cutting conditions, keeping diameter of the drill constant. Topography of drilled hole surfaces were analyzed through scanning electron microscope micrographs. From the experimental investigation, it was found that plain/neat composites produced better quality holes compared to particulate-filled composites when machined with HSS drills. Carbide drills produced better quality holes in particulate-filled composites, especially with graphite-filled ones.

9.1   INTRODUCTION

Glass-fabric-reinforced epoxy (G–E) composite materials are nowadays being commonly used in aircraft space, automotive industry, and others due to their high specific strength, hardness, better corrosion, and heat resistance compared to the other engineering materials.[1,2] During assembly of composite components, drilling operation is commonly used. As composites are nonhomogeneous and anisotropic, during drilling the tool encounters alternately matrix and reinforcement, resulting in force fluctuation and several types of surface damages such as matrix cracking, fiber pullout, fuzzing and delamination. Delamination particularly deleterious, since it creates a potential point of origin for failure during service.[3,4] Due to these damages occurring lot of mechanical parts are being rejected. For instance in the aircraft industry, 60% of the parts are being refused only because of these damages.[5] In the drilling of composite materials, some unconventional methods (water jet, laser, electro erosion) have been tested but because of the heat and mechanical properties of the material, conventional methods have been preferred in the economical drilling of holes with less damage.[6] Several researchers in their studies on the drilling of composite materials specified that the damages occurring at the drilling operations were due to cutting parameters and tool geometry.[7–11] Davim et al.[12] studied the cutting parameters (cutting velocity and feed) and the influence of the matrix under specific cutting force (K_c), delamination factor (F_d), and surface roughness (R_a) in two types of matrix (Viapal VUP 9731 and ATLAC 382-05). Abrao et al.,[13] Tsao et al.,[14] and Velayudham et al.[15] established that cutting parameters and tool geometry have a considerable influence on the drilling of composite materials. Mohan et al.[16] found that the damage factor decreased at low feed rates and high cutting speeds. Rubio et al.[17] carried out high speed machining to realize high performance drilling of glass-fiber-reinforced plastics (GFRP) with reduced damage. Based on the literature cited above, an experimental study on drilling of G–E composites filled with microfillers has been carried out to evaluate the performance of high speed steel (HSS) and carbide drills.

9.2 EXPERIMENTAL

9.2.1 MATERIALS

Woven glass plain weave fabrics made of 360 g/m², containing E-glass fibers of diameter of about 12 μm, have been employed. The epoxy resin (LAPOX L-12 with density 1.16 g/m²) was mixed with the hardener (K-6, supplied by ATUL India Ltd., Gujarat, India) in the ratio 10:1.2 by weight. Twenty-four layers of fabrics were used to obtain approximately laminates of thickness 10 mm. The microfillers chosen were aluminum oxide and graphite. The average particle size of microfillers is about 10 μm. A hand lay-up technique was used to fabricate G–E composites followed by compression molding. The weight percent of the glass fiber and microfillers in the composite is 60 and 5, respectively. The quality of all laminates was assessed by ultrasonic C-scanning prior to the test. The detailed compositions along with the designation are presented in Table 9.1.

TABLE 9.1 Composite Material Designation and Its Composition.

Material designation	Material composition		
	Epoxy (wt%)	Glass fibers (wt%)	Microfiller (wt%)
Neat/plain	40	60	0
Alumina	35	60	5
Graphite	35	60	5

9.2.2 TESTING PROCEDURE

The dimension of composites prepared for drill test is $250 \times 60 \times 10$ mm. Drilling trials were performed on a Praga column drilling machine by using two flutes HSS and carbide twist drills of 6 mm diameter with 118° point angle. The cutting force components such as thrust force F (N) and torque T (N m) for different cutting conditions were measured online using drill tool dynamometer. The details of cutting parameters used were presented in Table 9.2.

TABLE 9.2 Cutting Parameters.

Cutting speed (V_c) (m/min)	Feed rate (f) (mm/rev)
14.3	0.11
17.3	0.16
21.1	0.24

The delamination occurring on the side of the hole entrance was established by the determination of damage factor (F_d),[11] which was established by Mitutoyo Co-ordinate measuring machine, which gives the value directly. The value of specific cutting pressure K_c (MPa) was calculated by using an equation.[12] Surface roughness of the wall of the drilled hole was measured offline using Mitutoyo surface roughness measuring machine. Drilled surfaces were examined using scanning electron microscope (SEM) and analyzed.

9.3 RESULTS AND DISCUSSION

9.3.1 EFFECT OF CUTTING PARAMETERS ON THE THRUST FORCE

Figure 9.1 shows the parametric variation of thrust force with HSS and carbide drills for plain, alumina, and graphite-filler-filled G–E composites under same cutting conditions. It can be observed that the thrust force increased rapidly while drilling with HSS drill. This can be attributed to increased wear of HSS drills with increased cutting conditions. With carbides, the force values remained less and variation is also marginal. This suggests the better performance of carbide drills while machining G–E composites.

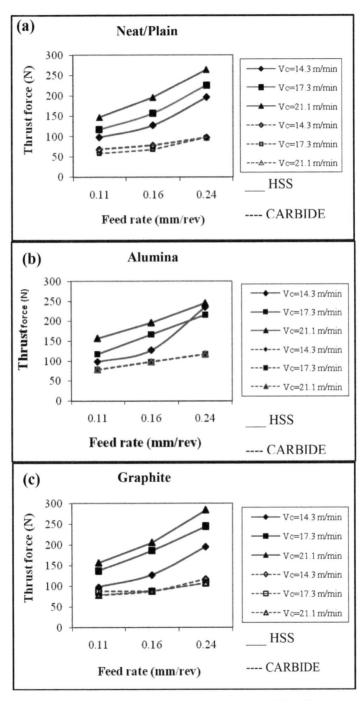

FIGURE 9.1 Typical variation in thrust force with HSS and carbide drills.

9.3.2 EFFECT OF CUTTING PARAMETERS ON THE TORQUE

Figure 9.2 shows the parametric variation of torque with HSS and carbide drills for plain, alumina, and graphite-filler-filled G–E composite under same cutting conditions. It can be observed that torque variation does not show any definite trend for both the drill materials and all the composites. It was also found that carbide drill presents lower torque value compared to HSS drill for all the G–E composites.

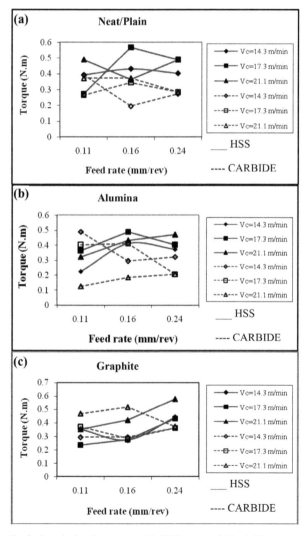

FIGURE 9.2   Typical variation in torque with HSS and carbide drills.

9.3.3 EFFECT OF CUTTING PARAMETERS ON THE DELAMINATION

The parametric variation of delamination factor (F_d) with HSS and carbide drills for plain, alumina, and graphite-filler-filled G–E composites under same cutting conditions as shown in Figure 9.3. It can be observed that the delamination factor registered a higher value compared to carbide drill. When drilled with HSS tool, a marginal increase in delamination factor for

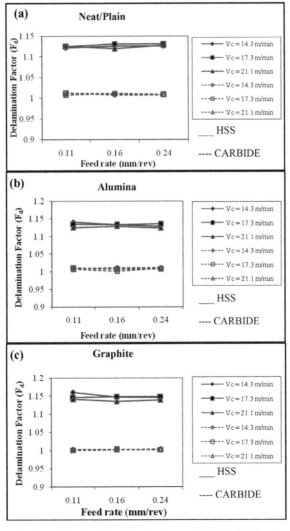

FIGURE 9.3 Typical variation in delamination factor (F_d) with HSS and carbide drills.

plain composites and a small decrease for particulate-filled composites was observed. With carbide drill, delamination factor decreased marginally with feed for all the composites. These suggest better performance of carbide drills while machining G–E composites.

9.3.4 EFFECT OF CUTTING PARAMETERS ON THE SPECIFIC CUTTING PRESSURE

A typical observation on influence of drill material on specific cutting pressure during drilling of G–E composites is illustrated in Figure 9.4. It can be seen from Figure 9.4 that the specific cutting pressure (K_c) decreased with significant change in the value of the same for both the drills while drilling all the G–E composites. It can also be observed that under the same cutting conditions, carbide drill registered a lower value of specific cutting pressure for all the G–E composites suggesting a better performance.

9.3.5 EFFECT OF CUTTING PARAMETERS ON THE SURFACE ROUGHNESS

Figure 9.5 shows the typical variation of surface roughness value (R_a) with HSS and carbide drills for plain, alumina, and graphite-filler-filled G–E composites under same cutting conditions. It can be observed that the surface roughness value increased with feed while drilling with both HSS and carbide drills. It can also be observed that under higher speed and feed conditions, carbide drill registers a lower value of surface roughness (R_a) for all the G–E composites.

9.3.6 DRILLED HOLE SURFACE TOPOGRAPHY

Topography of drilled surface of G–E produced by HSS and carbide drills under higher cutting conditions is shown in Figures 9.6 and 9.7. The thrust force caused visible cracking of surface layer, which resulted in deterioration of the surface. Figure 9.6b and c shows that particulate-filled G–E have more fiber fragmentation, matrix debris, large fiber breakage, inclined fiber fracture, severe debonding at fiber matrix interface, and the like compared to neat/plain. Similarly, surface roughness is higher which depends on the nature of the tool material which is responsible for increased cracking on

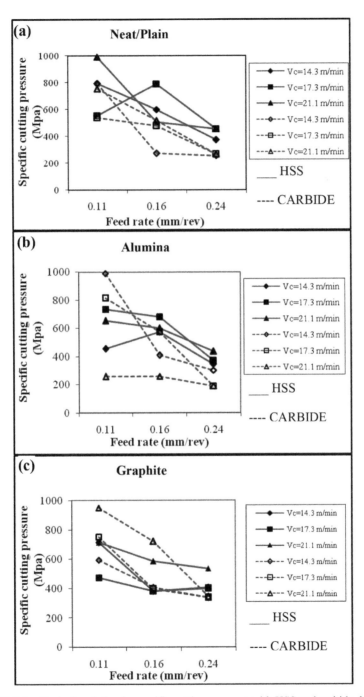

FIGURE 9.4 Typical variation in specific cutting pressure with HSS and carbide drills.

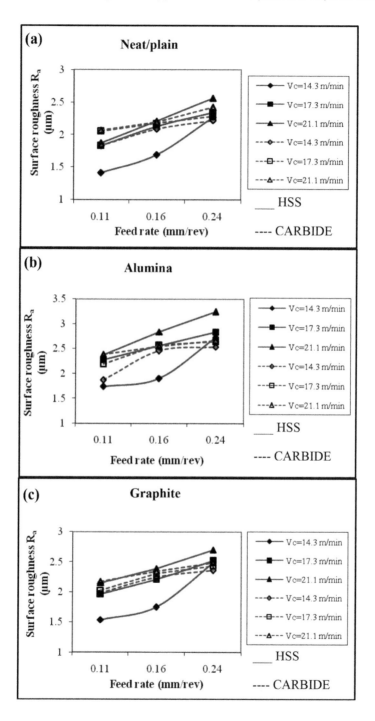

FIGURE 9.5   Typical variation in surface roughness (R_a) with HSS and carbide drills.

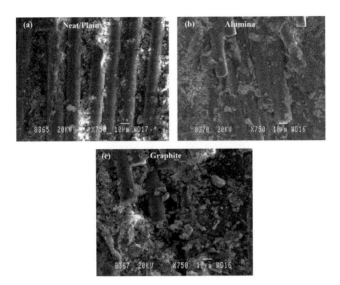

FIGURE 9.6 Drilled hole surface Topography by HSS at 750× at 21.1 m/min and 0.24 mm/rev.

surface. Figure 9.7a shows that neat/plain exhibits higher longitudinal fiber breakage, matrix cracking, more debris formation, inclined fiber fracture, plowing, cohesive resin fracture, and so on compared to particulate-filled composites (Fig. 9.7b and c). It suggests that carbide drills produces minimal material damage for particulate-filled G–E composites.

FIGURE 9.7 Drilled hole surface topography by carbide at 750× at 21.1 m/min and 0.24 mm/rev.

9.4 CONCLUSIONS

Based on the analysis of experimental results and findings, the following conclusions can be drawn: The present work showed that thrust force increased with feed rate for all composite systems. Addition of particulate filler showed increased thrust force (F) compared with the neat/plain composite under the same cutting conditions for both HSS and carbide drills. Observation on variation of torque suggested an existence of critical feed rate of 0.16 mm/rev at which a change in trend is seen for both the drill materials. From the delamination study, it is seen that HSS drills produced holes of better quality on neat/plain composites, whereas carbide drills produced better quality holes on particulate-filled composites, especially with graphite-filled one. The specific cutting pressure decreased with the feed rate and slightly with the cutting speed. Measurement of surface roughness values suggests that the surface deteriorated with increase in feed and speed. Addition of particulate filler showed increased surface roughness (R_a) compared with the neat/plain composite under the same cutting conditions for both HSS and carbide drill. Alumina-filled G–E composites registered higher surface roughness value. SEM micrographs show that in drilling of particulate-filled G–E composites by carbide drill bit minimal material damage was observed.

ACKNOWLEDGMENTS

The authors are grateful to management of St. Joseph Engineering College, Mangalore, Karnataka for their financial support. The authors also thankfully acknowledge the NIE, Mysore, Karnataka for providing the facilities to conduct the work.

KEYWORDS

- **glass-fabric-reinforced epoxy**
- **microfillers**
- **drilling**
- **topography**
- **glass fiber**

REFERENCES

1. El-Sonbaty, I.; Khashaba, U. A.; Machaly, T. *Compos. Struct.* **2004**, *63*, 329–338.
2. Enemuoh, U.; Sherif El-Gizawy, E.; Chukwujekwu, A.; Okafor, A. *Int. J. Mach. Tools Manuf.* **2001**, *41*, 1795–1814.
3. Ramkumar, J.; Malhotra, S. K.; Krishnamurthy, R. *J. Mater. Process. Technol.* **2004**, *152*, 329–332.
4. Singh, I.; Bhatnagar, N.; Viswanath, P. *Mater. Des.* **2008**, *29*, 546–553.
5. Khashaba, U. A. *Compos. Struct.* **2004**, *63*, 313–327.
6. Konig, W.; Wulf, C.; Grap, B.; Willerscheid, H. *Ann. CIRP* **1985**, *34*, 537–548.
7. Piquet, R.; Ferret, B.; Lachaud, F.; Swider, P. *Composites, A: Appl. Sci. Manufac.* **2000**, *31*, 1107–1115.
8. Zhang, H.; Chen, W.; Chen, D.; Zhang, L. *Precis. Mach. Adv. Mater.* **2001**, *196*, 43–52.
9. Rubio, J. C.; Abroa, A. M.; Correia, A. E.; Davim, J. P. *Int. J. Mach. Tools Manuf.* **2008**, *48*; 715–720.
10. Tsao, C. C.; Hocheng, H. *Int. J. Mach. Tools Manuf.* **2004**, *44*, 1085–1090.
11. Davim, J. P.; Reis, P.; Antonio, C. C. *Compos. Sci. Technol.* **2004**, *64*, 289–297.
12. Davim, J. P.; Reis, P.; Antonio, C. C. *J. Mater. Process. Technol.* **2004**, *156*, 1828–1833.
13. Abrao, A. M.; Faria, P. E.; Rubio, J. C.; Davim, J. P. *Mater. Des.* **2008**, *29*, 508–513.
14. Tsao, C. C. *Int. J. Adv. Manuf. Technol.* **2008**, *37*, 23–28.
15. Velayudham, A.; Krishnamuthy, R. *J. Mater. Process. Technol.* **2007**, *185*, 204–209.
16. Mohan, N. S.; Ramachandra, A.; Kulkarni, S. M. *J. Mater. Process. Technol.* **2007**, *186*, 265–271.
17. Campos Rubio, J. C.; Abrao, A. M.; Faria, P. E.; Reis, P.; Davim, J. P. *J. Mater. Process. Technol.* **2007**, *186*, 1–7.

CHAPTER 10

LUMINESCENCE PROPERTIES OF Eu^{3+}-ACTIVATED BiYInNbO$_7$ AS A POTENTIAL PHOSPHOR FOR WLEDS

G. JYOTHI[1,2], L. SANDHYA KUMARI[1], and
K. G. GOPCHANDRAN[1,*]

[1]*Department of Optoelectronics, University of Kerala, Thiruvananthapuram 695581, India*

[2]*Department of Physics, M. S. M. College, Kayamkulam 690502, India*

[*]*Corresponding author. E-mail: gopchandran@yahoo.com*

CONTENTS

ABSTRACT

New Eu^{3+} activated $BiYInNbO_7$ phosphors were prepared by conventional solid-state reaction method. The effect of Eu^{3+} substitution on photo luminescence properties was elucidated using X-ray diffraction, Raman spectroscopy, UV–visible spectroscopy, scanning electron microscopy, and photoluminescence spectroscopy. The Raman shifts observed at 328 cm^{-1} due to vibrational modes F_{2g} and E_g and at 533 cm^{-1} due to mode A_{1g} indicate cubic pyrochlore phase of samples. Intense red emission peaked at 611 nm with color coordinates (0.65, 0.35) and color purity 96% was successfully obtained while exciting with UV light (395 nm).

10.1   INTRODUCTION

The huge propulsion in developing phosphors of bright emission with color purity arises due to the growing demand of lighting devices having high energy efficiency and lumen equivalent. Among the various phosphor applications such as field emission displays, plasma display panels and light emitting diodes, phosphor-converted WLEDs have peak priority in solid-state lighting. Phosphors in various structures had been developed for deriving red luminescence with low correlated color temperature (CCT) and better color rendering index with UV or blue irradiation. Many studies have been conducted to find host materials as well as activators with high performance for red phosphor applications,[1,2] and the emission efficiency and intensity of red phosphors could be improved by both a new host material and synthetic techniques.[3,4] In effect, the luminescence properties of phosphors are strongly dependent on the crystal structure and grain size of the host materials. It has been reported that a fine grain size could enhance the efficiency and intensity of emission from phosphors.[5–7] Studies on Eu^{3+} luminescence gained much importance because of its sharp simple emission spectrum and the hypersensitive asymmetry probe, the f–f transition ($^5D_0$–$^7F_2$).[7–9] For trivalent europium ions, a given optical center in different host lattices will exhibit different optical properties due to the changes of the surroundings.[10] Development of new inorganic phosphors with high chemical stability and luminous efficiency is essential for white light emitting diodes.

   Pyrochlore compounds with general formula $A_2B_2O_7$ have high thermal and chemical stability, ability to accommodate large variety of cations of different valence at A and B sites and can tolerate vacancies at A and O sites to a certain extent. Pyrochlores include wide spectrum of compounds with electrical, magnetic, optical, catalytic, dielectric, and luminescence properties produced by the chemical substitution of combinations of different valance ions at A and B sites. The unit cell is face centered cubic structure with space group $Fd3m$. A unit cell contains eight formula units ($Z = 8$). Larger A site cations are in eight-fold coordination and B site cations in six-fold coordination. Red emitting $CaLaSnNbO_7$: Eu^{3+} and pyrochlore compounds like Bi_2InNbO_7, $Bi_2M^{III}NbO_7$ (M = In and Sc) have been investigated.[11,12] In the present work, the structural, morphological, and luminescence properties of $BiYInNbO_7$ which exhibit intense red emission with Eu substitution are discussed in detail.

10.2 EXPERIMENTAL

$BiY_{1-x}InNbO_7$: xEu^{3+} (x = 0.0, 0.05, 0.10, 0.15) pyrochlore oxides were synthesized by conventional solid-state method at a temperature of 1100°C using Bi_2O_3, In_2O_3, Y_2O_3, Nb_2O_5, Eu_2O_3 (Aldrich, 99.9% purity) as the starting materials.

The crystal structure and the phase purity of the calcined samples were identified by recording the powder X-ray diffraction (XRD) patterns using a powder X-ray diffractometer (D8 Advanced Bruker) with Ni filtered Cu $K\alpha$ radiation (λ = 1.541 Å). The data were recorded over the 2θ range of 10–80°. Room temperature micro-Raman spectra were recorded by using an integrated laser Raman system (Horiba Jobin Yvon LABRAM-HR800) with a confocal microscope and a thermoelectrically cooled multichannel charge-coupled device detection system. The morphology of the synthesized samples was recorded on a scanning electron microscope (Nova Nano SEM 450). The UV–visible absorption of the samples was measured in a spectrophotometer (JASCO V-550). Excitation and emission spectra were recorded using spectroflourimeter (Jobin Yvon Fluoromax) with 450-W xenon flash lamp as the exciting source.

10.3 RESULTS AND DISCUSSION

Figure 10.1 shows the powder X-ray diffraction pattern of $BiY_{1-x}InNbO_7$: xEu^{3+} (x = 0.0, 0.05, 0.10, and 0.15). All the peaks can be indexed to a cubic pyrochlore type structure with space group $Fd3m$ implying the nominal $Bi_2M(III)NbO_7$ stoichiometry.[12] In pyrochlore, Bi^{3+} and Y^{3+} occupy the A site with higher ionic radius (~1 Å), and In^{3+} and Nb^{5+} occupy the B site with smaller ionic radius. So, Eu^{3+} (~1.066 Å) is expected to occupy the Yittrium site (~1.019 Å), based on the ionic size and charge neutrality. The calculated lattice parameters of the compound are given in Table 10.1.

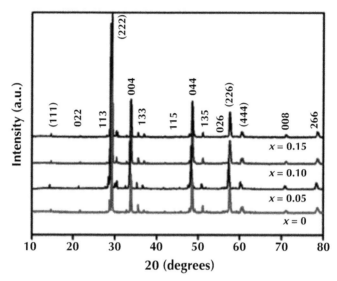

FIGURE 10.1   Powder XRD pattern of $BiYInNbO_7$: xEu^{3+} (x = 0.0, 0.05, 0.10, and 0.15) phosphors.

TABLE 10.1   The Lattice Parameter of the Eu^{3+}-doped $BiYInNbO_7$: xEu^{3+} (x = 0.0, 0.05, 0.10, and 0.15) Phosphors.

Value of x	Lattice parameter (Å)
0	10.5659
0.05	10.5675
0.1	10.5684
0.15	10.5698

Raman spectrum is a tool to analyze local disorder and site symmetry in the crystal structure. The cubic pyrochlore structure belongs to $Fd3m$ (O_h^7) space group. The site symmetry for A and B ions are ions are D_{3d}, and for O and O′ are C_{2v} and T_d, respectively. According to factor group analysis, 26 possible zone center vibrational modes are possible. They are expressed by irreducible representations as

$$\Gamma = A_{1g} + E_g + 4F_{2g} + 8F_{1u} + 3A_{2u} + 3E_u + 2F_{1g} + 4F_{2u} \qquad (10.1)$$

where A_{1g}, E_g, and F_{2g} are Raman active phonon modes. $8F_{1u}$ is infrared active, and the rests are optically inactive vibrational modes. The cations A and B do not take part in Raman active vibrations since they are situated at the inversion center and the six Raman active modes involve the movement of oxygen atoms (O and O′) only. Raman spectra of $BiYInNbO_7$: xEu^{3+} (x = 0.0, 0.05, 0.10, 0.15) phosphors is shown in Figure 10.2. Considering the spectra between 300 cm⁻¹ and 1000 cm⁻¹, low frequency bands arise from A–O stretching (F_{2g}) and BO_6 bending (E_g), mid-range bands from O–B–O bending (A_{1g} and F_{2g}) and high frequency bands from B–O stretching (F_{2g}).[13] The most intense band 328 cm⁻¹ is assigned to two modes F_{2g} and E_g, and the broad band at 533 cm⁻¹ is assigned to A_{1g}.[14,15] The peak at 809 cm⁻¹ may be due to local short range disorder at the B site. The lower wave number bands at 133, 212, and 235 cm¹ are second-order excitations originated due to over-tones of A_{1g} and A_{1g} + F_{2g} phonon modes.[16–18] Raman spectra of samples do not show any difference due to Eu³⁺ doping and this evidences that all the samples crystallize in the same symmetry.

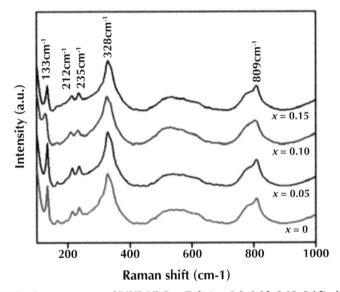

FIGURE 10.2   Raman spectra of $BiYInNbO_7$: xEu^{3+} (x = 0.0, 0.05, 0.10, 0.15) phosphors.

Figure 10.3 shows the FESEM micrographs of the samples $BiYInNbO_7$: xEu^{3+} (x = 0.0, 0.05, 0.10, 0.15). The powdered samples appear to be highly crystalline. The particles are in the range 2–5 µm with a homogeneous nature. As the Eu³⁺ concentration increases, the shape of the particle changes significantly. The particle size decreases with increase in concentration of Eu^{3+} and the particles become more aggregate and are irregular.

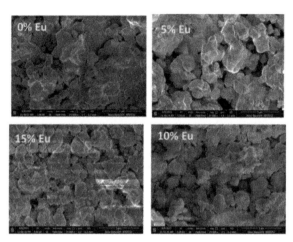

FIGURE 10.3 FESEM micrographs of $BiYInNbO_7$: xEu^{3+} (x = 0.0, 0.05, 0.10, 0.15) phosphors.

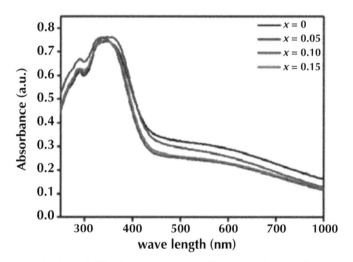

FIGURE 10.4 The UV–visible absorption spectra of $BiY_{1-x}InNbO_7$: xEu^{3+} phosphors.

The UV–visible absorption spectra of the host material and the Eu^{3+}-doped samples are shown in Figure 10.4. All these samples exhibit a strong absorption near 343 nm with a shoulder band at ~290 nm. These bands can be attributed to the charge transfer from the oxygen ligand to the central metal ions, bismuth, niobium, and europium in the metal–oxygen polyhedron. The host material has absorption edge around 410 nm, and as the concentration of Eu^{3+} increases, the absorption edge shifts to the high energy region.

Figure 10.5a shows the excitation spectra of 5 mol% Eu^{3+}-substituted $BiYInNbO_7$ phosphors for the emission monitored at 611 nm. The spectrum mainly consists of a series of lines ascribed to the typical intraconfigurational $4f–4f$ transitions of the dopant Eu^{3+} ions, the peaks at 382 nm ($^7F_0$–$^6G_{2-4}$), 395 nm ($^7F_0$–$^5L_6$), 417 nm ($^7F_0$–$^4D_6$), and 464 nm ($^7F_0$–$^5D_2$). Intense peaks at 395 and 465 nm facilitate the prepared phosphors to be used in commercial phosphor-converted WLEDs that can be excited by near UV or blue LEDs.

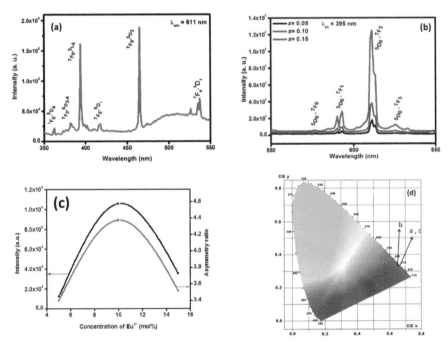

FIGURE 10.5   (a) Excitation spectrum of $BiYInNbO_7$: 0.1 Eu^{3+} phosphor monitored at 611 nm, (b) emission spectrum of $BiY_{1-x}InNbO_7$: xEu^{3+} phosphors excited at 395 nm, (c) variation of asymmetric ratio and emission intensity at 611 nm as a function of concentration of Eu^{3+} ions, (d) CIE chromaticity diagram of $BiYInNbO_7$: xEu^{3+} phosphors: $x = 0.05$ (a), $x = 0.10$ (b), and $x = 0.15$(c).

Room temperature emission spectra of $BiYInNbO_7$: $xEu^{3+}(x = 0.05, 0.10,$ and 0.15) phosphors at the excitation wavelength of 395 nm is demonstrated in Figure 10.5b. Photoluminescence (PL) spectra contain sharp intense lines in the range 580 to 640 nm. The peaks observed at 577, 593, 611, and 624 nm are assigned as $^5D_0 \rightarrow ^7F_J$, (J = 0, 1, 2, and 3) f–f transitions of Eu^{3+}, respectively. $^5D_0 \rightarrow ^7F_2$ electric-dipole transition is highly intense, while $^5D_0 \rightarrow ^7F_1$ (magnetic-dipole transition) emission is reduced and split to two lines at 589 nm and 593 nm. The electric dipole transition is hyper sensitive to site symmetry of Eu^{3+}, while the magnetic dipole transition in independent of symmetry of location of dopant ion. The symmetry forbidden, high intense ED transition at 611 nm evidences that Eu^{3+} ion occupies a site with noninversion symmetry in the host lattice.[19]

The PL emission intensity increases with concentration of Eu^{3+} and concentration quenching was observed at the concentration of 10 mol% Eu^{3+}. Usually, as the Europium ion concentration increases, probability of resonant energy transfers between excited Eu^{3+} ion and ground state Eu^{3+} ion increases. This leads to concentration quenching.[20] This nonradiative energy transfer can be due to exchange interaction or multipolar interaction. Obviously, the relative intensity of every wavelength peak dramatically depends on the concentration of the Eu^{3+} ions. As we know, excessive impurities in the host materials can lead to the formation of the deep trap center of the doped ion and this can bring about fluorescence quenching. No distinct diversifications of the emission spectra shapes and positions occurred when the concentration of Eu^{3+} varied. Yet, nearly no emission from the host could be observed because of the strong luminescence of Eu^{3+}, indicating an efficient energy transfer from the host to Eu^{3+} in the phosphors. The ratio of integrated emission intensity, $^5D_0 \rightarrow ^7F_2 / ^5D_0 \rightarrow ^7F_1$, known as asymmetric ratio is a measure of crystal field asymmetry around Eu^{3+} ion. Variation of asymmetric ratio and emission intensity with concentration of Eu^{3+} is shown in Figure 10.5c. High value of asymmetric ratio is an indication of large deviation from inversion symmetry. Based on Commission Internationale de l'Eclairage (CIE) 1931 diagram, emission color of phosphor materials can be represented in (x, y) color coordinates. Chromaticity coordinates of photo luminescence of the prepared samples calculated using CIE 1931 color matching functions are shown in Table 10.2, and corresponding chromaticity diagram is shown in Figure 10.5d.[21] These values indicate that emission in red region. Color purity of dominant wavelength in the emission can be calculated as

$$\text{Color purity} = \{[(x - x_i)^2 + (y - y_i)^2]/[(x_d - x_i)^2 + (y_d - y_i)^2]\}^{1/2} \times 100\% \qquad (10.2)$$

where (x_i, y_i) and (x_d, y_d) are the chromaticity coordinates of white illuminant (0.333, 0.333) and of dominant wavelength in the emission, respectively.[22,23] Color purity of emission from these phosphors is 96–97%. CCT is very important specification of an illuminating light, and its value represents the quality of light source. CCT were determined for the prepared phosphors (Table 10.2), using McCamy method as

$$CCT = 449n^3 + 3525n^2 + 6823.3n + 5520.33 \qquad (10.3)$$

where $n = (x - x_e)/(y_e - y)$ is the inverse slope line, and the epicenter (x_e, y_e) is (0.3320, 0.1858).[24]

TABLE 10.2 Color Coordinates, CCT, and Color Purity of BiYInNbO₇: xEu³⁺ Phosphors.

Value of x	Color coordinates		CCT (K)	Color purity (%)
	x	y		
0.05	0.66	0.34	2634	97.10
0.10	0.65	0.35	2266	96.30
0.15	0.66	0.34	2634	97.10

10.4 CONCLUSIONS

Highly luminescent new niobium-based pyrochlore type BiYInNbO₇: xEu³⁺ red phosphors having high color purity were successfully synthesized. These phosphors can be efficiently excited by near UV (395 nm) and blue light (465 nm). The dominating electric dipole transition at 611 nm confirms that Eu³⁺ occupies a site with noninversion symmetry. A concentration quenching was observed at 10 mol% concentration of Eu³⁺. XRD and Raman analysis revealed that up to 15 mol% of Eu³⁺ doping, the system maintained the cubic pyrochlore structure. FESEM images affirmed the high crystallinity of samples. In view of these results, we propose BiYInNbO₇: xEu³⁺ as promising red phosphors for WLEDs.

KEYWORDS

- photoluminescence
- pyrochlore
- red emission
- phosphor

REFERENCES

1. Li, Y. C.; Chang, Y. H.; Lin, Y. F.; Lin, Y. J.; Chang, Y. S. *App. Phys. Lett.* **2006**, *89*, 081110.
2. Chang, Y. S. *J. Electron. Mater.* **2008**, *37*(7), 1024.
3. Zhang, X. M.; Seo, H. J. *J. Alloys Compd.* **2010**, *503*, L14.
4. Koo, H. Y.; Ko, D. R.; Lee, S. H.; Lee, S. M.; Kang, Y. C.; Lee, J. H. *J. Alloys Compd.* **2010**, *503*, 260.
5. Wakefield, G.; Keron, H. A.; Dobson, P. J.; Hutchison, J. L. *J. Colloid Interface Sci.* **1999**, *215*, 179.
6. Williams, D. K.; Bihari, B.; Tissue, B. M.; McHale, J. M. *J. Phys. Chem. B* **1998**, *102*, 916.
7. Ekambaram, S.; Patil, K. C.; Maaza, M. *J. Alloys Compd.* **2005**, *393*, 81.
8. Rao, R. P. *Solid State Commun.* **1996**, *99*, 439.
9. Schmechel, R.; Kennedy, M.; Seggern, H. V.; Winkler, H.; Kolbe, M.; Fischer, R. A.; Li, X. M.; Benker, A.; Winterer, M.; Hahn, H. *J. Appl. Phys.* **2001**, *89*, 1679.
10. Chang, Y.-S.; Shi, Z.-R.; Tsai, Y.-Y.; Wu, S.; Chen, H.-L. *Opt. Mater.* **2011**, *33*, 375–380.
11. Zou, Z.; Ye, J.; Abe, R. *Catal. Lett.* **2000**, *68*(3), 235–239.
12. Liu, Y.; Withers, R. L.; Nguyen, H. B.; Elliott, K.; Ren, Q.; Chen, Z. *J. Solid State Chem.* **2009**, *182*, P2748–P2755.
13. Radhakrishnan, A. N.; Prabhakar Rao, P.; Sibi, K. S.; Deepa, M.; Koshy, P. *J. Solid State Chem.* **2009**, *182*, 2312–2318.
14. Sellami, N.; Sattonnay, G.; Grygiel, C.; Monnet, I.; Debelle, A.; Legros, C.; Menut, D.; Miro, S.; Simon, P.; Bechade, J. L.; Thome, L. *Nucl. Instrum. Methods Phys. Res. Sect. B* **2015**, *365*, 371–375.
15. Guje, R.; Ravi, G.; Palla, S.; Nageshwar Rao, K.; Vithal, M. *Mater. Sci. Eng. B* **2015**, *198*, 1–9.
16. Jana, Y. M.; Halder, P.; Ali Biswas, A.; Jana, R.; Mukherjee, G. D. *Vib. Spectrosc.* **2016**, *84*, 74–82.
17. Glerup, M.; Faurskov Nielsen, O.; Willy Poulsen, F. *J. Solid State Chem.* **2001**, *160*, 25–32.
18. Fischer, M.; Malcherek, T.; Bismayer, U.; Blaha, P.; Schwarz, K. *Phys. Rev. B* **2008**, *78*, 014108.
19. Lin, H.; Tanabe, S.; Lin, L.; Yang, D. L.; Liu, K.; Wong, W. H.; Yu, J. Y.; Pun, E. Y. B. *Phys. Lett. A* **2006**, *358*, 474.
20. Blasse, G.; Grabmaier, B. C. *Luminescent Materials*; Springer: Berlin, 1994.
21. Hunt, R. W. G. *Measuring Color: Applied Science and Industrial Technology*, 2nd ed.; Ellis Horwood: New York, 1991.
22. Dillip, G. R.; Mallikarjuna, K.; Dhoble, S. J.; Deva Prasad Raju, B. *J. Phys. Chem. Solids* **2014**, *75*, 8.
23. Fang, Y. C.; Chu, S. Y.; Kao, P. C.; Chuang, Y. M.; Zeng, Z. L. *J. Electrochem. Soc.* **2011**, *158*, J1.
24. McCamy, C. S. *Color Res. Appl.* **1992**, *17*, 142.

CHAPTER 11

THz GENERATION, DETECTION BY NONLINEAR CRYSTALS AND PHOTOCONDUCTIVE ANTENNAS FOR THE DESIGNING OF SPECTROPHOTOMETER AND ITS APPLICATION IN EXPLOSIVE DETECTION

M. VENKATESH and A. K. CHAUDHARY[*]

Advanced Centre of Research in High Energy Materials, University of Hyderabad, Hyderabad 500046, India

[*]*Corresponding author. E-mail: anilphys@yahoo.com; akcphys@gmail.com*

CONTENTS

ABSTRACT

This chapter presents a review on comparative study on efficient terahertz (THz) generation and detection techniques using LTGaAs photoconductive (PC) antennas and ZnTe (1 1 0) nonlinear crystal by employing 800 nm wavelength at 140-fs laser pulses. The obtained results from electrooptic and PC sampling detection techniques are evaluated in terms of their detection efficiency. Further, the developed THz spectrophotometer is employed for the recording of fingerprint spectra of RDX and TNT explosives in Teflon matrix. In addition, the value of absorption coefficients of RDX in different concentration is also ascertained.

11.1 INTRODUCTION

The electromagnetic (EM) waves having frequencies between 0.1×10^{12} (i.e., 3 mm) and 10×10^{12} Hz (i.e., 0.03 mm) range is defined as terahertz (THz) radiation. It lies between microwaves and infrared regions of EM spectrum. THz waves are also called as T-rays, submillimeter, or far-infrared radiation. It can interact strongly with polar materials (i.e., HCl, SO_2, etc.); moreover, it can also penetrate through nonpolar and nonmetallic materials (i.e., cloths, paper, wood, plastic, ceramics, etc.).[1,2] It has high reflection at metal surfaces due to high electrical conductivity of metals. THz radiation is less affected by Mie scattering than visible and infrared radiation due to their longer wavelengths.[3,4] THz radiation has photon energy less than ionization energy of most of the molecular systems, and hence, it is known as nonionizing radiation.[5] In addition, it has wide spectral features associated with vibrations in solids, intraband transitions in semiconductors, vibrational, and rotational transitions of organic molecules. Due to aforesaid properties, it has wide spread applications and opportunities in various fields such as homeland security, astronomy, biomedical science, paleontology, space applications telecommunications, and so on.[6–14]

THz region in EM spectrum is known as THz gap up to the 1980s due to lack of efficient, compact, and coherent emitters and detectors.[2,15] However, the advancement of semiconductor technology and invention of ultrafast laser systems fulfilled the mentioned technological gap. THz sources are divided into two classes (1) incoherent and (2) coherent. The mercury arc lamp, silicon carbide (SiC) globar lamp, and many more are treated as incoherent sources, which emit continuous (CW) THz radiation on heating.[16–19] The emitted THz power from these sources is weak as compared to coherent sources.

Coherent sources are classified into two types depending on THz emission mode and its operating frequency. The two types are (1) CW and (2) pulsed sources. The CW source emits THz radiation in CW or quasi-CW mode (i.e., modulation up to GHz frequencies) at single frequency. These sources emit narrowband THz radiation with high spectral resolution (~0.1 GHz) and high average power as compared to pulsed sources. CW sources are less affected by water vapor absorption if the generated THz frequency lies in air windows.[4,20] Free-electron lasers (FEL), Gunn oscillators, IMPATT diodes, backward wave oscillators (BWOs), far-infrared gas lasers, P-type germanium lasers, parametric oscillators, and quantum cascade lasers (QCL) are some of the well-known CW THz sources.[21–33] The photoconductive antennas (PCAs) and nonlinear (NL) crystals also emit CW THz radiation, when they are irradiated with multimode lasers, mixing of two frequency-offset lasers and photo-mixing of two CW laser sources.[34–36] Pulsed THz sources emit broadband THz radiation within sub-ps pulse duration. The illumination of PCA, NL crystals, bare semiconductor surfaces, metals, superconductors, semiconductor quantum wells, super-lattices, dye-molecules, and magnetic materials, and so on with ultrashort laser pulses generates pulsed THz radiation with broad bandwidth.[37–46] In addition, laser-induced air plasma also generates pulsed THz radiation.[47,48] Moreover, the BWO, QCL, and FEL devices are able to emit THz radiation in pulsed mode.[4,20] However, PCA and NL crystals are some of the widely used sources of THz radiation.

The detection techniques are divided into two categories, that is, coherent and incoherent. The basic difference between these two schemes is based on to retrieve the information of THz radiation. The coherent system provides information about the amplitude and phase, whereas incoherent techniques offer only the THz intensity information. Incoherent technique employs thermal detectors, which are capable of detecting THz broad spectral range. Some of the widely used incoherent detectors are bolometer, golay cell, and pyroelectric detectors.[10,49–51] These detectors require reference frequency to achieve high sensitivity and to surpass background noise. The PCA, NL crystals, Schottky diodes, and superconducting heterodyne receivers are some of basic coherent detectors employed for THz detection.[2,37,52–54] The PCA and NL crystals are efficient and widely used detectors of THz detection.

In this review, we focused on different aspects of THz generation, detection, and its enhancement from PCA, zinc telluride NL crystal. The comparative study of THz detection using electrooptic (EO) sampling (EOS) (using NL crystals) and photoconductive (PC) sampling (PCS) techniques reveals advantages of each technique. In addition, we have demonstrated the use of our THz spectrophotometer for the recording of time domain characteristic spectra of TNT, RDX explosive mixed with Teflon powder in different proportions.

11.2 THz GENERATION AND DETECTION

11.2.1 THz GENERATION

Since, the THz radiation is emitted from different sources by various mechanisms,[37–47] it is important to select these techniques judiciously based on intensity, bandwidth, cost, and others. Among them, PC switching in antennas and optical rectification in NL crystals are important mechanisms which are responsible for THz emission under the influence of ultrafast optical beam.

11.2.1.1 PC SWITCHING

In PC switching technique, antennas are fabricated on a semiconductor substrate that is used for THz generation. The ultrashort laser pulses are used to illuminate the semiconductor substrate for photo excitation, which generate ultrashort lifetime free charge carriers in semiconductor. These charge carriers are accelerated by applied field across the antenna, which creates a transient current in substrate. This transient current is responsible for THz generation. For photo excitation, the incident laser pulses must have photon energy greater than the semiconductor band gap. In this process, the semiconductor material flip-flops its state between nonconduction to conducting phases which depends on incident laser pulse width, carrier lifetime of material, respectively.[4] Figure 11.1 illustrates the schematic of THz generation from PCA.

Smith et al. first reported the THz generation from dipole antennas made on radiation damaged silicon on sapphire (RD-SOS) semiconductor substrate by 120 fs laser pulses.[37] Apart from RD-SOS, the semiconductors, such as amorphous silicon, low-temperature gallium arsenide (LT-GaAs), semiinsulating gallium arsenide (SI-GaAs), indium phosphide, and so on, are also used for THz generation.[55] Among them, LT and SI-GaAs are widely used semiconductors as antenna substrates. Bow tie, strip-line, spiral, slot, log-periodic, interdigital, fractal, dipole, and gammadion are some of the antenna designs used for THz generation and detection.[56–60] Among these designs, stripline, bow-tie, and dipole are popular and basic antenna geometries used for THz generation.

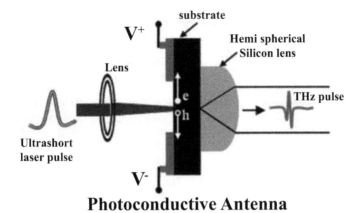

Photoconductive Antenna

FIGURE 11.1 Schematic of THz generation from photoconductive antenna.

THz pulse emitted by the Hertzian dipole antennas is governed by following equation:[1,61]

$$E_{\text{THz}} \propto \frac{\partial J}{\partial t} \propto \frac{\partial(\text{ne}V)}{\partial t} \propto \frac{\partial n}{\mu e E_{\text{bias}}/dt} \qquad (11.1)$$

where n, V, and μ represent the density, drift velocity, and mobility of free charge carriers, respectively. The electrical charge and applied bias voltage are represented by e, E_{bias}, respectively. From eq 11.1, the generated THz electric field is proportional to the mobility of antenna substrate materials, applied bias field, and rate of density of free charge carriers. Therefore, the PC material selected as substrate of antenna must have high carrier mobility and high break down field or voltage in addition to low carrier lifetime for high intensity THz emission.[1,62]

In addition to substrate's properties, antenna design (pattern) also determines the emission intensity and bandwidth of generated THz radiation.[63-65] We have fabricated dipole antennas for generation and detection of THz radiation. These designs are fabricated on LT and SI-GaAs semiconductor wafers. SI and LT-GaAs semiconductor wafers of 3 in. diameter were purchased from New Way Semiconductor Co. Ltd. The resistivity and thickness of SI-GaAs and LT-GaAs wafer were 50 MΩ, 625 μm, and 1 MΩ, 625 μm, respectively. The PC antennas of desired design and type were fabricated on semiconductors using conventional photolithography technique. The fabricated dipole PC antennas consist of transmission lines along with pad for wire bonding as shown in Figure 11.2. The gap (G), width (W), and length (L) of the PC dipole antenna were 5, 20, and 30 μm, respectively.

The pad size of 250 × 250 μm is utilized for electrical connections or wire bonding. The antenna embedded in PCB was transferred to commercially available antenna holder (i.e., Menlo-T8-H2 alignment package) for characterization and THz measurements.

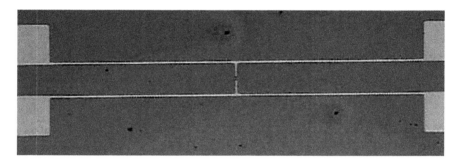

FIGURE 11.2   Picture of fabricated dipole PC antennas.

The intensity and bandwidth of the generated THz radiation can be controlled by modifying the design of PC antennas, substrate parameters as well as the laser parameters.[37,64,66–78] Zhang et al. studied the generation of THz radiation in photo-mixer antennas in the 780–860-nm wavelength range and found that the THz amplitude varies inversely with central wavelength of pump pulse.[79] We have also enhanced THz radiation by tuning the incident laser wavelengths incident on SI-GaAs and LT-GaAs-based dipole antennas. The obtained results revealed that generated THz radiation is high at particular wavelength were the optical absorption was maximum.[80,81] By using the setup mentioned[85,86] in Ref. [80], we have studied the effect of incident laser pulse durations on generated THz radiation. The temporal profiles and bandwidth of THz pulses generated from antennas at laser pulse widths 140 and 15 fs at 20-V fixed bias voltage and 7-mW laser power are illustrated in Figure 11.3. Figure 11.3a and c shows the nature of generated THz waveform. The amplitude of generated THz radiation with 15-fs laser pulse is high as compared to 140-fs laser pulses. So, the THz pulses obtained from 15 fs are stronger than 140 fs. Figure 11.3b and d shows the frequency spectra of generated THz pulses using 140-fs and 15-fs laser pulses are extended to 4 and 6 THz, respectively. Therefore, the generated THz amplitude and bandwidth are more for laser pulses having lesser pulse duration. Previously, Budiarto et al. measured energy and width of THz pulse obtained from large aperture PC antennas with respect to incident laser duration.[82] They demonstrated that THz pulse energy is inversely related to laser pulse width. In addition, the THz pulse width is increasing with an increase in laser pulse duration.

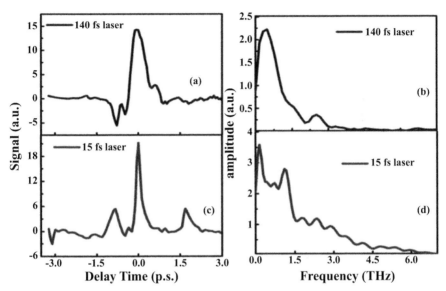

FIGURE 11.3   THz radiation waveforms and its corresponding spectrum: (a) THz temporal profile at 15 fs, (b) Fourier transform of waveform shown in 'a,' (c) THz temporal profile at 140 fs, and (d) Fourier transform of waveform shown in 'c.'

11.2.1.2   OPTICAL RECTIFICATION

Optical rectification is a second-order NL process, which develops dc polarization in NL material, when it is subjected to intense femtosecond laser pulses.[83] The duration of dc polarization is of the order subpicoseconds when NL crystal is illuminated with ultrashort laser pulses. The generated time dependent (transient) polarization creates an EM field in THz domain. In other words, the optical rectification is a special case of difference frequency mixing, where spectral components present in single ultrashort laser pulse are mixed.[84] The THz generation from NL crystal by optical rectification is illustrated in Figure 11.4. The THz radiation from NL crystals by means of optical rectification is given by the following equation:

$$E_{\text{THz}}(t) \propto \frac{\partial^2 P(t)}{\partial t^2} \tag{11.2}$$

where E_{THz} is the electric field of generated THz radiation, and $P(t)$ is the NL polarization induced in the crystal (in temporal domain). From eq 11.2, it is clear that the amplitude of THz electric field depends on the rate of change of induced nonlinear polarization in crystal. The induced NL polarization in the crystals is written as[85,86]

$$P(t) = \hat{f}\{P(\omega)\} = \varepsilon_0 \chi^2 \int_0^\infty \int_0^\infty E\omega_1 * E\omega_2 \exp[-i(\omega = \omega_2 - \omega_1)t]d\omega_1 d\omega_2$$

$$= \varepsilon_0 \chi^2 E^*(t)E(t) \propto \chi^2 I_{opt}(t) \tag{11.3}$$

where $P(\omega) = \varepsilon_0 \chi^2 (\omega = \omega_2 - \omega_1)E_{\omega_1}E_{\omega_2}$

where χ^2 is the second-order NL susceptibility, ω is the frequency of generated EM radiation, $P(\omega)$ is the NL polarization in spectral domain, and $I_{opt}(t)$ is intensity of incident laser beam. From above equations, the strength of THz electric field depends on χ^2 (depends on crystal structure) and $I_{opt}(t)$.

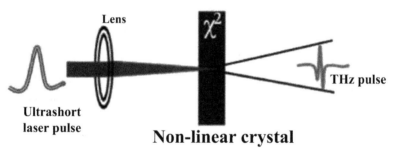

FIGURE 11.4   Schematic of THz generation from nonlinear crystal by optical rectification[1,38,87] process.

In optical rectification process, the THz generation depends on NL properties of material but not on generation of free charge carriers. Hence, this process generates THz radiation even with laser pulses having photon energy less than band gap of employed material. On the other hand, the NL properties of material are enhanced by illuminating with laser pulses of energy higher than bandgap, which can generate more THz power.[4,87,88] The NL crystals are extensively used for THz generation due to the simplicity in alignment, sustaining at high laser power, and generating broadband THz pulses. The amplitude and bandwidth of emitted THz pulses using this mechanism depend on incident laser pulse properties, NL material properties, crystal orientation, crystal thickness, crystal damage threshold, and THz and optical absorption in material.[89] In addition, the velocity matching condition is crucial to increase the signal strength and detection sensitivity.[90,91] The velocity matching condition is defined in terms of equality of optical group velocity and THz phase velocity of the medium.[84] The coherence length is used to measure the degree of velocity matching between THz phase velocity and group velocity of optic beam in the crystal. In other words, it is an effective interaction length between optical pulse and generated THz radiation. The equation for coherence length is given by[84]

$$L_c = \frac{1}{2v(1_{V_{gr}} - 1_{V_{THz}})} \Rightarrow L_c \frac{\lambda_{THz}}{2(n_{gr} - n_{THz})} \tag{11.4}$$

where $n_{gr} = c / V_{gr} = n_{opt} - \left[\left(\partial n_{opt} / \partial \lambda_{opt} \right) \lambda_{opt} \right]$

where c is the velocity of light. The n_{opt} and n_{THz} are optical group velocity and THz phase velocity, respectively. The n_{opt} is optical refractive index and n_{THz} is THz refractive index of material. Incident optical wavelength is denoted by λ_{opt} and the emitted THz wavelength is represented by λ_{THz}. The THz signal strength and detection sensitivity increase with crystal thickness, while the bandwidth of emitted and detected THz radiation decreases, within the coherence length of crystal.[84,85]

The NL crystals employed for THz generation are divided into three principal categories, namely, (1) inorganic (LiNbO$_3$, LiTaO$_3$, etc.), (2) semiconductor (ZnTe, GaSe, CdTe, GaP, GaAs, ZnGeP$_2$, etc.), and (3) organic (DAST, BNA, MBANP, HMQ-T, etc.).[92–103] The first optical rectification-based THz generation from LiNbO$_3$ crystal using picosecond pulses was reported by Yang et al. in 1971.[104,105,106] Here, we discuss THz generation from zinc telluride (ZnTe) crystal.

ZnTe is a cubic (zinc blende) red-colored crystal with chemical formula ZnTe. It is a p-type semiconductor material that belongs to II–VI group compounds with a direct band gap ~2.26 eV. ZnTe crystal comes under the category of semiconductor crystals, and its properties are given in Table 11.1.

Table 11.1 clearly shows that ZnTe and GaSe crystals have high EO, NL coefficients, and perfect phase matching at tunable wavelengths of Ti:sapphire lasers. GaSe has higher EO coefficient with low absorption and transparent over a broad range compare to ZnTe crystal. Therefore, it can generate broadband THz radiation with higher amplitude than ZnTe crystal. In addition, it generates and detects selective range of THz radiation by using suitable thicknesses of emitter and detector. However, the growth of pure GaSe single crystals is difficult and available only in <0 0 1> plane. In addition, GaSe is soft and delicate material which can even break during the process of growth itself.[107] On the other side, ZnTe single crystal is moisture resistant and harder than GaSe crystal. It can be grown in <1 0 0>, <1 1 0>, and <1 1 1> orientation. Therefore, ZnTe crystal is most suitable semiconductor material for THz generation and detection.

TABLE 11.1   Optical Properties of Semiconductor Crystals.

Material	EO coefficient (pm/V)	Phase-matching wavelength (nm)[+]	NL coefficient (pm/V)	Absorption coefficient @ 1 THz (cm^{-1})	Noise equivalent power (10^{-16} W/Hz$^{0.5}$)
ZnTe	4.0	~822	68.5	1.3	0.27
CdTe	4.5	~970	81.8	4.8	–
GaAs	1.43	~1405	65.6	0.5	0.89
GaSe	14.4	~780[*]	28	0.5	–
GaP	0.97	~1030	24.8	0.2	2.2

EO—electrooptic and NL—nonlinear.
[*]@ 53° emitter phase matching angle at 8 THz and 25 THz.[93,108]
[+]Phase matching at specific THz frequencies.

The first demonstration on THz generation and its detection from <1 1 0> oriented ZnTe crystals using 130 fs pulses at 800 nm was reported in 1999 by Nahata et al.[90] The intensity and bandwidth of emitted THz radiation from crystals can be realized by changing the crystal thickness,[92,109–121] azimuthal orientation, and laser parameters. The arbitrary THz waveforms were generated from ZnTe crystal using shaped ultrafast laser pulses.[122–125] The arbitrary shaped THz pulses have applications in the fields of remote sensing, wireless communications, signal processing, material characterization, and narrow band radiation for efficient imaging of objects. A few attempts have been made to study the effect of laser wavelength on generated THz pulse shape and spectra from ZnTe crystal.[109,125–128] Vander Valk et al. demonstrated the variations in emitted THz pulse shape and its spectrum by changing the crystal thickness and laser central wavelengths (i.e., 1080, 1148, and 1128 nm). He reported the period and amplitude of oscillations in THz spectrum increases as central wavelength of the laser decreases.[109] Li et al. achieved tuning in THz central frequency (i.e., 4.2–1.1 THz) obtained from ZnTe crystal as a function of laser wavelengths tuned between 700 and 900 nm (resolution 25 nm).[127] Later, Tu et al. predicated the change in THz pulse shape and periodic oscillations were due to the excitation of coherent-phonon polariton in ZnTe crystal.[128] Herein, we studied the effect of laser wavelengths tunable between 740 and 840 nm on generated THz frequencies from ZnTe crystal. The experimental schematic and laser used for this experiment is same as in Ref. [116].

The spectral amplitudes of obtained THz radiation from ZnTe crystal with laser wavelengths are shown in Figure 11.5. The obtained THz spectra consist of three distinct humps for 750 and 760-nm incident laser wavelengths. These spectral humps have peaks at 0.78, 1.82, and 2.54 THz for

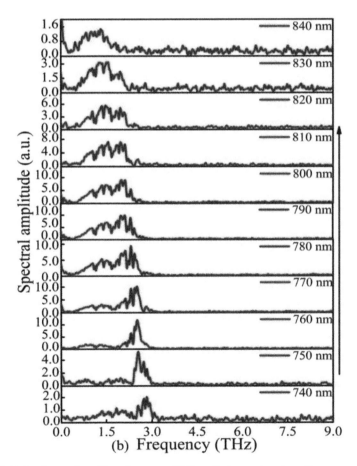

FIGURE 11.5 Spectral profiles of generated THz radiation as function of laser wavelengths.

750 nm while 0.79, 1.35, and 2.517 THz for 760 nm. The three spectral humps correspond to THz pulse, nonphase matched frequencies, and *lower disorder–order transition* (LDOT).[128] The spectra correspond to LDOT was due to the presence of excited coherent-phonon polariton (CPP) in ZnTe crystal. The effect of LDOT (i.e., CPP) on THz spectra is overcome by reducing by increasing the incident laser wavelength. The humps in THz spectra are overlapped (or merge) with each other as wavelength increases. In addition, the spectrum of emitted radiation is shifted to higher frequency side for shorter incident laser wavelengths. The emitted THz bandwidth is relatively narrow for shorter wavelengths as compared to longer wavelength. This is due to absorbance of THz frequencies near to transverse optical phonon modes in ZnTe crystal.

The generated THz peak frequency with respect to incident laser wavelength is depicted in Figure 11.6. As pump wavelength increases, the peak frequency of emitted THz radiation moves toward lower frequency side. It is due to velocity matching of pump pulse and THz lower frequency with increasing incident laser wavelength.[127]

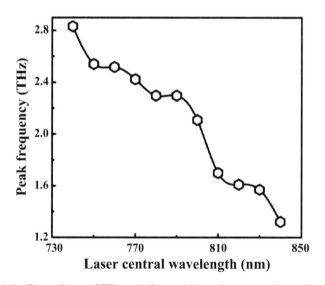

FIGURE 11.6 Dependence of THz peak frequencies on laser central wavelength.

11.2.2 DETECTION TECHNIQUES

The emitted THz radiation is detected using different detectors, which follows different mechanisms. The PCS and EOS are widely used coherent techniques for THz detection. The EOS technique is based on Pockels effect in NL crystals. The mechanism used for THz detection in PCS method is identical to the generation mechanism in PC antennas. Here, we discuss the THz detection using PC antennas and ZnTe crystals and compared results obtained using these techniques.

11.2.2.1 PC SAMPLING

In PCS, the generated THz pulses and laser pulses are focused in the gap of antenna. The focused laser pulses create free charge carriers and are accelerated by electric field of focused THz radiation. The accelerated charge carriers generate photocurrent. The induced current is proportional to field amplitude of focused THz radiation on the detector (i.e., antenna) and transient carrier density.[85] The induced photo current is measured using current meter (Pico ammeter + Lock-in amplifier). The temporal profile of THz radiation is obtained by measuring the induced current with respect to delay time between THz and laser pulse. The Fourier transform of measured THz profile gives the spectral amplitude and bandwidth of THz radiation. The schematic of the experimental THz detection using PC antennas is illustrated in Figure 11.7. The induced photocurrent in detector at delay time (t) is given in following equation:[84]

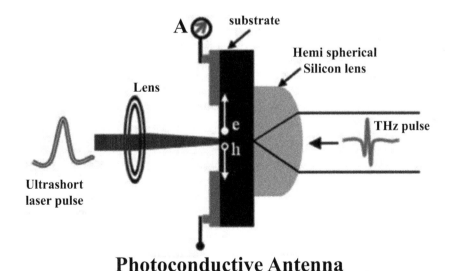

Photoconductive Antenna

FIGURE 11.7   Schematic of THz detection using photoconductive antenna (Ammeter-A).

$$J(t) \propto \int_{-\infty}^{t} \sigma(t - t^6) \, E_{THz} \, (t^6) \, dt^6 \tag{11.5}$$

where (σ) is the transient conductivity, and $E_{THz} \, (t)$ is the electric field of focused THz radiation on to the detector. From eq 11.5, the measured current exactly reflects the THz electric field when surface conductivity has Dirac delta function. The transient surface conductivity is written as

$$\sigma(t) \propto \int_{-\infty}^{t} \mu(t) n(t) I_{opt}(t) dt \tag{11.6}$$

According to Ref. [84] eq (11.6) becomes

$$\sigma_{(t)} \propto \int_{-\infty}^{t} e^{-t/\tau_{2p_e} \, -t/\tau_{c[1-es]} \, -t/\tau_{dt}} \tag{11.7}$$

From eq 11.7, the surface conductivity is restricted by laser pulse duration and carrier lifetime of detector. Hence, detector lifetime and incident laser duration is crucial for detecting the THz radiation.[129] In addition, the detector antenna structure also affects the efficiency of collected THz signal.[130,131] The collection efficiency depends on overlap between THz beam waist and antenna gap.

11.2.2.2   ELECTROOPTIC SAMPLING

EOS is another popular technique employed for detection of THz radiation which is first demonstrated by Valdmanis et al. in 1982.[132] It uses EO (pockels) phenomenon of the crystal for THz detection. By applying electric field, the optical properties of EO crystals can be changed. The applied electric field induces birefringence in crystals, which changes the polarization state of beam propagating through crystal. The EO crystals, such as GaSe, ZnTe, DAST, GaAs, GaP, EO polymer films, and others, are used for THz detection.[90,133–137] Among them, ZnTe crystal is most popular and widely used crystal for THz generation and detection due to its excellent phase matching at Ti:sapphire laser wavelengths, high EO coefficient, and high transparency at optical and THz frequencies.

The schematic of EO detection for THz radiation is illustrated in Figure 11.8. The EO crystal, $\lambda/4$ plate, Wollaston prism and photo-detectors are essential optics required for EO detection setup. THz pulse and laser pulses are focused on the EO crystal. The polarization state of laser pulses remain same until the propagated THz field and linearly polarized laser pulses are not overlapped both temporally and spatially in EO crystal. The combination of $\lambda/4$ plate and Wollaston prism converts the linear polarized pulses into circular and circular to two orthogonal polarized beams having same intensity. These polarization components were focused onto the balanced photo-diodes, which measures the intensity difference between inputs. The output of photo-diodes was zero in the absence of THz radiation. In the presence of THz field, polarization state of linear polarized pulse is changed to elliptically polarized pulse due to field-induced birefringence in EO crystal. The $\lambda/4$ plate induces $\pi/2$ phase retardation in elliptically polarized laser pulses, which reduces signal distortion. Wollaston prism splits the probe beam into two orthogonal polarization components, that is, I_x and I_y having different intensity. The intensity difference is measured by balanced photo-diodes, which is proportional to field amplitude of THz radiation.

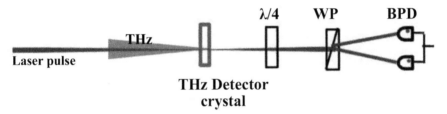

FIGURE 11.8 Schematic of THz detection using electrooptic sampling: BPD—balanced photo diodes, WP—Wollaston prism, and $\lambda/4$—quarter wave plate.

The output (i.e., intensity difference) of the balance photodiodes can be written as[84]

$$I_d = I_x - I_y = I_0 \Delta\phi = \frac{I_0 \omega L}{c} n^3 r E_{0.41\,\text{THz}} \qquad (11.8)$$

where I_o is the incident laser intensity on Wollaston prim, $\Delta\phi$ is the phase retardation of laser beam and $n_0^{\ 3}$ is the group refractive index of EO crystal at optical frequency (ω). The L, r_{41}, and E_{THz} are crystal length, EO coefficient, and THz electric field, respectively. From eq 11.8, the I_d is proportional to crystal thickness and EO coefficient that indicates thicker crystals with high EO coefficient is good for detection of THz signal. The actual resolution of this technique is limited by optical pulse duration, susceptibility dispersion, and velocity mismatch between laser pulse and THz radiation.[84,134,138–140]

The equation of EO signal which comprises all these factors is

$$E_{eos}(t) \propto \int_{-\infty}^{\infty} E_{THz}(\omega_{THz}) R_F(\omega, \omega_{THz}) e^{-i\omega_{THz}t} d\omega_{THz} \qquad (11.9)$$

where R_F is the detector response function which is the combination of velocity mismatch, laser duration, and susceptibility dispersion. The R_F is expressed as

$$R_F(\omega, \omega_{THz}) \propto I_{opt}(\omega) \chi \, eff^{2\Delta\varphi(\omega,\omega_{THz})}$$
$$\propto \Delta\varphi(\omega, \omega_{THz}) \qquad (11.10)$$
$$\propto e^{\dfrac{-i\Delta K(\omega,\omega_{THz})L-1}{i\Delta K(\omega,\omega_{THz})}}$$

where I_{opt}, χ^2, and $\Delta\varphi$ are laser intensity profile, NL susceptibility, and frequency dependent velocity mismatch, respectively. From eq 11.10, the EO temporal and spectral resolution decreases with velocity mismatch between the group velocity of optical pulse and phase velocity of THz radiation.

11.2.3   COMPARATIVE STUDY OF PCS AND EOS TECHNIQUES

Figure 11.9 illustrates the detected temporal profiles of emitted THz radiation from ZnTe crystal using PCS and EOS techniques. The average laser power used for illumination of emitter and detectors were ~600 and ~150 mW, respectively, at 800-nm incident laser wavelength. The employed PCA in PCS technique is dipole shaped with a gap ~5 μm and length ~20 μm.

The shape of the THz waveforms obtained by these techniques is different, and they are similar to the previously reported by Park et al.[141] The detected peak–peak amplitude of THz signal using PCS and EOS techniques are 1052 and 420 a.u., respectively. Our observation shows that detected peak–peak amplitude is more in PCS compared to EOS technique.

The signal-to-noise ratio (S/N ratio or SNR) of signal is defined as the ratio of obtained peak amplitude of THz radiation to noise floor.[142] The obtained peak amplitude and noise floor in EOS was ~230 and ~15 a.u., while for PCS method, it was ~540 and ~10 a.u., respectively. Therefore, S/N ratio of PCS was 3.5 times higher in magnitude as compared to EOS technique in our system. The higher SNR in PCS could be due to chopping of input laser beam at 1-kHz reference frequencies.[143] The triggered frequency employed for chopping the laser beam in our experiments was ~1.569 kHz. The EOS technique is sensitive to laser noise, low frequency acoustic, and mechanical disturbances at kHz chopping frequencies. Consequently, EOS technique has low S/N ratio at kHz chopping frequencies compared to PCS method. The S/N ratio in EOS method can be enhanced by chopping the laser beam at MHz frequencies that can overcome the noises to an extent.[143]

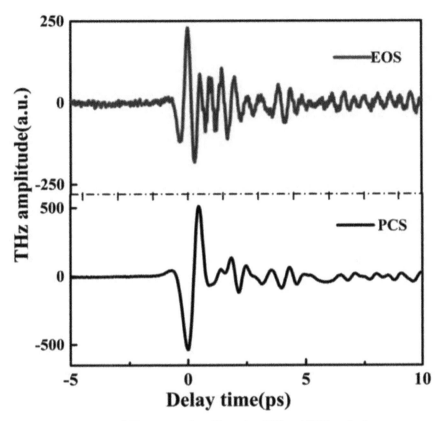

FIGURE 11.9   Detected THz temporal profiles using PCS and EOS methods.

Figure 11.10 and its inset show the dependence of THz temporal profiles and peak–peak amplitude as a function of incident laser power (175–600 mW) at 800 nm, respectively. Here, ZnTe crystal and PCA were employed for the emission and detection of THz radiation. The obtained shapes of THz waveforms were independent of incident laser power, while its peak amplitude was increasing with incident laser power. In addition, THz peak–peak amplitude is increasing linearly with laser power as shown in the inset of the figure. It is evident from eq 11.4, the emitted THz peak amplitude is proportional to input laser power/intensity.

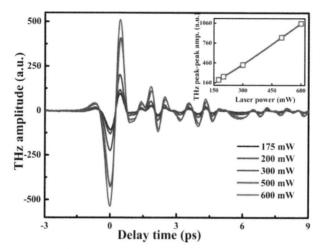

FIGURE 11.10 Dependence of THz temporal profiles and THz peak–peak amplitude (shown in inset) with respect to laser power.

The dependence of THz peak amplitude was measured as a function of azimuthal orientation of ZnTe (emitter) as illustrated in Figure 11.11. The angular dependence of THz signal from ZnTe follows three-fold rotation symmetry as shown in the figure. The maximum signal obtained in azimuthal rotation could be due to the parallel alignment of crystal polar axis, polarization of optical beam, and detector dipole axis. The change in THz peak amplitude and its polarity was due to change/dispersion of NL EO coefficients (i.e., NL susceptibility tensor d_{14}) of crystal.[96]

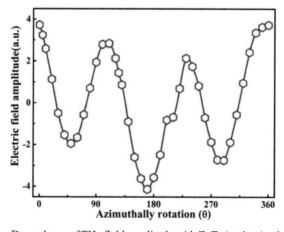

FIGURE 11.11 Dependence of THz field amplitude with ZnTe (emitter) azimuthally rotation.

11.3   DETECTION OF EXPLOSIVES USING THZ RADIATION

THz radiation has innumerable spectral features associated with rotational and vibrational spectroscopy. Nonionizing and noninvasive nature of THz radiation do not impose any safety measures in spectroscopy, material and personal scanning, and others. Therefore, THz spectroscopy has become one of the promising techniques to detect/characterize the explosives, biological samples, pharmaceutical materials, polymers, liquids, semiconductors, and gases as well as flames and flows.[7,66,144–159]

Usually, there are three types of THz spectroscopy experiments, namely, (1) THz time domain spectroscopy (THz-TDS), (2) time-resolved THz spectroscopy (TRTS), and (3) THz emission spectroscopy (THz-ES).[160,161] The THz-TDS provides static properties of sample, whereas TRTS provide dynamic evolving properties of material. In THz-ES, sample properties can be evaluated by analyzing the emitted temporal profile, amplitude, shape, and bandwidth of generated THz radiation. In this paper, the results obtained using THz-TDS of explosives will be discussed.

THz-TDS provides information about sample properties such as absorption coefficient, refractive index, transmission coefficient, and many more. This technique is noninvasive and nondestructive in nature. Hence, it can be applied directly at the production site for monitoring. In THz-TDS, the sample properties are directly measured without using the Karmers–Kronig relations. Other spectroscopic techniques such as Fourier transform infrared spectroscopy (FTIR) and Raman spectroscopy use Karmers–Kronig analysis to measure sample properties.[148,160,162–165] In addition, FTIR and Raman spectroscopy have experimental difficulties and require theoretical models to analyze the data. The comparison of THz-TDS and FTIR spectroscopic techniques are given in Table 11.2.

In THz-TDS, the time domain signal was measured in two steps. In the first step, signal was measured in the absence of the sample, which is taken as a reference signal. In a second step, the sample was inserted between the source and detector to record the transmittance of the signal. The obtained signal was used to ascertain the absorption and attenuation coefficient of the sample under consideration. Previously, several research groups estimated the refractive index, transmission, absorption, and attenuation coefficient of materials in THz domain using both coherent and incoherent spectroscopic techniques.[6,8,145–148,166–176] Here, the spectral fingerprints of RDX and TNT mixed with Teflon matrix are ascertained using THz-TDS.

TABLE 11.2 Comparison between FTIR and THz-TDS Techniques.

Name of the Technique	Advantages	Disadvantages
THz-TDS	Operation at room temperature It can be worked in coherent and incoherent detection Real-time monitoring can inspect materials under cover High selectivity No need to use Kramer–Kronig relations for analysis	Low frequency resolution compared to FTIR Less bandwidth compared to FTIR
FTIR	Frequency resolution is up to 10 cm^{-1} High sensitivity High sensitivity	Need to use Kramer–Kronig relations for analysis Require cryogenic cooling No real-time monitoring No coherent detection Cannot inspect materials under cover

The schematic of conventional THz time domain spectrometer is illustrated in Figure 11.12. The employed laser source is coherent chameleon ultra-II oscillator, which delivers transform limited pulses of duration 140 fs at 80-MHz repetition rate. The average power of laser is 3.7 W @ 800 nm wavelength. The ZnTe crystal and Baptop PCA are used as emitter and detector for the THz radiation, respectively. The ZnTe crystal has thickness 2 mm, diameter 10 mm, and orientation <1 1 0>. The length and gap of SI-GaAs PC dipole antenna were ~20 and ~5 μm, respectively. The laser power used for generation and detection of THz radiation was ~ 1 W and ~ 175 mW, respectively. The samples for spectroscopic studies were kept between the parabolic mirrors (PM$_1$ and PM$_2$), where collimated THz radiation was used for investigation of sample properties. The thickness and diameter of pellets are 3 and 12 mm, respectively. A pure pellet of PTFE (Teflon) was taken as a reference. The 50-mg RDX and TNT are mixed with 450 mg of Teflon pellet. Therefore, the overall quantity of sample pellet is 500 mg. It is assumed that explosive sample is homogeneously distributed in Teflon matrix.

The THz temporal profiles through Teflon reference and sample pellet are shown in Figure 11.13a. The reference and sample THz spectrum are obtained by the FFT of THz temporal profiles and are shown in Figure

11.13b. The formula used for the absorption coefficient (α) of sample pellet is expressed as[176]

$$\alpha = \frac{1}{l}\ln\frac{E_{\text{THz-sam}}(v)}{E_{\text{THz-Ref}}(v)} \tag{11.11}$$

where $E_{\text{THz-sam}}(v)$ and $E_{\text{THz-Ref}}(v)$ are obtained THz spectrum when radiation is transmitted through sample and reference. l is the effective thickness of pellet and ω is the THz frequency. The effective length of sample is given as

$$l = \frac{m \times 4}{\rho \pi D^2} \tag{11.12}$$

where m is the weight of the sample, ρ is the density of material, and D is the diameter of sample. Density (ρ) and the calculated effective thicknesses (l) of RDX and TNT explosives distributed in Teflon matrix are 1.82 g/cm³, 0.243 cm, and 1.65 g/cm³, 0.2681 cm, respectively.

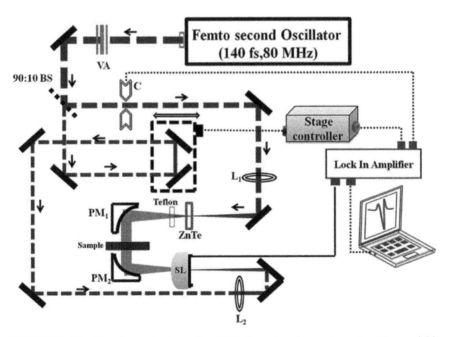

FIGURE 11.12 Experimental schematic of THz time domain spectrometer: VA—variable attenuator, SL—hyper hemispherical silicon lens, BS—beam splitter, L1 and L2—plano convex lens, ZnTe—zinc telluride crystal, C—mechanical chopper, and PM1 and PM2—parabolic mirrors.

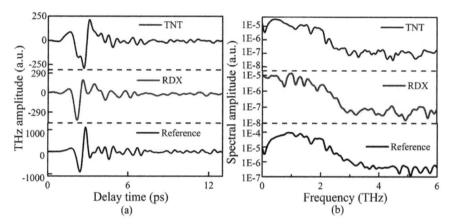

FIGURE 11.13 (a) Transmitted THz temporal profiles and (b) transmitted THz frequency spectra through samples.

The absorption coefficients of RDX and TNT using equation are illustrated in Figure 11.14. The measured absorption peaks or spectral fingerprints of RDX and TNT in our experiment are observed at 0.76, 1.43, 1.88 THz and 1.65, 2.18 THz, respectively. The obtained results reveal that there is slight shift in absorption peaks of RDX and TNT compared to previous results. This might be due to purity of the explosive sample mixed with Teflon matrix. In addition, we also measured the absorption coefficient RDX sample in THz domain which is mixed with Teflon powder in different proportions, that is, 100, 50, 30, and 10 mg of RDX is mixed with Teflon having quantity 400, 450, 470, and 490 mg, respectively. Figure 11.15 illustrates the absorption coefficient value of RDX in different proportions. It shows that the absorption coefficient is more for pellet having more amount of RDX sample.

11.4 CONCLUSIONS

We presented a thorough review on comparative study on THz generation and detection mechanisms of PCAs and NL crystal. The comparison between EOS and PCS detection techniques revealed that the PCS is good at kHz reference frequencies using femtoseconds oscillator pulses at 80 MHz repetition rate. In addition, the potential of the designed THz spectrophotometer is evaluated by recording the finger prints of premier explosives like RDX and TNT mixed in Teflon matrix. We recorded the absorption coefficients of RDX in various concentrations. The obtained results revealed that the absorption coefficient value changes with the concentration of explosive.

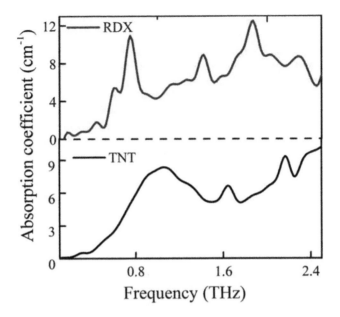

FIGURE 11.14   Absorption coefficient of RDX and TNT.

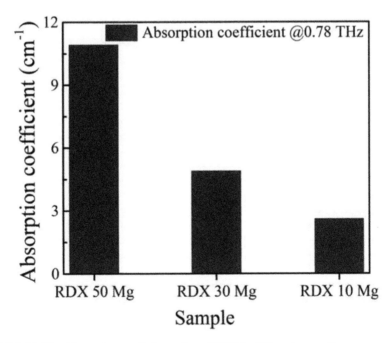

FIGURE 11.15   Absorption coefficient value of RDX in different proportions.

ACKNOWLEDGMENT

The authors gratefully acknowledge the financial support provided by the DRDO, Ministry of Defense, Govt. of India.

KEYWORDS

- **THz generation**
- **THz detection**
- **photoconductive switching**
- **optical rectification**
- **detection techniques**

REFERENCES

1. Dragoman, D.; Dragoman, M. Terahertz Fields and Applications. *Prog. Quantum Electron.* **2004,** *28,* 1–66.
2. Lee, Y.-S. Introduction. In *Principles of Terahertz Science and Technology*; Springer: US, 2009, pp 1–9.
3. Qiao, W. Liquids and Molecular Solids Investigated by THz Time-Domain Reflection and Parallel-Plate Waveguide Spectroscopy, 2012. http://nbn-resolving.de/urn:nbn:de:bsz:352-209972
4. Zhang, X.-C.; Xu, J. Terahertz Radiation. In *Introduction to THz Wave Photonics*; Springer: USA, 2010, pp 1–26.
5. Son, J.-H. *Terahertz Biomedical Science and Technology*; CRC Press: USA, 2014.
6. Shen, Y.; Lo, T.; Taday, P.; Cole, B.; Tribe, W.; Kemp, M. Detection and Identification of Explosives Using Terahertz Pulsed Spectroscopic Imaging. *Appl. Phys. Lett.* **2005,** *86,* 241116.
7. Shen, Y.-C. Terahertz Pulsed Spectroscopy and Imaging for Pharmaceutical Applications: A Review. *Int. J. Pharm.* **2011,** *417,* 48–60.
8. Federici, J. F.; Schulkin, B.; Huang, F.; Gary, D.; Barat, R.; Oliveira, F.; Zimdars, D. THz Imaging and Sensing for Security Applications—Explosives, Weapons and Drugs. *Semicond. Sci. Technol.* **2005,** *20,* S266.
9. Mittleman, D. M.; Jacobsen, R. H.; Nuss, M. C. T-Ray Imaging. *IEEE J. Sel. Top. Quantum Electron.* **1996,** *2,* 679–692.
10. Siegel, P. H. Terahertz Technology. *IEEE Trans. Microwave Theory Tech.* **2002,** *50,* 910–928.
11. Waters, J. W. Submillimeter-wavelength Heterodyne Spectroscopy and Remote Sensing of the Upper Atmosphere. *Proc. IEEE* **1992,** *80,* 1679–1701.

12. Federici, J.; Moeller, L. Review of Terahertz and Subterahertz Wireless Communications. *J. Appl. Phys.* **2010,** *107*, 111101.

13. Kleine-Ostmann, T.; Nagatsuma, T. A Review on Terahertz Communications Research. *J. Infrared Millim. Terahertz Waves* **2011,** *32*, 143–171.

14. Siegel, P. H. In *Terahertz Technology in Biology and Medicine*, Microwave Symposium Digest, 2004 IEEE MTT-S International, June 6–11, 2004, Vol. 3, pp 1575–1578.

15. Barnes, M. Terahertz Emission from Ultrafast Lateral Diffusion Currents within Semiconductor Devices. University of Southampton, 2014.

16. Shumyatsky, P.; Alfano, R. R. Terahertz Sources. *J. Biomed. Opt.* **2011,** *16*, 033001-033001-9.

17. Arenas, D.; Shim, D.; Koukis, D.; Seok, E.; Tanner, D. Characterization of Near-Terahertz Complementary Metal-Oxide Semiconductor Circuits Using a Fourier-Transform Interferometer. *Rev. Sci. Instrum.* **2011,** *82*, 103106.

18. Bründermann, E.; Hübers, H.-W.; Kimmitt, M. F. *Terahertz Techniques*; Springer: USA, 2012; Vol. 151.

19. Chakkittakandy, R. *Quasi-near Field Terahertz Spectroscopy*; TU Delft, Delft University of Technology, 2010.

20. Radhanpura, K. All-optical Terahertz Generation from Semiconductors: Materials and Mechanisms, 2012. http://ro.uow.edu.au/cgi/viewcontent.cgi?article=4848&context=theses

21. Williams, G. FAR-IR/THz Radiation from the Jefferson Laboratory, Energy Recovered Linac, Free Electron Laser. *Rev. Sci. Instrum.* **2002,** *73*, 1461–1463.

22. Eisele, H.; Kamoua, R. Submillimeter-wave InP Gunn Devices. *IEEE Trans. Microwave Theory* **2004,** *52*, 2371–2378.

23. Eisele, H. InP Gunn Devices for 400–425 GHz. *Electron. Lett.* **2006,** *42*, 358–359.

24. Ino, M.; Ishibashi, T.; Ohmori, M. C. W. Oscillation with p⁺-p-n⁺ Silicon IMPATT Diodes in 200 GHz and 300 GHz Bands. *Electron. Lett.* **1976,** *12*, 148–149.

25. Gorshunov, B.; Volkov, A.; Spektor, I.; Prokhorov, A.; Mukhin, A.; Dressel, M.; Uchida, S.; Loidl, A. Terahertz BWO-spectroscopy. *Int. J. Infrared Millimeter Waves* **2005,** *26*, 1217–1240.

26. Tells, E. M.; Viscovini, R.; Scalabrin, A.; Pereira, D. In *Far-Infrared Laser Lines from CH3 Rocking of* ¹³*CD*₃*OD*, SPIE's International Symposium on Optical Science, Engineering, and Instrumentation; International Society for Optics and Photonics, 1998; pp 60–63.

27. Ghoranneviss, M.; Kashani, M. A.; Hogabri, A.; Kohiyan, A.; Anvari, A. In *Design and Modification of the FIR HCN Laser*, SPIE's International Symposium on Optical Science, Engineering, and Instrumentation; International Society for Optics and Photonics, 1998; pp 85–87.

28. Bergner, A.; Heugen, U.; Bründermann, E.; Schwaa b, G.; Havenith, M.; Chamberlin, D. R.; Haller, E. E. New p-Ge THz Laser Spectrometer for the Study of Solutions: THz Absorption Spectroscopy of Water. *Rev. Sci. Instrum.* **2005,** *76*, 063110.

29. Bründermann, E.; Chamberlin, D. R.; Haller, E. E. High Duty Cycle and Continuous Terahertz Emission from Germanium. *Appl. Phys. Lett.* **2000,** *76*, 2991–2993.

30. Edwards, T. J.; Walsh, D.; Spurr, M. B.; Rae, C. F.; Dunn, M. H.; Browne, P. G. Compact Source of Continuously and Widely-tunable Terahertz Radiation. *Opt. Express* **2006,** *14*, 1582–1589.

31. Kawase, K.; Shikata, J.-I.; Imai, K.; Ito, H. Transform-limited, Narrow-linewidth, Terahertz-wave Parametric Generator. *Appl. Phys. Lett.* **2001,** *78*, 2819–2821.

32. Williams, B. S. Terahertz Quantum-Cascade lasers. *Nat. Photonics* **2007**, *1*, 517–525.

33. Wade, A.; Fedorov, G.; Smirnov, D.; Kumar, S.; Williams, B.; Hu, Q.; Reno, J. Magnetic-Field-Assisted Terahertz Quantum Cascade Laser Operating up to 225 K. *Nat. Photonics* **2009**, *3*, 41–45.

34. Brown, E.; McIntosh, K.; Nichols, K.; Dennis, C. Photomixing up to 3.8 THz in Low Temperature Grown GaAs. *Appl. Phys. Lett.* **1995**, *66*, 285–287.

35. Shi, W.; Ding, Y. J.; Fernelius, N.; Vodopyanov, K. Efficient, Tunable, and Coherent 0.18–5.27-THz Source Based on GaSe Crystal. *Opt. Lett.* **2002**, *27*, 1454–1456.

36. Morikawa, O.; Tonouchi, M.; Hangyo, M. Sub-THz Spectroscopic System Using a Multimode Laser Diode and Photoconductive Antenna. *Appl. Phys. Lett.* **1999**, *75*, 3772–3774.

37. Smith, P. R.; Auston, D. H.; Nuss, M. C. Subpicosecond Photoconducting Dipole Antennas. *IEEE J. Quant. Electron.* **1988**, *24*, 255–260.

38. Rice, A.; Jin, Y.; Ma, X.; Zhang, X. C.; Bliss, D.; Larkin, J.; Alexander, M. Terahertz Optical Rectification from <110> Zinc Blende Crystals. *Appl. Phys. Lett.* **1994**, *64*, 1324–1326.

39. Zhang, X. C.; Auston, D. Optoelectronic Measurement of Semiconductor Surfaces and Interfaces with Femtosecond Optics. *J. Appl. Phys.* **1992**, *71*, 326–338.

40. Welsh, G. H.; Hunt, N. T.; Wynne, K. Terahertz-pulse Emission through Laser Excitation of Surface Plasmons in a Metal Grating. *Phys. Rev. Lett.* **2007**, *98*, 026803.

41. Ozyuzer, L.; Koshelev, A.; Kurter, C.; Gopalsami, N.; Li, Q.; Tachiki, M.; Kadowaki, K.; Yamamoto, T.; Minami, H.; Yamaguchi, H. Emission of Coherent THz Radiation from Superconductors. *Science* **2007**, *318*, 1291–1293.

42. Planken, P. C.; Nuss, M. C.; Brener, I.; Goossen, K. W.; Luo, M. S.; Chuang, S. L. Terahertz Emission in Single Quantum Wells after Coherent Optical Excitation. *Phys. Rev. Lett.* **1992**, *69*, 3800–3803.

43. Waschke, C.; Roskos, H. G.; Schwedler, R.; Leo, K.; Kurz, H.; Köhler, K. Coherent Submillimeter-Wave Emission from Bloch Oscillations in a Semiconductor Superlattice. *Phys. Rev. Lett.* **1993**, *70*, 3319.

44. Beard, M. C.; Turner, G. M.; Schmuttenmaer, C. A. Measuring Intramolecular Charge Transfer via Coherent Generation of THz Radiation. *J. Phys. Chem. A.* **2002**, *106*, 878–883.

45. Hilton, D. J.; Averitt, R.; Meserole, C.; Fisher, G. L.; Funk, D. J.; Thompson, J. D.; Taylor, A. J. Terahertz Emission via Ultrashort-Pulse Excitation of Magnetic Metal Films. *Opt. Lett.* **2004**, *29*, 1805–1807.

46. Shen, J.; Fan, X.; Chen, Z.; DeCamp, M. F.; Zhang, H.; Xiao, J. Q. Damping Modulated Terahertz Emission of Ferromagnetic Films Excited by Ultrafast Laser Pulses. *Appl. Phys. Lett.* **2012**, *101*, 072401.

47. Cook, D.; Hochstrasser, R. Intense Terahertz Pulses by Four-Wave Rectification in Air. *Opt. Lett.* **2000**, *25*, 1210–1212.

48. Xie, X.; Dai, J.; Zhang, X. C. Coherent Control of THz Wave Generation in Ambient Air. *Phys. Rev. Lett.* **2006**, *96*, 075005.

49. Greene, B. I.; Federici, J. F.; Dykaar, D. R.; Jones, R. R.; Bucksbaum, P. H. Interferometric Characterization of 160 fs Far Infrared Light Pulses. *Appl. Phys. Lett.* **1991**, *59*, 893–895.

50. Daghestani, N.; Persheyev, S.; Cataluna, M.; Ross, G.; Rose, M. THz Generation from a Nanocrystalline Silicon-Based Photoconductive Device. *Semicond. Sci. Technol.* **2011**, *26*, 075015.

51. Turkoglu, F.; Koseoglu, H.; Demirhan, Y.; Ozyuzer, L.; Preu, S.; Malzer, S.; Simsek, Y.; Müller, P.; Yamamoto, T.; Kadowaki, K. Interferometer Measurements of Terahertz Waves from $Bi_2Sr_2CaCu_2O_{8+d}$ Mesas. *Supercond. Sci. Technol.* **2012,** *25,* 125004.

52. Wu, Q.; Zhang, X. C. Free Space Electro Optic Sampling of Terahertz Beams. *Appl. Phys. Lett.* **1995,** *67,* 3523–3525.

53. Liu, L.; Hesler, J. L.; Haiyong, X.; Lichtenberger, A. W.; Weikle, R. M. A Broadband Quasi-Optical Terahertz Detector Utilizing a Zero Bias Schottky Diode. *IEEE Microwave Wireless Compon. Lett.* **2010,** *20,* 504–506.

54. Hubers, H. W. Terahertz Heterodyne Receivers. *IEEE J. Sel. Top. Quantum Electron.* **2008,** *14,* 378–391.

55. Sakai, K.; Tani, M. Introduction to Terahertz Pulses. In *Terahertz Optoelectronics*; Springer: USA, 2005; pp 1–30.

56. Tani, M.; Matsuura, S.; Sakai, K.; Nakashima, S.-I. Emission Characteristics of Photoconductive Antennas Based on Low-temperature-grown GaAs and Semi-insulating GaAs. *Appl. Opt.* **1997,** *36,* 7853–7859.

57. Dykaar, D. R.; Greene, B. I.; Federici, J. F.; Levi, A. F. J.; Pfeiffer, L. N.; Kopf, R. F. Log Periodic Antennas for Pulsed Terahertz Radiation. *Appl. Phys. Lett.* **1991,** *59,* 262–264.

58. Cao, H.; Linke, R. A.; Nahata, A. Broadband Generation of Terahertz Radiation in a Waveguide. *Opt. Lett.* **2004,** *29,* 1751.

59. Suo, H.; Takano, K.; Ohno, S.; Kurosawa, H.; Nakayama, K.; Ishihara, T.; Hangyo, M. Polarization Property of Terahertz Wave Emission from Gammadion-type Photoconductive Antennas. *Appl. Phys. Lett.* **2013,** *103,* 111106.

60. Bartels, A.; Thoma, A.; Janke, C.; Dekorsy, T.; Dreyhaupt, A.; winner, S.; Helm, M. High-Resolution THz Spectrometer with kHz Scan Rates. *Opt. Express* **2006,** *14*(1), 430.

61. Zhang, X.-C.; Xu, J. *Introduction to THz Wave Photonics*; Springer: USA, 2010; p 28.

62. Sakai, K.; Kikō, J. T. K. *Terahertz Optoelectronics*; Springer: Berlin, 2005.

63. Heiman, D.; Hellwarth, R. W.; Levenson, M. D.; Martin, G. *Phys. Rev. Lett.* **1976,** *36,* 189.

64. Stone, M. R.; Naftaly, M.; Miles, R. E.; Fletcher, J. R.; Steenson, D. P. Electrical and Radiation Characteristics of Semilarge Photoconductive Terahertz Emitters. *IEEE Trans. Microwave Theory* **2004,** *52,* 2420–2429.

65. Miyamaru, F.; Saito, Y.; Takeda, M. W.; Liu, L.; Hou, B.; Wen, W.; Sheng, P. Emission of Terahertz Radiations from Fractal Antennas. *Appl. Phys. Lett.* **2009,** *95,* 221111.

66. Jepsen, P. U.; Jacobsen, R. H.; Keiding, S. R. Generation and Detection of Terahertz Pulses from Biased Semiconductor Antennas. *J. Opt. Soc. Am. B.* **1996,** *13,* 2424–2436.

67. Yano, R.; Gotoh, H.; Hirayama, Y.; Miyashita, S.; Kadoya, Y.; Hattori, T. Terahertz Wave Detection Performance of Photoconductive Antennas: Role of Antenna Structure and Gate Pulse Intensity. *J. Appl. Phys.* **2005,** *97,* 103103.

68. Miyamaru, F. Dependence of Emission of Terahertz Radiation on Geometrical Parameters of Dipole Photoconductive Antennas. *Appl. Phys. Lett.* **2010,** *96,* 211104.

69. Upadhya, P. C.; Fan, W.; Burnett, A.; Cunningham, J.; Davies, A. G.; Linfield, E. H.; Lloyd-Hughes, J.; Castro-Camus, E.; Johnston, M. B.; Beere, H. Excitation-Density-Dependent Generation of Broadband Terahertz Radiation in an Asymmetrically Excited Photoconductive Antenna. *Opt. Lett.* **2007,** *32,* 2297–2299.

70. Winnerl, S.; Peter, F.; Nitsche, S.; Dreyhaupt, A.; Zimmermann, B.; Wagner, M.; Schneider, H.; Helm, M.; Kohler, K. Generation and Detection of THz Radiation with

Scalable Antennas Based on GaAs Substrates with Different Carrier Lifetimes. *IEEE J. Sel. Top. Quantum Electron.* **2008**, *14*, 449–457.

71. Awad, M.; Nagel, M.; Kurz, H.; Herfort, J.; Ploog, K. Characterization of Low Temperature GaAs Antenna Array Terahertz Emitters. *Appl. Phys. Lett.* **2007**, *91*, 181124.

72. Berry, C. W.; Wang, N.; Hashemi, M. R.; Unlu, M.; Jarrahi, M. Significant Performance Enhancement in Photoconductive Terahertz Optoelectronics By Incorporating Plasmonic Contact Electrodes. *Nat. Commun.* **2013**, *4*, 1622.

73. Singh, A.; Prabhu, S. S. Microlensless Interdigitated Photoconductive Terahertz Emitters. *Opt. Express* **2015**, *23*, 1529–1535.

74. Rihani, S.; Faulks, R.; Beere, H.; Page, H.; Gregory, I.; Evans, M.; Ritchie, D. A.; Pepper, M. Effect of Defect Saturation on Terahertz Emission and Detection Properties of Low Temperature GaAs Photoconductive Switches. *Appl. Phys. Lett.* **2009**, *95*, 051106.

75. Madéo, J.; Margiolakis, A.; Zhao, Z.-Y.; Hale, P. J.; Man, M. K. L.; Zhao, Q.-Z.; Peng, W.; Shi, W.-Z.; Dani, K. M. Ultrafast Properties of Femtosecond-Laser-Ablated GaAs and Its Application to Terahertz Optoelectronics. *Opt. Lett.* **2015**, *40*, 3388–3391.

76. Singh, A.; Pal, S.; Surdi, H.; Prabhu, S. S.; Nanal, V.; Pillay, R. G. Highly Efficient and Electrically Robust Carbon Irradiated Semi-insulating GaAs Based Photoconductive Terahertz Emitters. *Appl. Phys. Lett.* **2014**, *104*, 063501.

77. Duvillaret, L.; Garet, F.; Roux, J.-F.; Coutaz, J.-L. Analytical Modeling and Optimization of Terahertz Time-Domain Spectroscopy Experiments, Using Photoswitches as Antennas. *IEEE J. Sel. Top. Quantum Electron.* **2001**, *7*, 615–623.

78. Sun, F. Analysis of Terahertz Pulse Measurement with a Chirped Probe Beam. *Appl. Phys. Lett.* **1998**, *73*, 2233.

79. Zhang, J.; Hong, Y.; Braunstein, S.; Shore, K. Terahertz Pulse Generation and Detection with LT-GaAs Photoconductive Antenna. *IEE Proc. Optoelectron.* **2004**, *151*, 98–101.

80. Venkatesh, M.; Rao, K. S.; Abhilash, T. S.; Tewari, S. P.; Chaudhary, A. K. Optical Characterization of GaAs Photoconductive Antennas for Efficient Generation and Detection of Terahertz Radiation. *Opt. Mater.* **2014**, *36*, 596–601.

81. Venkatesh, M.; Chaudhary, A. K. In *Generation of THz Radiation from Low Temperature Gallium Arsenide (LT-GaAs) Photoconductive (PC) Antennas Using Tunable Femtosecond Oscillator*, 12th International Conference on Fiber Optics and Photonics, Kharagpur, December 13, 2014; Optical Society of America: Kharagpur, 2014; p S5A.33.

82. Budiarto, E.; Margolies, J.; Jeong, S.; Son, J.; Bokor, J. High-intensity Terahertz Pulses at 1-kHz Repetition Rate. *IEEE J. Quant. Electron.* **1996**, *32*, 1839–1846.

83. Bass, M.; Franken, P.; Ward, J.; Weinreich, G. Optical Rectification. *Phys. Rev. Lett.* **1962**, *9*, 446.

84. Lee, Y.-S. Generation and Detection of Broadband Terahertz Pulses. In *Principles of Terahertz Science and Technology*; Springer: US, 2009; pp 1–66.

85. Kono, S.; Tani, M.; Sakai, K. Generation and Detection of Broadband Pulsed Terahertz Radiation. In *Terahertz Optoelectronics*; Sakai, K., Ed.; Springer: Berlin, Heidelberg, 2005; Vol. 97, pp 31–62.

86. Zhang, X.-C.; Xu, J. Generation and Detection of THz Waves. In *Introduction to THz Wave Photonics*; Springer: USA, 2010; pp 27–48.

87. Khurgin, J. B. Optical Rectification and Terahertz Emission in Semiconductors Excited Above the Band Gap. *J. Opt. Soc. Am. B.* **1994**, *11*, 2492–2501.

88. Xiaodong, M.; Zotova, I. B.; Ding, Y. J. Power Scaling on Efficient Generation of Ultrafast Terahertz Pulses. *IEEE J. Sel. Top. Quantum Electron.* **2008**, *14*, 315–332.

89. Wilke, I.; Sengupta, S. Nonlinear Optical Techniques for Terahertz Pulse Generation and Detection—Optical Rectification and Electrooptic Sampling. *Terahertz Spectrosc.: Principles Appl. Opt. Sci. Eng.* **2008**, *131*, 41.

90. Nahata, A.; Weling, A. S.; Heinz, T. F. A Wideband Coherent Terahertz Spectroscopy System Using Optical Rectification and Electro Optic Sampling. *Appl. Phys. Lett.* **1996**, *69*, 2321–2323.

91. Schneider, A.; Biaggio, I.; Günter, P. Optimized Generation of THz Pulses via Optical Rectification in the Organic Salt DAST. *Opt. Commun.* **2003**, *224*, 337–341.

92. Blanchard, F.; Razzari, L.; Bandulet, H.; Sharma, G.; Morandotti, R.; Kieffer, J.; Ozaki, T.; Reid, M.; Tiedje, H.; Haugen, H. Generation of 1.5 µJ Single-cycle Terahertz Pulses by Optical Rectification from a Large Aperture ZnTe Crystal. *Opt. Express* **2007**, *15*, 13212–13220.

93. Huber, R.; Brodschelm, A.; Tauser, F.; Leitenstorfer, A. Generation and Field-Resolved Detection of Femtosecond Electromagnetic Pulses Tunable up to 41 THz. *Appl. Phys. Lett.* **2000**, *76*, 3191–3193.

94. Xie, X.; Xu, J.; Zhang, X.-C. Terahertz Wave Generation and Detection from a CdTe Crystal Characterized by Different Excitation Wavelengths. *Opt. Lett.* **2006**, *31*, 978–980.

95. Chang, G.; Divin, C. J.; Liu, C.-H.; Williamson, S. L.; Galvanauskas, A.; Norris, T. B. Power Scalable Compact THz System based on an Ultrafast Yb-Doped Fiber Amplifier. *Opt. Express* **2006**, *14*, 7909–7913.

96. Rice, A.; Jin, Y.; Ma, X. F.; Zhang, X. C.; Bliss, D.; Larkin, J.; Alexander, M. Terahertz Optical Rectification from <1 1 0> Zinc-Blende Crystals. *Appl. Phys. Lett.* **1994**, *64*, 1324–1326.

97. Rowley, J. D.; Bas, D. A.; Zawilski, K. T.; Schunemann, P. G.; Bristow, A. D. Terahertz Emission from $ZnGeP_2$: Phase-Matching, Intensity, and Length Scalability. *J. Opt. Soc. Am. B.* **2013**, *30*, 2882–2888.

98. Winnewisser, C.; Jepsen, P. U.; Schall, M.; Schyja, V.; Helm, H. Electro-Optic Detection of THz Radiation in $LiTaO_3$, $LiNbO_3$ and ZnTe. *Appl. Phys. Lett.* **1997**, *70*, 3069–3071.

99. Carrig, T. J.; Rodriguez, G.; Clement, T. S.; Taylor, A. J.; Stewart, K. R. Scaling of Terahertz Radiation via Optical Rectification in Electro-optic Crystals. *Appl. Phys. Lett.* **1995**, *66*, 121–123.

100. Schneider, A.; Neis, M.; Stillhart, M.; Ruiz, B.; Khan, R. U.; Günter, P. Generation of Terahertz Pulses Through Optical Rectification in Organic DAST Crystals: Theory and Experiment. *J. Opt. Soc. Am. B* **2006**, *23*, 1822–1835.

101. Kuroyanagi, K.; Fujiwara, M.; Hashimoto, H.; Takahashi, H.; Aoshima, S.-I.; Tsuchiya, Y. All Organic Terahertz Electromagnetic Wave Emission and Detection Using Highly Purified *N*-Benzyl-2-Methyl-4-Nitroaniline Crystals. *Jpn. J. Appl. Phys.* **2006**, *45*, 4068.

102. Carey, J. J.; Bailey, R. T.; Pugh, D.; Sherwood, J.; Cruickshank, F.; Wynne, K. Terahertz Pulse Generation in an Organic Crystal by Optical Rectification and Resonant Excitation of Molecular Charge Transfer. *Appl. Phys. Lett.* **2002**, *81*, 4335–4337.

103. Kang, B. J.; Baek, I. H.; Jeong, J.-H.; Kim, J.-S.; Lee, S.-H.; Kwon, O. P.; Rotermund, F. Characteristics of Efficient Few-cycle Terahertz Radiation Generated in As-grown Nonlinear Organic Single Crystals. *Curr. Appl. Phys.* **2014**, *14*, 403–406.

104. Yang, K. H.; Richards, P. L.; Shen, Y. R. Generation of Far-infrared Radiation by Picosecond Light Pulses in $LiNbO_3$. *Appl. Phys. Lett.* **1971**, *19*, 320–323.

105. Hebling, J.; Yeh, K.-L.; Hoffmann, M. C.; Bartal, B.; Nelson, K. A. Generation of High-power Terahertz Pulses by Tilted-pulse-front Excitation and Their Application Possibilities. *J. Opt. Soc. Am. B* **2008**, *25*, B6–B19.

106. Wu, Q.; Zhang, X. C. Ultrafast Electro Optic Field Sensors. *Appl. Phys. Lett.* **1996**, *68*, 1604–1606.

107. Ingrid, W.; Suranjana, S. Nonlinear Optical Techniques for Terahertz Pulse Generation and Detection—Optical Rectification and Electrooptic Sampling. In *Terahertz Spectroscopy*; CRC Press: USA, 2007; pp 41–72.

108. Kübler, C.; Huber, R.; Leitenstorfer, A. Ultra Broadband Terahertz Pulses: Generation and Field-resolved Detection. *Semicond. Sci. Technol.* **2005**, *20*, S128.

109. van der Valk, N. C. J.; Planken, P. C. M.; Buijserd, A. N.; Bakker, H. J. Influence of Pump Wavelength and Crystal Length on the Phase Matching of Optical Rectification. *J. Opt. Soc. Am. B* **2005**, *22*, 1714–1718.

110. Planken, P. C. M.; Nienhuys, H.-K.; Bakker, H. J.; Wenckebach, T. Measurement and Calculation of the Orientation Dependence of Terahertz Pulse Detection in ZnTe. *J. Opt. Soc. Am. B* **2001**, *18*, 313–317.

111. Dakovski, G. L.; Kubera, B.; Shan, J. Localized Terahertz Generation via Optical Rectification in ZnTe. *J. Opt. Soc. Am. B* **2005**, *22*, 1667–1670.

112. Kampfrath, T.; Notzold, J.; Wolf, M. Sampling of Broadband Terahertz Pulses with Thick Electro-optic Crystals. *Appl. Phys. Lett.* **2007**, *90*, 231113.

113. Ahmed, S.; Savolainen, J.; Hamm, P. Detectivity Enhancement in THz Electrooptical Sampling. *Rev. Sci. Instrum.* **2014**, *85*, 013114.

114. Sajadi, M.; Wolf, M.; Kampfrath, T. Terahertz Field Enhancement via Coherent Superposition of the Pulse Sequences after a Single Optical-Rectification Crystal. *Appl. Phys. Lett.* **2014**, *104*, 091118.

115. Vidal, S.; Degert, J.; Tondusson, M.; Freysz, E.; Oberlé, J. Optimized Terahertz Generation via Optical Rectification in ZnTe Crystals. *J. Opt. Soc. Am. B* **2014**, *31*, 149–153.

116. Venkatesh, M.; Chaudhary, A. K. In *Generation of Temporally Shaped Terahertz (THz) Pulses from Zinc Telluride (ZnTe) Crystal Using Tunable Femtosecond Laser Wavelengths*, 13th International Conference on Fiber Optics and Photonics, Kanpur, December 4, 2016; Optical Society of America: Kanpur, 2016; p Tu4A.17.

117. Venkatesh, M.; Thirupugalmani, K.; Rao, K. S.; Brahadeeswaran, S.; Chaudhary, A. K. Generation of Efficient THz Radiation by Optical Rectification in DAST Crystal Using Tunable Femtosecond Laser Pulses. *Indian J. Phys.* **2016**, 1–8.

118. Venkatesh, M.; Rao, K. S.; Chaudhary, A. K.; Ghosh, S. K.; Dubey, S.; Yadav, N. Evaluation of DAST and Zinc Telluride Nonlinear Crystals for Efficient Terahertz Generation. *AIP Conf. Proc.* **2015**, *1670*, 020005.

119. Rao, K. S.; Chaudhary, A. K.; Venkatesh, M.; Thirupugalmani, K.; Brahadeeswaran, S. DAST Crystal Based Terahertz Generation and Recording of Time Resolved Photoacoustic Spectra of N_2O Gas at 0.5 and 1.5 THz Bands. *Curr. Appl. Phys.* **2016**, *16*, 777–783.

120. Kumari, A.; Venkatesh, M.; Chaudhary, A. K. In *Terahertz Generation from Cadmium Telluride Crystal Using Tunable Oscillator Laser Pulses*, 13th International Conference on Fiber Optics and Photonics, Kanpur, December 4, 2016; Optical Society of America: Kanpur, 2016; p W3A.6.

121. Konda, S. R.; Venkatesh, M.; Thirupugalmani, K.; Brahadeeswaran, S.; Chaudhary, A. K. In *Optical Parametric Amplifier Based Efficient Terahertz Generation in DAST*

Crystal Using Optical Rectification, 12th International Conference on Fiber Optics and Photonics, Kharagpur, December 13, 2014; Optical Society of America: Kharagpur, 2014; p S5A.28.

122. Kohli, K. K.; Vaupel, A.; Chatterjee, S.; Rühle, W. W. Adaptive Shaping of THz-pulses Generated in <1 1 0> ZnTe Crystals. *J. Opt. Soc. Am. B* **2009,** *26,* A74–A78.

123. Vidal, S.; Degert, J.; Oberlé, J.; Freysz, E. Femtosecond Optical Pulse Shaping for Tunable Terahertz Pulse Generation. *J. Opt. Soc. Am. B* **2010,** *27,* 1044–1050.

124. Lu, C.; Zhang, S.; Jia, T.; Qiu, J.; Sun, Z. Manipulation of Terahertz Pulse Generation in ZnTe Crystal by Shaping Femtosecond Laser Pulses with a Square Phase Modulation. *Opt. Commun.* **2014,** *310,* 90–93.

125. Ahn, J.; Efimov, A.; Averitt, R.; Taylor, A. Terahertz Waveform Synthesis via Optical Rectification of Shaped Ultrafast Laser Pulses. *Opt. Express* **2003,** *11,* 2486–2496.

126. Prabhu, S. S.; Vengurlekar, A. S. In *Study of Excitation Wavelength Dependence of THz Emission from ZnTe*, Physics of Semiconductor Devices, 2007. IWPSD 2007. International Workshop on, December 16–20, 2007; pp 389–390.

127. Li, D.; Ma, G. Pump-wavelength Dependence of Terahertz Radiation via Optical Rectification in (1 1 0)-Oriented ZnTe Crystal. *J. Appl. Phys.* **2008,** *103,* 123101.

128. Tu, C. M.; Ku, S. A.; Chu, W. C.; Luo, C. W.; Chen, J. C.; Chi, C. C. Pulsed Terahertz Radiation due to Coherent Phonon–Polariton Excitation in <1 1 0> ZnTe Crystal. *J. Appl. Phys.* **2012,** *112,* 093110-5.

129. Singh, A.; Pal, S.; Surdi, H.; Prabhu, S. S.; Mathimalar, S.; Nanal, V.; Pillay, R. G.; Döhler, G. H. Carbon Irradiated Semi Insulating GaAs for Photoconductive Terahertz Pulse Detection. *Opt. Express* **2015,** *23,* 6656–6661.

130. Jepsen, P. U.; Jacobsen, R. H.; Keiding, S. Generation and Detection of Terahertz Pulses from Biased Semiconductor Antennas. *J. Opt. Soc. Am. B* **1996,** *13,* 2424–2436.

131. Kono, S.; Tani, M.; Sakai, K. Ultrabroadband Photoconductive Detection: Comparison with Free-space Electro-Optic Sampling. *Appl. Phys. Lett.* **2001,** *79,* 898–900.

132. Valdmanis, J. A.; Mourou, G.; Gabel, C. W. Picosecond Electro Optic Sampling System. *Appl. Phys. Lett.* **1982,** *41,* 211–212.

133. Kübler, C.; Huber, R.; Tübel, S.; Leitenstorfer, A. Ultrabroadband Detection of Multi-Terahertz Field Transients with GaSe Electro-optic Sensors: Approaching the Near Infrared. *Appl. Phys. Lett.* **2004,** *85,* 3360.

134. Wu, Q.; Zhang, X.-C. 7 Terahertz Broadband GaP Electro-Optic Sensor. *Appl. Phys. Lett.* **1997,** *70,* 1784–1786.

135. Vossebürger, M.; Brucherseifer, M.; Cho, G. C.; Roskos, H. G.; Kurz, H. Propagation Effects in Electro-Optic Sampling of Terahertz Pulses in GaAs. *Appl. Opt.* **1998,** *37,* 3368–3371.

136. Sinyukov, A. M.; Hayden, L. M. Generation and Detection of Terahertz Radiation with Multilayered Electro-Optic Polymer Films. *Opt. Lett.* **2002,** *27,* 55–57.

137. Han, P.; Tani, M.; Pan, F.; Zhang, X.-C. Use of the Organic Crystal DAST for Terahertz Beam Applications. *Opt. Lett.* **2000,** *25,* 675–677.

138. Zheng, X.; McLaughlin, C. V.; Cunningham, P.; Hayden, L. M. Organic Broadband Terahertz Sources and Sensors. *J. Nanoelectron. Optoelectron.* **2007,** *2,* 58–76.

139. Gallot, G.; Grischkowsky, D. Electro-optic Detection of Terahertz Radiation. *J. Opt. Soc. Am. B* **1999,** *16,* 1204–1212.

140. Gallot, G.; Zhang, J.; McGowan, R.; Jeon, T.-I.; Grischkowsky, D. Measurements of the THz Absorption and Dispersion of ZnTe and Their Relevance to the Electro-Optic Detection of THz Radiation. *Appl. Phys. Lett.* **1999,** *74,* 3450–3452.

141. Park, S.-G.; Melloch, M.; Weiner, A. Comparison of Terahertz Waveforms Measured by Electro-Optic and Photoconductive Sampling. *Appl. Phys. Lett.* **1998,** *73,* 3184–3186.

142. Shen, Y. C.; Upadhya, P. C.; Beere, H. E.; Linfield, E. H.; Davies, A. G.; Gregory, I. S.; Baker, C.; Tribe, W. R.; Evans, M. J. Generation and Detection of Ultrabroadband Terahertz Radiation Using Photoconductive Emitters and Receivers. *Appl. Phys. Lett.* **2004,** *85,* 164–166.

143. Cai, Y.; Brener, I.; Lopata, J.; Wynn, J.; Pfeiffer, L.; Stark, J.; Wu, Q.; Zhang, X.; Federici, J. Coherent Terahertz Radiation Detection: Direct Comparison Between Free-Space Electro-Optic Sampling and Antenna Detection. *Appl. Phys. Lett.* **1998,** *73,* 444–446.

144. Van Exter, M.; Grischkowsky, D. Carrier Dynamics of Electrons and Holes in Moderately Doped Silicon. *Phys. Rev. B* **1990,** *41,* 12140.

145. Dai, J.; Zhang, J.; Zhang, W.; Grischkowsky, D. Terahertz Time-Domain Spectroscopy Characterization of the Far-Infrared Absorption and Index of Refraction of High-Resistivity, Float-Zone Silicon. *J. Opt. Soc. Am. B* **2004,** *21,* 1379–1386.

146. Leahy-Hoppa, M.; Fitch, M.; Zheng, X.; Hayden, L.; Osiander, R. Wideband Terahertz Spectroscopy of Explosives. *Chem. Phys. Lett.* **2007,** *434,* 227–230.

147. Davies, A. G.; Burnett, A. D.; Fan, W.; Linfield, E. H.; Cunningham, J. E. Terahertz Spectroscopy of Explosives and Drugs. *Mater. Today* **2008,** *11,* 18–26.

148. Chen, J.; Chen, Y.; Zhao, H.; Bastiaans, G. J.; Zhang, X.-C. Absorption Coefficients of Selected Explosives and Related Compounds in the Range of 0.1–2.8 THz. *Opt. Express* **2007,** *15,* 12060–12067.

149. Wang, Y.; Zhao, Z.; Chen, Z.; Kang, K.; Feng, B.; Zhang, Y. Terahertz Absorbance Spectrum Fitting Method for Quantitative Detection of Concealed Contraband. *J. Appl. Phys.* **2007,** *102,* 113108.

150. Globus, T.; Woolard, D.; Khromova, T.; Crowe, T.; Bykhovskaia, M.; Gelmont, B.; Hesler, J.; Samuels, A. THz-Spectroscopy of Biological Molecules. *J. Biol. Phys.* **2003,** *29,* 89–100.

151. Siegel, P. H. Terahertz Technology in Biology and Medicine. *IEEE Trans. Microwave Theory.* **2004,** *52,* 2438–2447.

152. Walther, M.; Plochocka, P.; Fischer, B.; Helm, H.; Uhd Jepsen, P. Collective Vibrational Modes in Biological Molecules Investigated by Terahertz Time Domain Spectroscopy. *Biopolymers.* **2002,** *67,* 310–313.

153. D'Angelo, F.; Mics, Z.; Bonn, M.; Turchinovich, D. Ultra-Broadband THz Time-Domain Spectroscopy of Common Polymers Using THz Air Photonics. *Opt. Express* **2014,** *22,* 12475–12485.

154. Van Exter, M.; Fattinger, C.; Grischkowsky, D. Terahertz Time-domain Spectroscopy of Water Vapor. *Opt. Lett.* **1989,** *14,* 1128–1130.

155. Chrzanowski, L. S. v.; Beckmann, J.; Marchetti, B.; Ewert, U.; Schade, U. Terahertz Time Domain Spectroscopy for Non-Destructive Testing of Hazardous Liquids. *Mater. Test* **2012,** *54,* 444–450.

156. Kindt, J.; Schmuttenmaer, C. Far-Infrared Dielectric Properties of Polar Liquids Probed by Femtosecond Terahertz Pulse Spectroscopy. *J. Phys. Chem.* **1996,** *100,* 10373–10379.

157. Mittleman, D. M.; Jacobsen, R. H.; Neelamani, R.; Baraniuk, R. G.; Nuss, M. C. Gas Sensing Using Terahertz Time-Domain Spectroscopy. *Appl. Phys. B: Lasers Opt.* **1998,** *67,* 379–390.

158. 158. Harde, H.; Grischkowsky, D. Coherent Transients Excited by Subpicosecond Pulses of Terahertz Radiation. *J. Opt. Soc. Am. B* **1991,** *8,* 1642–1651.

159. Gowen, A.; O'Sullivan, C.; O'Donnell, C. Terahertz Time Domain Spectroscopy and Imaging: Emerging Techniques for Food Process Monitoring and Quality Control. *Trends Food Sci. Technol.* **2012**, *25*, 40–46.

160. Dexheimer, S. L. *Terahertz Spectroscopy: Principles and Applications.* CRC Press: USA, 2007.

161. Beard, M. C.; Turner, G. M.; Schmuttenmaer, C. A. Terahertz Spectroscopy. *J. Phys. Chem. B.* **2002**, *106*, 7146–7159.

162. Neugebauer, J.; Hess, B. A. Resonance Raman Spectra of Uracil Based on Kramers–Kronig Relations Using Time-Dependent Density Functional Calculations and Multi-reference Perturbation Theory. *J. Chem. Phys.* **2004**, *120*, 11564–11577.

163. Krivokhvost, O. Conventional and Nonconventional Kramers–Kronig Analysis in Optical Spectroscopy, 2014. http://citeseerx.ist.psu.edu/viewdoc/download?doi=10.1.1.1020.1177&rep=rep1&type=pdf

164. Zhang, X.-C.; Xu, J. *Introduction to THz Wave Photonics*; Springer: USA, 2010.

165. Zhang, X.-C. *The Terahertz Wave Ebook*; Zomega THz Cooperation, 2012.

166. Morikawa, O.; Tonouchi, M.; Hangyo, M. A Cross-Correlation Spectroscopy in Subtera-hertz Region Using an Incoherent Light Source. *Appl. Phys. Lett.* **2000**, *76*, 1519–1521.

167. Cunningham, P. D.; Valdes, N. N.; Vallejo, F. A.; Hayden, L. M.; Polishak, B.; Zhou, X.-H.; Luo, J.; Jen, A. K.-Y.; Williams, J. C.; Twieg, R. J. Broadband Terahertz Characterization of the Refractive Index and Absorption of Some Important Polymeric and Organic Electro-optic Materials. *J. Appl. Phys.* **2011**, *109*, 043505–043505-5.

168. Hai-Bo, L.; Zhong, H.; Karpowicz, N.; Chen, Y.; Xi-Cheng, Z. Terahertz Spectroscopy and Imaging for Defense and Security Applications. *Proc. IEEE* **2007**, *95*, 1514–1527.

169. Hirakawa, Y.; Ohno, Y.; Gondoh, T.; Mori, T.; Takeya, K.; Tonouchi, M.; Ohtake, H.; Hirosumi, T. Nondestructive Evaluation of Rubber Compounds by Terahertz Time-Domain Spectroscopy. *J Infrared Millim. Terahertz Waves* **2011**, *32*, 1457–1463.

170. Du Bosq, T.; Peale, R.; Boreman, G. Terahertz/Millimeter Wave Characterizations of Soils for Mine Detection: Transmission and Scattering. *Int. J. Infrared Millim. Waves* **2008**, *29*, 769–781.

171. Bjarnason, J. E.; Chan, T. L. J.; Lee, A. W. M.; Celis, M. A.; Brown, E. R. Millimeter-Wave, Terahertz, and Mid-Infrared Transmission through Common Clothing. *Appl. Phys. Lett.* **2004**, *85*, 519–521.

172. Huang, F.; Federici, J.; Gary, D.; Barat, R.; Zimdars, D. Noninvasive Study of Explosive Materials by Time Domain Spectroscopy and FTIR. *AIP Conf. Proc.* **2005**, *760*, 578–585.

173. Dean, P.; Saat, N. K.; Khanna, S. P.; Salih, M.; Burnett, A.; Cunningham, J.; Linfield, E. H.; Davies, A. G. Dual-frequency Imaging Using an Electrically Tunable Terahertz Quantum Cascade Laser. *Opt. Express* **2009**, *17*, 20631–20641.

174. Fitzgerald, A. J.; Berry, E.; Zinov'ev, N. N.; Homer-Vanniasinkam, S.; Miles, R. E.; Chamberlain, J. M.; Smith, M. A. Catalogue of Human Tissue Optical Properties at Terahertz Frequencies. *J. Biol. Phys.* **2003**, *29*, 123–128.

175. Mottamchetty, V.; Chaudhary, A. K. Improvised Design of THz Spectrophotometer Using LT-GaAs Photoconductive Antennas, Pyroelectric Detector and Band-pass Filters. *Indian J. Phys.* **2015**, 1–6.

176. Trzcinski, T.; Palka, N.; Szustakowski, M. THz Spectroscopy of Explosive-Related Simulants and Oxidizers. *Polska Akademia Nauk. Bull. Pol. Acad. Sci.* **2011**, *59*, 445.

CHAPTER 12

WATER VAPOR PERMEABILITY, MECHANICAL, OPTICAL, AND SENSORIAL PROPERTIES OF PLASTICIZED GUAR GUM EDIBLE FILMS

XOCHITL RUELAS CHACON[1,2,*],
JUAN CARLOS CONTRERAS ESQUIVEL[1], JULIO MONTACEZ[3],
ANTONIO FRANCISCO AGUILERA CARBO[4],
MARIA DE LA LUZ REYES VEGA[1,*],
RENE DARIO PERALTA RODRIGUEZ[5], and
GABRIELA SANCHEZ BRAMBILA[6]

[1]*Department of Food Research, Faculty of Chemistry, Universidad Autonoma de Coahuila, Blvd. V. Carranza, Colonia Republica Oriente, Saltillo 25280, Coahuila, Mexico*

[2]*Department of Food Science and Technology, Universidad Autonoma Agraria Antonio Narro, Calzada Antonio Narro 1923, Colonia Buenavista, Saltillo 25315, Coahuila, Mexico*

[3]*Department of Chemical Engineering, Faculty of Chemistry, Universidad Autonoma de Coahuila, Blvd. V. Carranza, Colonia Republica Oriente, Saltillo 25280, Coahuila, Mexico*

[4]*Department of Animal Nutrition, Universidad Autonoma Agraria Antonio Narro, Calzada Antonio Narro 1923, Colonia Buenavista, Saltillo 25315, Coahuila, Mexico*

[5]*Research Center for Applied Chemistry, Blvd. Enrique Reyna Hermosillo No. 140, Saltillo 25253, Coahuila, Mexico*

[6]*Russell Research Center-ARS, Quality and Safety Assessment Research Unit USDA, 950 College Station Road, Athens 30605, GA, USA*

Corresponding author. E-mail: xochitl.ruelas@uaaan.mx, xruelas@yahoo.com; mlrv20@yahoo.com

CONTENTS

ABSTRACT

Edible films were prepared by casting method using guar gum (1.0%, 1.5%, and 2.0%) and glycerol (20%, 30%, and 40%, w/v) in different ratios. The water vapor permeability (WVP), mechanical properties, thickness, optical properties, solubility, moisture content (MC), and sensory acceptability were investigated. As the plasticizer concentration increased, the MC, solubility, and WVP of the films increased significantly ($p < 0.05$). The tensile strength (TS) decreased as levels of glycerol increased, and the elongation at break increased as polyol and guar gum levels increased. Thickness and optical properties were affected significantly by guar gum and glycerol concentrations ($p < 0.05$). Sensory properties of films showed that taste and overall acceptability were not significantly different, while color and stickiness were significantly different ($p < 0.05$). This study provides basic information on the properties of these biodegradable and flexible films, which are made with a natural biopolymer and represent an attractive option for future trends in food applications.

12.1 INTRODUCTION

The deterioration of packaged foodstuffs largely depends on the mass and heat transfer that may occur between the internal and the external environments of the packaged food.[1,2] The use of natural biopolymers as edible films has increased the shelf life of food by modifying these environment exchanges[3–5] and offered better means of availability by allowing to preserve food for longer time. The properties of these films and their applications will depend on the composition as well as their conditions during their preparation.[6] Biopolymers used to prepare packaging materials include polysaccharides, lipids, proteins, and their derivatives.[7] The polysaccharides, such as cellulose derivatives, chitosan, starch, alginate, carrageenan, and pectin, are the most preferred because of their high film-forming ability and mechanical properties.[7–9] On the other hand, the fact that they come from a natural resource meets the trends in consumer preferences.

In order to apply edible films to foods, it is necessary to study their transport properties (water vapor permeability; WVP), mechanical properties, and sensory characteristics. WVP and the mechanical properties are the most important characteristics of edible films. The WVP is related to the property that films have as barriers against gases (O_2 and CO_2), water vapor, or oil, thus influencing shelf life and improving quality and handling management of food products.[10] Elongation (stretchability) and toughness (film strength × elongation at break) of the film determine their application as food wrap or coating.[11] The edible films should stand mechanical stress and strain to such an extent that they do not break easily under various mechanical forces. Other important elements for edible films and coatings are those related to sensory evaluation such as acceptable color, odor, taste, flavor, and texture (stickiness), which are of main importance to the acceptability for consumers.[12–14]

Edible films can be plasticized by low molecular weight carbohydrates, for example, polyols.[10] Plasticizers such as glycerol, polyethylene glycol, and sorbitol increase the flexibility of the films due to their capacity to reduce the internal hydrogen links among the polymer chains as they increase the molecular space.[15] The plasticizers most often used and recommended in film formulations are sorbitol and glycerol.[15–17] The incorporation of plasticizers is necessary to reduce polymer intermolecular forces, increasing the mobility of the polymeric chains, and improving the mechanical characteristics of the film, for example, extensibility.[18,19] Also, plasticizers affect the water barrier property of the films, since they have a great affinity for water.[16]

Guar gum, galactomannan-rich flour, is a water-soluble polysaccharide obtained from the leguminous Indian cluster bean *Cyamopsis tetragonoloba* (L.) *Taub.* The backbone of this hydrocolloid is a linear chain of D-mannopyranosyl units connected to each other by β-1,4-bonds linked to galactose residues by 1,6-bonds forming short side-branches.[20–22] Guar gum is one of the most important thickeners and a versatile material used for many food applications due to its different physicochemical properties as well as its high availability, low cost, and biodegradability. The guar gum is an excellent nontoxic stiffener used in the textile, pharmaceutical, biomedical, cosmetic, and food industries.[23,24] In addition, this galactomannan exhibits surface, interfacial, and emulsification activities.[20,24,25] Films based on galactomannans can be used to reduce water vapor, oxygen, lipid, and flavor migration between components of multi-component food products, and between food and its surroundings.[8] There is a little published information about the WVP, mechanical, optical, and sensorial properties of plasticized guar gum films. Therefore the aim of this research was to develop edible films based on guar gum biopolymer and to evaluate the effect of glycerol as plasticizer on the WVP, mechanical, optical, and sensory acceptability properties of these films.

12.2 MATERIALS AND METHODS

12.2.1 REAGENTS

Guar gum (G4129-500G) was purchased from Sigma-Aldrich (St. Louis, MO, USA). Glycerol, anhydrous $CaCl_2$, NaCl, and KCl were purchased from Jalmek Co. (Monterrey, Nuevo Leon, Mexico).

12.2.2 FILM PREPARATION

A total of nine different formulations of films were prepared. Guar gum and glycerol were used; the two components were dissolved in distilled water at 60°C with constant agitation (500 rpm) for 40 min. A series of blends were prepared with varying concentration (%) of guar gum and plasticizer (1.0/20, 1.0/30, 1.0/40, 1.5/20, 1.5/30, 1.5/40, 2.0/20, 2.0/30, and 2.0/40). Film-forming solutions were homogenized using a hot plate/stirrer (Talboys, Thorofare, NJ, USA) at 700 rpm. The polymer films were prepared by a casting method, 20 mL of recently prepared suspensions were immediately poured on polyethylene Petri dishes of 8 cm in diameter, resting on a level surface. The suspensions in the Petri dishes were then dried at 50°C in a ventilated oven (Quincy Lab Inc., Chicago, IL, USA) during 10 h. The dried films were peeled from the casting surface and stored in desiccators at 0% RH with silica gel at 25°C.

12.2.3 THICKNESS MEASUREMENT

Thickness was determined using a dial thickness gauge micrometer (Mitutoyo Manufacturing Co. Ltd., Tokyo, Japan) by generating the mean of five measurements at randomized positions on the film. Samples with air bubbles and nicks or tears were excluded from analysis.

12.2.4 OPTICAL PROPERTIES

Optical properties of light transmittance and opacity of the films were obtained using a Genesis 10 UV spectrophotometer (Thermo Electron Corporation, Madison, WI, USA). For each film specimen, a sample of a rectangular piece (1×3 cm) was placed directly in a spectrophotometer test cell, and measurements were performed using air as reference. The light transmittance of the films was scanned from wavelength of 400–800 nm using the spectrophotometer. The determination was done by triplicate and, from these spectra; the average transparency at 600 nm (T600) was calculated. The T600 was obtained from the following eq 12.1:[26]

$$T600 = \log \frac{\%T}{b} \tag{12.1}$$

where %T is the percentage transmittance and b is the film thickness (mm).

The opacity of the films was calculated by the following eq 12.2 according to the method described by Gontard et al.:[27]

$$\text{Opacity} = \text{AU500nm} * b \tag{12.2}$$

where AU_{500nm} is absorbance units at 500 nm and b is the film thickness (mm).

12.2.5   COLOR MEASUREMENT

Color of the films was assessed using a colorimeter (Minolta CR-400, Tokyo, Japan). A white standard color plate ($L = 97.75$, $a = 0.49$, $b = 1.96$, supplied by Minolta Co.) for instrument calibration was used as a background for color measurements of the films. The system provides the values of three-color components; L* (black-white component, luminosity) and chromaticness coordinates, a* (+red to −green component) and b* (+yellow to −blue component) were recorded by triplicates for each sample. Total color change from standard (ΔE), the yellowness (YI), and whiteness (WI) indexes of samples were calculated following Ahmadi et al.[15] and Bolin et al.[28] recommended eqs 12.3–12.5. The samples were analyzed at five random positions.

$$\Delta E = \sqrt{(L'-L)^2 + (a'-a)^2 + (b'-b)^2} \tag{12.3}$$

$$YI = \frac{(142.86 * b)}{L} \tag{12.4}$$

$$WI = 100 - \sqrt{(100-L)^2 + a^2 + b^2} \tag{12.5}$$

12.2.6 MOISTURE DETERMINATION

Percent MC of each film was determined according to the method reported by Mei et al.[29] The films were cut in squares of 2 cm × 2 cm and were placed into a previously dried and cooled aluminum dishes. The films along with aluminum dishes were dried inside a laboratory oven (Quincy Lab Inc., Chicago, IL, USA) at 100°C for 24 h. Weights of the film samples were taken before and after drying using a digital balance (Adventurer Ohaus Corp., Pine Brook, NJ, USA) with an accuracy of 0.0001 g. The MC was determined according to the eq 12.6. Three replications of each film were measured for MC values.

$$\text{Moisture content}(\%) = \frac{(\text{Initial dry weight} - \text{Final dry weight})*100}{(\text{Initial dry weight})} \quad (12.6)$$

12.2.7 SOLUBILITY MEASUREMENT

The solubility of the films was carried out following the methodology reported by Romero-Bastida et al.[30] Dried film samples were placed in a glass beaker with 80 mL of distilled water under constant agitation of 300 rpm at 25°C during 60 min. After this time, the samples were removed and dried for 24 h at 60°C to achieve constant weight. Samples weight was determined by using a digital balance (Adventurer Ohaus Corp., Pine Brook, NJ, USA) with an accuracy of 0.0001 g. Films solubility was calculated using the eq 12.7. The samples were analyzed and the average values were reported.

$$\text{Solubility}(\%) = \frac{(\text{Initial dry weight} - \text{Final dry weight})*100}{(\text{Initial dry weight})} \quad (12.7)$$

12.2.8 WATER VAPOR PERMEABILITY

The WVP of the prepared films was measured following the methodology reported by ASTMA E 96.[31] Granular anhydrous $CaCl_2$ (approximately 3.0 g) was used as a desiccant in the acrylic permeability cell covered with the studied film. Distance between the surface of desiccant and film was less than 6 mm as suggested by ASTM E 96.[31] Thickness of each film was measured with a micrometer (Mitutoyo, Japan with an accuracy of 0.01 mm) at five randomly selected points. The cells were placed in desiccators at different relative humidity values (RH) 75%, 85%, and 100% at $25 \pm 1.5°C$.[32] Weight gain due to water vapor permeation was determined gravimetrically as a function of time during 10 h. When the relationship between weight gain (Δw) and time (Δt) is linear, the slope of the plot is used to calculate the water vapor transmission rate (WVTR) and WVP.[1] The slope is obtained by linear regression. WVTR was calculated from the slope ($\Delta w/\Delta t$) divided by the test area (A) ($g*m^{-2}*d^{-1}$), with eq 12.8:

$$WVTR = \left(\frac{\Delta w}{\Delta t}\right) \div A \qquad (12.8)$$

12.2.9 MECHANICAL PROPERTIES

Where $\Delta w/\Delta t$ = transfer rate, amount of moisture loss per unit time ($g\ d^{-1}$); A = area exposed to moisture transfer (m^2).

WVP is calculated as following eq 12.9:

$$WVP = \frac{WVTR * L}{\Delta p} \qquad (12.9)$$

where WVP is ($g\ mm\ m^{-2}\ d^{-1}\ kPa^{-1}$), WVTR is WVTR ($gm^{-2}d^{-1}$), L is thickness of film (mm) and Δp is water vapor pressure difference at both sides of the film (kPa).[1] All determinations were evaluated in triplicate.

The films were cut in 1 cm × 9 cm strips and conditioned in a desiccator for 5 days at 57 % RH.[33] Tensile strength (TS) and elongation-at-break (E) were determined from a stress–strain curve using a texture analyzer instrument (TA-XT2; Texture Technologies Corp., Scarsdale, NY, USA) following the procedure outlined in ASTM method D 82-91.[33] Initial grip distance and crosshead speed were 5 cm and 1.00 mm*min[-1], respectively. TS was calculated by dividing the peak load by the cross-section area (thickness × 1 cm) of the initial specimen. E was expressed as the percentage of change in the length of the specimen to the original length between the grips (5 cm). TS and E values were obtained from 10 replications.

12.2.10 SENSORY EVALUATION

Sensory tests were carried out with a sensory panel of 30-trained members who were instructed to give a subjective evaluation of the films. The sensory characteristics such as color, taste, stickiness and overall acceptability of each of the samples were rated on a nine-point hedonic scale (where 1 = disliked extremely and 9 = liked extremely).[14] The samples consisted of pieces of 2.0 × 2.0 cm^2 of each film samples which were presented to the panelists in coded disposable plates and in randomized order presentation. Panelists were asked to evaluate color followed by tasting and chewing the samples to score the taste, stickiness, and overall acceptability, respectively. Cups with water and unsalted crackers were provided to the panelists to clean palate during tasting.[14]

12.2.11   STATISTICAL ANALYSIS

Differences between the variables were tested for significance by factorial analysis of variance ANOVA with Tukey's posttest using JMP software (Version 5.01; SAS Institute Inc., Cary, NC, USA). Differences at $p < 0.05$ were considered to be significant.

12.3   RESULTS AND DISCUSSION

12.3.1   FILM APPEARANCE

Homogeneous, thin, and flexible films were obtained from guar gum and glycerol solutions. Films could be easily removed from the acrylic plates. The films did not roll over or break. Visually, all the films were colorless and translucent similar to findings reported by Pereda et al.[34] with chitosan–gelatin films and by Matta Fakhoury et al.[35] with blends of manioc starch and gelatin films.

12.3.2   FILM THICKNESS

Thickness of the films showed significant differences ($p < 0.05$) depending on the percentage of guar gum and glycerol used in the formulation (Table 12.1). This has to be done with the plasticizer capacity to reduce the internal hydrogen links among the polymer chains as they increase the molecular space as reported by Farahnaky et al.[16] and Mali et al.[17] and with the amount of galactomannan used as reported by Cerqueira et al.[36]

TABLE 12.1 Effect of Guar Gum and Glycerol Concentration on the Film Thickness, Opacity, Percentage of Solubility, and Moisture Content.

Guar gum (%)	Glycerol (%, v/v)	Thickness (mm)	Opacity 500 nm (UA*mm)	Solubility (%)	Moisture content (%)
1.0	20	0.038 ± 0.003^e	0.0032 ± 0.0002^a	85.53 ± 1.05^{bc}	16.99 ± 0.99^{ab}
1.5	20	0.087 ± 0.009^a	0.0111 ± 0.0029^{ab}	78.85 ± 1.05^e	13.32 ± 1.09^c
2.0	20	0.088 ± 0.001^a	0.0121 ± 0.0020^{ab}	65.41 ± 0.79^g	6.90 ± 0.91^e
1.0	30	0.045 ± 0.002^{de}	0.0038 ± 0.0018^a	88.41 ± 1.03^{ab}	18.16 ± 0.86^a
1.5	30	0.071 ± 0.002^{bc}	0.0090 ± 0.0003^{ab}	81.04 ± 0.94^{de}	12.90 ± 0.50^c
2.0	30	0.079 ± 0.005^{ab}	0.0162 ± 0.0049^{bc}	66.44 ± 0.88^g	8.57 ± 0.50^{de}
1.0	40	0.058 ± 0.003^{cd}	0.0059 ± 0.0004^a	90.34 ± 1.84^a	19.41 ± 1.01^a
1.5	40	0.063 ± 0.007^c	0.0062 ± 0.0010^a	82.70 ± 1.01^{cd}	14.54 ± 0.98^{bc}
2.0	40	0.085 ± 0.007^{ab}	0.0251 ± 0.0017^c	69.86 ± 0.98^f	9.36 ± 0.83^d

Means and standard deviation values followed by different superscripts letters within the same column were significantly different ($p < 0.05$). Three replications of each film were analyzed.

12.3.3 TRANSPARENCY AND OPACITY

Addition of guar gum and glycerol in various levels of the formulation led to changes in transparency and opacity of films (Table 12.1 and Fig. 12.1). It can be seen that, opacity and transparency are inversely correlated. There is a significant difference ($p < 0.05$) on transparency of guar gum films since 1% (w/v) guar gum film showed highest transparency followed by films with 1.5% and 2% (w/v) as the concentration of this galactomannan increased and as the glycerol content increased the films were least transparent (Fig. 12.1). The interaction of water molecules with glycerol modifies the refractive index of guar gum affecting the film's transparency (Fig. 12.1). The opacities of the films also varied with guar gum concentration. Guar gum films with 2% (w/v) of this galactomannan were most opaque, followed by guar gum films with 1.5% and 1% (w/v) and the influence of glycerol concentrations at 20%, 30%, and 40% (v/v) (Table 12.1) affected opacity. There were no significant differences ($p < 0.05$) on films with 20% of glycerol, meanwhile there were significant differences ($p < 0.05$) on films with 30% and 40% of glycerol (Table 12.1) due to the nature and structure of the edible film as Mu et al.[37] and Zhang et al.[38] mentioned on their investigations. These findings are important since film transparency and opacity are critical properties in various film applications, particularly if the films will be used as food coatings or to improve the product's appearance.[39]

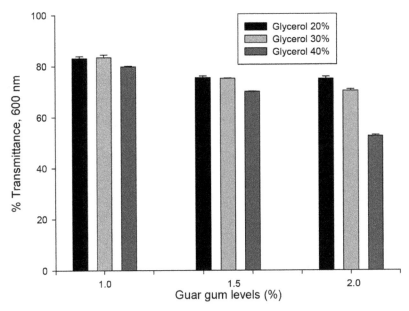

FIGURE 12.1 Effect of guar gum films with different levels of glycerol on transmittance at 600 nm.

12.3.4 SOLUBILITY OF FILMS

The solubility of the films increased basically ($p < 0.05$) at high concentrations of glycerol and this effect is shown in Table 12.1. This can be attributed to the flexibility of the polymer structure due to glycerol and that guar gum, as a polysaccharide, is hydrophilic so the films made with it are highly soluble in water; increasing the potential elasticity (E) and decreasing guar gum resistance (TS) at all dry material concentrations (Table 12.3). Since solubility is an important property for films because it strongly affects the shelf life of packed foods,[40,41] an alternative to decrease the rate of solubility in water is to modify the dry material or the plasticizer concentration used on edible films.[39] Similar investigations agree with these results, for example, Farahnaky et al.[16] and Ramos et al.[42] mentioned solubility (%) values ranging from 6.51% to 27.66% and from 63.91% to 84.22%, respectively. Farahnaky et al.[16] worked with wheat starch edible films plasticized with glycerol at 20% reporting a solubility of 6.51%. Mali et al.[43] worked with yam starch edible films plasticized with glycerol and the solubilities observed ranged from 23.63% to 27.72%. Kurt et al.[40] reported a solubility of 75.94% for an edible film made with 1.5% of guar gum and 10% of glycerol content.

A higher solubility is directly related to a high level of glycerol[44] and a low content of biopolymer.[45] However, in this study, using guar gum concentrations of 1.5% and 2.0% with 40% of glycerol content, the solubility did not show the effect mentioned earlier, this could be attributed to the high concentration of dry material that probably causes an inappropriate spreading of the film solution and weakening of solubility of the film in water (Table 12.1).

12.3.5 MOISTURE CONTENT

The MC of guar gum films prepared with different glycerol concentrations is shown in Table 12.1. The MC of the films increased significantly ($p < 0.05$) at high levels of plasticizer content; however, the MC decreased with high levels of guar gum content and the values ranged from 6.90% to 19.41%. These results may be attributed to the fact that water acts as a plasticizer in most hydrophilic films and is not only associated with the galactomannan film's structure, but also with the glycerol hydrophilic nature that retains water in the matrix.[46] Similar findings were mentioned by Osés et al.[47] whose composite films based on whey protein isolate and mesquite gum tended to become more hydrophilic with high levels of plasticizer content and the MC values ranged from 9.20% to 11.00% at 50% RH and from 15.00% to 16.80% at 75% RH. Our findings reveal that glycerol-plasticized guar gum films had greater MC when glycerol was at 40%, the values ranged from 9.36% to 19.41%. These results may be interpreted as follows: films with high concentration of glycerol have higher MC; this fact is probably due to the hydrophilic properties of guar gum and glycerol which has higher affinity for binding water thus increases the plasticizing activity and greatly influences mechanical properties of films.[48]

12.3.6 COLOR ATTRIBUTES

Color attributes are of prime importance because they directly influence consumer acceptability. Table 12.2 shows the measured color parameters including L (black-white), a* (green-red), and b* (blue-yellow) of guar gum films. Several experiments showed that all of the color parameters of guar gum films were significantly ($p < 0.05$) altered when glycerol and guar gum concentrations increased. Increasing of glycerol concentration provoke an increase in luminosity (L) while b* (blue-yellow) and a* (green-red) parameters decreased.

TABLE 12.2 Effect of Guar Gum and Glycerol Concentration on L, a*, b* Parameters, ΔE, Yellowness (YI), and Whiteness (WI) Indexes.

Guar gum (%)	Glycerol (%)	L	a*	b*	ΔE	YI	WI
1.0	20	90.91 ± 0.06[abc]	0.39 ± 0.06[e]	3.03 ± 0.11[bc]	8.01 ± 0.21[abc]	4.77 ± 0.16[b]	90.41 ± 0.05[bc]
1.0	30	90.68 ± 0.29[abc]	0.23 ± 0.05[e]	2.68 ± 0.04[b]	8.99 ± 0.17[a]	5.24 ± 0.06[b]	90.11 ± 0.15[cd]
1.0	40	91.33 ± 0.18[ab]	0.92 ± 0.02[cd]	1.05 ± 0.06[e]	7.11 ± 0.19[cd]	2.03 ± 0.48[d]	91.12 ± 0.22[ab]
1.5	20	91.03 ± 0.66[a]	1.04 ± 0.09[bc]	0.66 ± 0.16[f]	8.91 ± 0.53[a]	0.72 ± 0.24[e]	91.35 ± 0.28[bcd]
1.5	30	90.43 ± 0.53[bc]	0.90 ± 0.06[cd]	3.48 ± 0.33[e]	7.53 ± 0.59[bcd]	2.54 ± 0.51[d]	90.40 ± 0.26[bcd]
1.5	40	91.38 ± 0.49[d]	0.89 ± 0.07[f]	2.17 ± 0.26[a]	6.62 ± 0.54[d]	11.41 ± 0.43[a]	91.07 ± 0.52[ab]
2.0	20	90.50 ± 0.41[cd]	1.18 ± 0.09[ab]	2.19 ± 0.32[cd]	8.47 ± 0.65[ab]	4.10 ± 0.52[bc]	89.83 ± 0.19[d]
2.0	30	90.86 ± 0.06[abc]	1.39 ± 0.13[a]	1.88 ± 0.19[e]	7.70 ± 0.30[abcd]	2.81 ± 0.29[d]	90.64 ± 0.07[abc]
2.0	40	91.01 ± 0.72[abc]	0.74 ± 0.03[d]	1.87 ± 0.48[de]	7.03 ± 0.61[cd]	3.02 ± 0.74[cd]	90.70 ± 0.42[abc]

Means and standard deviation values followed by different superscripts letters within the same column were significantly different ($p < 0.05$). Three replications of each film were analyzed.

The ΔE indicates the degree of total color difference from the standard color plate, YI indicates the degree of yellowness, and WI indicates the degree of whiteness which can be more described as a result of increasing glycerol concentration in guar gum films (Table 12.2).

While WI increased significantly ($p < 0.05$) as a result of greater glycerol content, YI and ΔE decreased considerably. Hence, guar gum films became very slightly greenish and yellowish, but there were still transparent. Furthermore, visual observation confirmed this fact.

Comparing the results with Rao et al.[2] values for L* (43.83–49.18) are higher probably because the chitosan they used makes the films less luminescent. And for a* (−2.37 to −4.72) and b* (4.91 to 7.62) the values for the films of chitosan-guar gum are negative and positive but leading to the greenish zone, and the results obtained by Ekrami et al.[39] working with a salep-based edible film mentioned that increasing solid material content did not cause changes in L (62.07–69.37); however, b* (−0.36 to −2.73) increased, a* (0.43 to 0.85) decreased significantly ($p < 0.05$), and increasing the glycerol content results in a new bond formation, which altered the color index. The film color tends toward yellow, and an increase in YI confirms this apparent change.

Kurt et al.[40] report L values for guar gum films such as 81.82, a* values 5.69 and b* values 1.28, this values are somewhat different as the ones found in this research (Table 12.2). The difference could be due to several factors such as the concentration of glycerol used, the amount of solution cast in acrylic plates and the kind of guar gum used. Even though the differences among the results the guar gum films formulated are between the range of the values reported by others.[2,39-41]

12.3.7 MECHANICAL PROPERTIES

The mechanical nature of edible films is influenced by the formulation and concentration of ingredients and by changes in plasticizer and biopolymers. An important factor that influences the evaluation of the mechanical properties is the RH because water acts as softener; when RH increases in the film, mechanical resistance decreases and elasticity increases.[39] In this study, prior to assay, the samples were conditioned at 57% RH for 5 days (120 h). The TS and elongation at break (E) of plasticized guar gum films are shown in Table 12.3. There were significant differences ($p < 0.05$) between the TS and E of the film, the elongation at break of the guar gum films increased as the guar gum content increased. By increasing the glycerol content, small molecules of this plasticizer enter the polymer structure, so film solubility and chain mobility increases, which can led to improve the film elasticity (Table 12.3), similar findings are reported by other researchers.[44,49] The incorporation of plasticizers is necessary to reduce polymer intermolecular forces, increasing the mobility of the polymeric chains, and improving the mechanical characteristics of the film, such as film extensibility, since they have a great affinity for water.[11,19,50] Similar results have been reported by Leceta et al.,[52] who worked with chitosan and glycerol films, the TS values ranged from 31.89 to 61.82 MPa while the E values ranged from 4.59% to 30.51%; Kurt et al.[40] prepared films with salep gum, locust gum, and guar gum using 10% of glycerol and the range values for TS were 42.89–71.41 MPa and for E the range values were 16.1–37.2%; and Banegas et al.[51] have characterized cross-linked guar gum films, the TS range values were 16.00–43.80 MPa and for E the values were 2.00–2.90%. Other researchers using different bases for the film and using glycerol as a plasticizer obtained range values for TS from 0.57 to 9.87 MPa and for E from 17.11% to 250.00%.[19,44,52,53]

TABLE 12.3 Effect of Guar Gum and Glycerol Concentration on Tensile Strength (TS) and Elongation at Break (E) of Edible Films.

Guar gum (%)	Glycerol (%)	TS (MPa)	E (%)
1.0	20	32.68 ± 0.42[bcd]	8.97 ± 2.64[e]
1.0	30	23.97 ± 3.89[d]	18.60 ± 2.55[b]
1.0	40	30.63 ± 3.12[cd]	14.51 ± 0.14[cd]
1.5	20	32.41 ± 2.03[bcd]	12.40 ± 0.35[de]
1.5	30	23.11 ± 1.57[d]	19.30 ± 0.72[c]
1.5	40	27.88 ± 2.87[d]	17.43 ± 1.33[cd]
2.0	20	48.26 ± 3.49[a]	18.50 ± 0.85[b]
2.0	30	38.66 ± 0.19[abc]	21.15 ± 2.05[c]
2.0	40	41.08 ± 1.04[ab]	39.60 ± 1.94[a]

Means and standard deviation values followed by different superscripts letters within the same column were significantly different ($p < 0.05$). Ten replications of each film were analyzed.

12.3.8 WATER VAPOR PERMEABILITY

WVP of a film is an important property that greatly influences the usefulness of the film in foods.[45] Our results indicate that low WVP on films was related to the concentration of guar gum, the levels of glycerol used and the RH where the films were kept in a conditioning desiccator. WVP of guar gum films increased as glycerol and guar gum concentration increased ($p < 0.05$). WVP of films is dependent on both solubility and diffusion rate of water vapor.[44] Figure 12.2A–C illustrates that WVP increased as plasticizer levels increased. Hydrogen bonding between guar gum and glycerol is disrupted by water sorption therefore the intensity of the guar gum–water and the polyol–water interactions increases resulting in swelling of the film and an increased in water diffusion through the film.[44] This is due to increases in free volume and chain movement, reducing the rigidity and increasing the molecular mobility of films, thus allowing higher water vapor diffusion through their structure and therefore increasing WVP as stated by Cerqueira et al.[36] Plasticizers may also promote water vapor diffusivity through the polymeric structure by increasing interchain spacing between polymer chains; consequently, accelerating the water vapor transmission.[19] These results of the permeability of the films are in agreement with other works where the increase of plasticizer concentrations has increased the values of WVP.[19,36,40,41] WVP of films with good water vapor barrier properties (low or no water permeation and diffusion through the film) should not increase or increase very little with increasing relative vapor pressure.[10]

12.3.9   SENSORY EVALUATION

Table 12.4 illustrates results scored by the panelists of the sensory acceptability of the guar gum and glycerol edible films. Color was rated by the trained sensory panelists in a range from 5.13 to 6.17 of hedonic scale. The film with the highest color acceptability was the one formulated using 1.5% of guar gum and 30% of glycerol. There was a significantly difference ($p < 0.05$) in sensory panelists rating on the color attribute of the edible films which is an important factor that determines the general appearance and the consumer acceptability of the biofilms.[14] Other important sensory attributes are stickiness and taste because they provide fundamental information on the applicability of edible films and coatings on food surfaces as protective layers.[54] The rating given by the sensory panelists for the stickiness of each film ranged from 5.50 to 6.86 of hedonic scale. The increase in stickiness may be attributed to the water holding capacity of glycerol and likewise depends on its high water absorption capacity.[14,40,41,55] The sensory scores of taste from guar gum films ranged from 5.61 to 6.90 and were not significantly different ($p < 0.05$). The overall acceptability of edible films

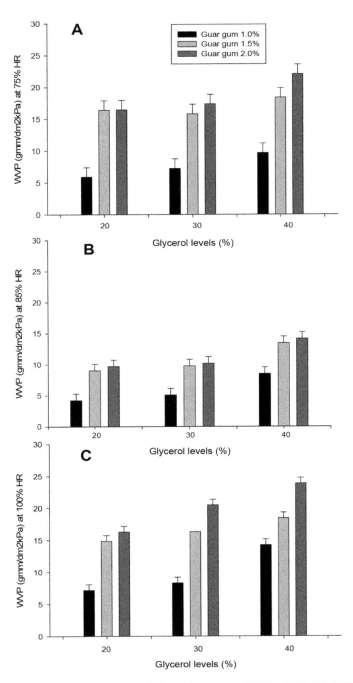

FIGURE 12.2 Effects of guar gum and glycerol content on WVP at 75% (A), 85% (B), and 100% (C) of relative humidity.

ranged from 5.42 to 7.16. The film samples were not significantly different ($p < 0.05$) in terms of overall acceptability even though other authors report that some factors like transparency, flexibility, brittleness, and elasticity of films influence on this attribute.[14,56,57] This is an important result since all formulations of the films could be eventually accepted by consumers.

TABLE 12.4   Sensory Acceptability of Guar Gum Films.

Guar gum (%)	Glycerol (%, v/v)	Color	Taste	Stickiness	Overall acceptability
1.0	20	5.88 ± 1.03^{ab}	6.45 ± 1.12^a	6.03 ± 1.05^{abc}	6.21 ± 0.88^a
1.5	20	5.99 ± 1.13^{ab}	6.65 ± 0.89^a	6.16 ± 1.05^{abc}	6.35 ± 1.17^a
2.0	20	5.54 ± 0.91^b	6.90 ± 0.72^a	6.21 ± 0.79^{abc}	5.42 ± 1.22^a
1.0	30	6.05 ± 0.82^{ab}	6.35 ± 0.68^a	6.11 ± 1.03^{abc}	6.19 ± 0.86^a
1.5	30	6.17 ± 1.02^{ab}	6.25 ± 1.18^a	6.65 ± 0.94^{ab}	7.16 ± 1.24^a
2.0	30	5.41 ± 1.05^b	5.90 ± 0.83^a	6.74 ± 0.88^{ab}	5.57 ± 1.16^a
1.0	40	5.86 ± 0.79^{ab}	5.80 ± 0.74^a	5.50 ± 1.64^c	5.85 ± 1.01^a
1.5	40	6.13 ± 0.97^{ab}	5.84 ± 1.23^a	6.24 ± 1.01^{abc}	5.54 ± 0.98^a
2.0	40	5.13 ± 1.07^a	5.61 ± 0.78^a	6.86 ± 0.98^a	5.51 ± 0.83^a

Means and standard deviation of 30 panelists score. Means values followed by different superscripts letters within the same column were significantly different ($p < 0.05$).

12.4   CONCLUSIONS

Elastic and flexible guar gum-based films plasticized with different levels of glycerol were prepared successfully. The concentration of the components in the film affected the optical, barrier, mechanical and sensorial properties of the films to various extents. Optical properties such as opacity and color (L, a*, and b*) were affected by the concentration of guar gum and glycerol used. Solubility of films depended on the interaction of guar gum and glycerol content. WVP and elongation at break increased as the concentration of glycerol increased, on the other hand TS decreased as the level of glycerol and guar gum increased. Overall acceptability of films was independent of formulation composition. Sensory acceptability of color and stickiness was affected by the concentration of guar gum and glycerol. Further studies should consider films prepared with guar gum at 1.5% and 30% of glycerol to determine their application in real food systems.

12.5 ACKNOWLEDGMENTS

The authors are grateful to the National Council of Science and Technology (CONACyT) for the financial support through a Ph.D. Scholarship for M.Sc. Ruelas-Chacon and to the Universidad Autonoma Agraria Antonio Narro (UAAAN) for the permission granted to MSc. Ruelas-Chacon to pursue Ph.D. studies.

KEYWORDS

- *Cyamopsis tetragonoloba* (L.) *Taub.*
- edible gum
- optical characteristics
- water vapor permeability
- mechanical characteristics
- glycerol

REFERENCES

1. Arevalo-Niño, K.; Aleman-Huerta, M. E.; Rojas-Verde, M. G.; Morales-Rodriguez, L. A. Peliculas Biodegradables a Partir de Residuos de Cítricos: Propuesta de Empaques Activos. *Rev. Latinoam. Biotecnol. Amb. Algal.* **2010,** *1*, 124–134.
2. Rao, M. S.; Kanatt, S. R.; Chala, S. P.; Sharma, A. Chitosan and Guar Gum Composite Films: Preparation, Physical, Mechanical and Antimicrobial Properties. *Carbohydr. Polym.* **2010,** *82*, 1243–1247.
3. Tefera, A.; Seyoum, T.; Woldetsadik, K. Effect of Disinfection, Packaging and Storage Environment on the Shelf Life of Mango. *Biosyst. Eng.* **2007,** *96*, 201–212.
4. Saucedo-Pompa, S.; Rojas-Molina, R.; Aguilera-Carbó, A. F.; Sáenz-Galindo, A.; De la Garza, H.; Jasso-Cantú, D.; Aguilar, C. N. Edible Film Based on Candelilla Wax to Improve the Shelf Life and Quality of Avocado. *Food Res. Int.* **2009,** *42*, 511–515.
5. Pushkala, R.; Raghuram, P. K.; Srividya, N. Chitosan Based Powder Coating Technique to Enhance Phytochemicals and Shelf Life Quality of Radish Shreds. *Postharvest Biol. Technol.* **2013,** *86*, 402–408.
6. Ettelaie, R.; Tasker, A.; Chen, J.; Alevisopoulos, S. Kinetics of Food Biopolymer Film Dehydration: Experimental Studies and Mathematical Modeling. *Ind. Eng. Chem. Res.* **2013,** *52*, 7391–7402.
7. Adeodato Vieria, M. G.; Altenhofen da Silva, M.; Oliviera dos Santos, L.; Beppu, M. M. Natural-Based Plasticizers and Biopolymer Films: A Review. *Eur. Polym. J.* **2011,** *47*, 254–263.

8. Hendrix, K. M.; Morra, M. J.; Lee, H. B. Min, S. C. Defatted Mustard Seed Meal-Based Biopolymer Film Development. *Food Hydrocoll.* **2012**, *26*, 118–125.

9. Wang, Y.; Dong, L.; Wang, L.; Yang, L.; Özkan, N. Dynamic Mechanical Properties of Flaxseed Gum Based Edible Films. *Carbohydr. Polym.* **2011**, *86*, 499–504.

10. Talja, R. A.; Helen, H.; Roos, Y. H.; Jouppila, K. Effect of Various Polyols and Polyol Contents on Physical and Mechanical Properties of Potato Starch-Based Films. *Carbohydr. Polym.* **2007**, *67*, 288–295.

11. Sothornvit, R.; Krochta, J. M. Plasticizer Effect on Mechanical Properties of β-Lactoglobulin Films. *J. Food Eng.* **2001**, *50*, 149–155.

12. Krogars, K.; Heinamaki, J.; Karjalainen, M.; Niskanen, A.; Leskela, M.; Yliruusi, J. Enhanced Stability of Rubbery Amylose-Rich Maize Starch Films Plasticized with a Combination of Sorbitol and Glycerol. *Int. J. Pharm.* **2003**, *251*, 205–208.

13. Mali, S.; Grossmann, M. V. E.; García, M. A.; Martino, M. N.; Zaritzky, N. E. Effects of Controlled Storage on Thermal, Mechanical and Barrier Properties of Plasticized Films from Different Starch Sources. *J. Food Eng.* **2006**, *75*, 453–460.

14. Chinma, C. E.; Ariahu, C. C.; Abu, J. O. Development and Characterization of Cassava Starch and Soy Protein Concentrate Based Edible Films. *Int. J. Food Sci. Technol.* **2012**, *47*, 383–389.

15. Ahmadi, R.; Kalbasi-Ashtari, A.; Oromiehie, A.; Yarmand, M. S.; Jahandideh, F. Development and Characterization of a Novel Biodegradable Edible Film Obtained from Psyllium Seed (*Plantago ovata* Forsk). *J. Food Eng.* **2012**, *109*, 745–751.

16. Farahnaky, A.; Saberi, B.; Majzoobi, M. Effect of Glycerol on Physical and Mechanical Properties of Wheat Starch Edible Films. *J. Texture Stud.* **2013**, *44*, 176–186.

17. Mali, S.; Grossmann, M. V. E.; Garcia, M. A.; Martino, M. N.; Zaritzky, N. E. Mechanical and Thermal Properties of Yam Starch Films. *Food Hydrocoll.* **2005**, *19*, 157–164.

18. Krochta, J. M. Proteins as Raw Materials for Films and Coatings: Definitions, Current Status, and Opportunities. In *Protein-Based Films and Coatings;* Gennadios, A., Ed.; CRC Press LCC: Boca Raton, FL, 2002; pp 1–41.

19. Rezvani, E.; Schleining, G.; Sümen, G.; Taherian, A. R. Assessment of Physical and Mechanical Properties of Sodium Caseinate and Stearic Acid Based Film-Forming Emulsions and Edible Films. *J. Food Eng.* **2013**, *116*, 598–605.

20. Heyman, B.; De Vos, W. H.; Depypere, F.; Van der Meeren, P.; Dewettinck, K. Guar and Xanthan Gum Differentially Affect Shear Induced Breakdown of Native Waxy Maize Starch. *Food Hydrocoll.* **2014**, *35*, 546–556.

21. Moser, P.; Lopes Cornelio, M.; Nicoletti Telis, V. R. Influence of the Concentration of Polyols on the Rheological and Spectral Characteristics of Guar Gum. *LWT Food Sci. Technol.* **2013**, *53*, 29–36.

22. Roberts, K. T. The Physiological and Rheological Effects of Foods Supplemented with Guar Gum. *Food Res. Int.* **2011**, *44*, 1109–1114.

23. Srivastava, M.; Kapoor, V. P. Seed Galactomannans: An Overview. *Chem. Biodivers.* **2005**, *2*, 295–317.

24. Cerqueira, M. A.; Souza, B. W. S.; Simões, J.; Teixeira, J. A.; Domigues, M. R. M.; Coimbra, M. A.; Vicente, A. A. Structural and Thermal Characterization of Galactomannans from Non-Conventional Sources. *Carbohydr. Polym.* **2011**, *83*, 179–185.

25. Cui, W.; Eskin, M. A. M.; Wu, Y.; Ding, S. Synergisms between Yellow Mustard Mucilage and Galactomannans and Applications in Food Products—A Mini Review. *Adv. Colloid Interface Sci.* **2006**, *128–130*, 249–256.

26. Han, J. H.; Floros, J. D. Casting Antimicrobial Packaging Films and Measuring Their Physical Properties and Antimicrobial Activity. *J. Plast. Film Sheeting* **1997**, *13*, 287–298.

27. Gontard, N.; Guilbert, S. Biopackaging: Technology and Properties of Edible and/or Biodegradable Material of Agricultural Origin. In *Food Packaging and Preservation;* Mathlouthi, M., Ed.; Blackie Academic & Professional: New York, NY, 1994; pp 159–181.

28. Bolin, H. R.; Huxsoll, C. C. Control of Minimally Processed Carrot (*Dascus carota*) Surface Discoloration Caused by Abrasion Peeling. *J. Food Sci.* **1991**, *56*, 416–418.

29. Mei, Y.; Zhao, Y. Barrier and Mechanical Properties of Milk Protein-Based Edible Films Containing Nutraceuticals. *J. Agric. Food Chem.* **2003**, *51*, 1914–1918.

30. Romero-Bastida, C. A.; Bello-Pérez, L. A.; García, M. A.; Martino, M. N.; Solorza-Feria, J.; Zaritzky, N. E. Physicochemical and Microstructural Characterization of Films Prepared by Termal and Cold Gelatinization from Non-Conventional Sources of Starches. *Carbohydr. Polym.* **2005**, *60*, 235–244.

31. ASTM. Standard Methods of Test for Water Vapor Transmission of Materials in Sheet form (E 96-00). In *Annual Book of ASTM Standards*, American Society for Testing and Material: Philadelphia, PA, 2001; pp 1048–1053.

32. Labuza, T. P.; Kaanane, A.; Chen, J. Y. Effect of Temperature on the Moisture Sorption Isotherms and Water Activity Shift of Two Dehydrated Foods. *J. Food Sci.* **1985**, *5*, 385–391.

33. ASTM D 82–91. Standard Test Methods for Tensile Properties of Thin Plastic Sheeting. In *Annual Book of ASTM Standards;* American Society for Testing & Materials: Philadelphia, PA, 1991.

34. Pereda, M.; Ponce, A. G.; Marcovich, N. E.; Ruseckaite, R. A.; Martucci, J. F. Chitosan-Gelatin Composites and Bilayer Films with Potential Antimicrobial Activity. *Food Hydrocoll.* **2011**, *25*, 1372–1381.

35. Matta Fakhoury, F.; Martelli, S. M.; Canhadas Bertan, L.; Yamashita, F.; Innocentini Mei, L. H.; Collares Queiroz, F. P. Edible Films Made from Blends of Manioc Starch and Gelatin-influence of Different Types of Plasticizer and Different Levels of Macromolecules on Their Properties. *LWT Food Sci. Technol.* **2012**, *49*, 149–154.

36. Cerqueira, M. A.; Souza, B. W. S.; Teixeira, J. A.; Vicente, A. A. Effect of Glycerol and Corn Oil on Physicochemical Properties of Polysaccharide Films—A Comparative Study. *Food Hydrocoll.* **2012**, *27*, 175–184.

37. Mu, C.; Guo, J.; Li, Z.; Lin, W.; Li, D. Preparation and Properties of Dialdehyde Carboximethyl Cellulose Crosslinked Gelatin Edible Films. *Food Hydrocoll.* **2012**, *27*, 22–29.

38. Zhang, Y.; Han, J. Plasticization of Pea Starch Films with Monosaccharides and Polyols. *J. Food Sci.* **2006**, *71*, E253–E261.

39. Ekrami, M.; Emam-Djomeh, Z. Water Vapor Permeability, Optical and Mechanical Properties of Salep-Based Edible Film. *J. Food Process. Pres.* **2013**, *38*, 1812–1820.

40. Kurt, A.; Kahyaoglu, T. Characterization of a New Biodegradable Edible Film Made from Salep Glucomannan. *Carbohydr. Polym.* **2014**, *104*, 50–58.

41. Jouki, M.; Khazaei, N.; Ghasemlou, M.; HadiNezhad, M. Effect of Glycerol Concentration on Edible Production from Cress Seed Carbohydrate Gum. *Carbohydr. Polym.* **2013**, *96*, 39–46.

42. Ramos, O. L.; Reinas, I.; Silva, S. I.; Fernandes, J. C.; Cerqueira, M. A.; Pereira, R. N.; Vicente, A. A.; Fatima Pocas, M.; Pintado, M. E.; Xavier Malcata, F. Effect of Whey

Protein Purity and Glycerol Content upon Physical Properties of Edible Films Manufactured Therefrom. *Food Hydrocoll.* **2013**, *30*, 110–122.

43. Mali, S.; Grossmann, M. V. E.; Garcia, M. A.; Martino, M. N.; Zaritzky, N. E. Barrier, Mechanical and Optical Properties of Plasticized Yam Starch Films. *Carbohydr. Polym.* **2004**, *56*, 129–135.

44. Aguirre, A.; Borneo, R.; León, A. E. Properties of Triticale Protein Films and Their Relation to Plasticizing-antiplasticinzing Effects of Glycerol and Sorbitol. *Ind. Crops Prod.* **2013**, *50*, 297–303.

45. Viña, S. Z.; Mugridge, A.; García, M. A.; Ferreyra, R. M.; Martino, M. N.; Chaves, A. R.; Zaritzky, N. E. Effects of Polyvinylchloride Films and Edible Starch Coatings on Quality Aspects of Refrigerated Brussels Sprouts. *Food Chem.* **2007**, *103*, 701–709.

46. Reddy, N.; Yang, Y. Completely Biodegradable Soyprotein-jute Biocomposites Developed Using Water Without any Chemicals as Plasticizer. *Ind. Crops Prod.* **2011**, *33*, 35–41.

47. Osés, J.; Fabregat-Vázquez, M.; Pedroza-Islas, R.; Tomás, S. A.; Cruz-Orea, A.; Maté, J. I. Development and Characterization of Composite Edible Films Based on Whey Protein Isolate and Mesquite Gum. *J. Food Eng.* **2009**, *92*, 56–62.

48. Cubero, N.; Monferrer, A.; Villalta, J. *Food Additives. Food Technology Collection.* Mundi-Prensa: Madrid, Spain, 2002; pp 133–134.

49. Al-Hassan, A. A.; Norziah, M. H. Starch-Gelatin Edible Films: Water Vapor Permeability and Mechanical Properties as Affected by Plasticizers. *Food Hydrocoll.* **2012**, *26*, 108–117.

50. Mali, S.; Grossmann, M. V. E.; Garcia, M. A.; Martino, M. N.; Zaritzky, N. E. Microstructural Characterization of Yam Starch Films. *Carbohydr. Polym.* **2002**, *50*, 379–386.

51. Banegas, R. S.; Zornio, C. F.; Borges, A. M. G.; Porto, L. C.; Soldi, V. Preparation, Characterization and Properties of Films Obtained from Cross-Linked Guar Gum. *Polimeros* **2013**, *23*, 182–188.

52. Leceta, I.; Guerrero, P.; Ibarburu, I.; Dueñas, M. T.; de la Caba, K. Characterization and Microbial Analysis of Chitosan-Based Films. *J. Food Eng.* **2013**, *116*, 889–899.

53. Lazaridou, A.; Biliaderis, C. G. Thermophysical Properties of Chitosan, Chitosan–Starch and Chitosan–Pullulan Films Near the Glass Transition. *Carbohydr. Polym.* **2002**, *48*, 179–190.

54. Kester, J. J.; Fennema, O. R. Edible Films and Coatings: A Review. *Food Technol.* **1986**, *40*, 47–59.

55. Srichuwong, S.; Snuarti, T. C.; Mishima Isonoa, T. N.; Hisamatsu, M. Starches from Different Botanical Sources II: Contribution of Starch Structures to Swelling and Pasting Properties. *Carbohydr. Polym.* **2005**, *62*, 25–34.

56. Srinivasa, P. C.; Ramesh, M. N.; Tharanathan, R. N. Food Hydrocolloid. Properties and Sorption Study on Chitosan–polyvinyl Alcohol Blends Films. *Carbohydr. Polym.* **2003**, *52*, 431–438.

57. Perez-Gallardo, A.; Bello-Perez, L. A.; Garcia-Almendarez, B.; Montejano-Gaitan, G.; Barbosa-Canovas, G.; Regalado, C. Effect of Structural Characteristics of Modified Waxy Corn Starches on Rheological Properties, Film-forming Solutions, and on Water Vapor Permeability, Solubility and Opacity of Films. *Starch Starke.* **2012**, *64*, 27–36.

CHAPTER 13

PHOTOELECTROCATALYTIC DEGRADATION OF METHYLENE BLUE AND INACTIVATION OF *Escherichia Coli* BY SPRAY-DEPOSITED Au:ZnO THIN FILMS

R. T. SAPKAL*

Nano-Materials Laboratory, Department of Physics, Tuljaram Chaturchand College, Baramati 413103, India

Corresponding author. E-mail: rt_sapkal@yahoo.co.in

CONTENTS

ABSTRACT

Spray-deposited Au-doped ZnO thin films were successfully prepared by using zinc acetate as precursor onto the glass and fluorine-doped tin oxide (FTO)-coated glass substrates at 400°C. In this study, the polycrystalline Au:ZnO films were prepared with the different Au concentration in the starting solution was varied from 0 to 4 at%. The optimized Au doping concentration was 3 at%. The direct optical band gap of the Au:ZnO film (3 at% Au doping) was 3.10 eV. The photoelectrocatalytic activity of the prepared thin films was evaluated by measuring the photoelectrocatalytic degradation of methylene blue. 94% degradation of MB with rate constant $k = 0.0555/s$. Inactivation studies of suspensions of *Escherichia coli* in a parallel plate reactor showed that the bacterial count can be reduced by a factor of 100 by direct UVA illumination, by a factor of 10^7 with a ZnO electrode with applied external bias of 1.5 V vs. a stainless steel counter electrode and by a factor of 10^{12} under a photocurrent of 18 mA across the Au:ZnO electrode with applied external bias of 1.5 V vs. a stainless steel counter electrode. The high antibacterial activity in the latter case is ascribed to the suppression of charge carrier recombination and auxiliary radical reactions occurring at the surface of bacteria adsorbed on the Au:ZnO electrode. The Au:ZnO improved photoelectrocatalytic degradation of MB and inactivation efficiency of *E. coli*.

13.1   INTRODUCTION

Zinc oxide (ZnO) is II–VI semiconductor of wurtzite structure and has a large direct band gap of 3.3 eV.[1] ZnO-based materials have been widely used in a variety of applications, such as, optical waveguides,[2] ultraviolet (UV) light emitters,[3] transparent conductive oxides,[4] chemical and gas sensors,[5] and photocatalysts.[6] Among these applications, photocatalysis of ZnO has been intensively studied due to its environmental eco-friendly applications. Illumination with UVA light ($\lambda < 385$ nm) will excite electrons from valence band to the conduction band, producing an e^--h^+ pair. If the adsorbed couples are considered to be water and dissolved oxygen, then water gets oxidized by positive holes and it splits into ·HO and H^+. As oxygen is easily reducible, reduction of oxygen by photoelectrons of the conduction band results in generation of superoxide radical anions (·O_2^-), which in turn reacts with H^+ to generate ·HO_2. On subsequent reaction with electron to produce hydrogen peroxide (HO_2^-), and then with hydrogen ion, a molecule of H_2O_2

is produced.[7] However, the photocatalytic efficiency of ZnO is always affected by the quick recombination of the photo excited electrons and holes. To retard the recombination of the excited electrons and holes, one possible strategy is to dope ZnO with metal.[8,9] The excited electrons in ZnO are supposed to transfer to the metal if the interface between ZnO and metal is optimized and the work function of the metal is greater than that of ZnO. The electron transfer will greatly suppress the electron–hole recombination and then improve the catalytic efficiency of ZnO. Until now, many kinds of noble metals have been combined with semiconductors, such as Au–TiN,[10] Comet-like superstructures of Au–ZnO,[11] Au-doped iron oxide thin films,[12] and Au-doped mesostructured SnO_2.[13]

Different methods have been utilizing to obtain pure and doped ZnO. These methods include sol–gel,[14] atom beam cosputtering,[15] RF magnetron sputtering,[16] microwave irradiation,[17] pulsed-laser deposition,[18] hydrothermal growth process,[19] chemical spray pyrolysis (CSP).[20] Of these, CSP technique has received considerable attention because of its simplicity, easy to incorporate the impurity and possibility of large area commercial deposition. Therefore, we have tuned our attention to study the feasibility of CSP technique for the synthesis of Au:ZnO thin films.

Several research groups have been attempted synthesis of catalytic materials and their catalytic efficiency.[21,22] However, the ZnO–Au thin films obtained in these previous works are not well designed for photoelectrocatalysis but for chemical sensors.[23–27] For instance, Sarkar et al. reported the photo-dependent excitonic mechanism and the charge migration kinetics in a colloidal ZnO–Au NC system.[28] Also, photocatalysis has been used for photochemical sterilization of water, in which microorganisms could be inactivated. *Shigella flexneri*, *Listeria monocytogenes*, *Vibrio parahaemolyticus*, *Staphylococcus aureus*, *Streptococcus pyogenes*, and *Acinetobacter baumannii* were killed by visible-light-illuminated undoped and nitrogen-doped TiO_2 thin films.[29] The antibacterial activity of suspensions of ZnO nanofluids against *Escherichia coli* (*E. coli*) has been evaluated by Jalal et al.[30] ZnO have notable bactericidal activity against a biofilm of *Staphylococcus epidermidis* under UVA light illumination.[31] The nanopowered ZnO seems to show a fungal growth inhibition stronger than the $ZnTiO_3$.[32] Ag/ZnO nanocomposites have potential applications in inhibition both Gram-negative and Gram-positive bacteria.[33]

In this work, the Au:ZnO thin films were prepared by spray pyrolosis method. Then, the prepared samples were structurally characterized by X-ray diffraction (XRD), scanning electron microscope (SEM), optical properties, and photoelectrochemical (PEC) properties. Finally, their photoelectrocatalytic performances in degradation of MB and inactivation of *E. coli* were studied.

13.2 EXPERIMENTAL

ZnO thin films were deposited onto ultrasonically cleaned preheated Corning glass and fluorine-doped-tin-oxide (FTO)-coated substrates using the CSP technique. Spraying solution (0.2 M) was prepared by mixing the appropriate volumes of zinc acetate (Zn(CH3COO)$_2$·2(H$_2$O), AR grade, 99.9% pure, Merck) and Auric chloride (HAuClO$_4$, AR grade, 99.9% pure, Merck) in a mixture of the solvents formed by double distilled water, acetic acid and methanol (25:10:65).[34] In order to dope Au in ZnO thin films, four different concentrations (1, 2, 3, and 4 at% measured as atomic weight percentage) were selected. The other deposition parameters such as spray rate (5 cm^3/min), nozzle to substrate distance (30 cm), and carrier gas pressure (2 atm) were kept at their fixed values. The resulting solution (100 cm^3) was sprayed at an optimized substrate temperature of 400°C. The structural properties were studied by a Philips X-ray diffractometer PW-1710 (λ, 1.5405 Å) using Cu-$K\alpha$ radiation in the span of 10–100°. Surface morphology of the thin film was studied with JEOL JSM-6360 SEM. Optical absorption study was carried out in the wavelength range 300–1100 nm using spectrometer Systronic Model-119. PEC characterization was performed using a potentiostatic set-up. The PEC cell comprised glass/FTO/Au:ZnO as working electrode (with an active surface area of 1 cm^2), a graphite as counter-electrode and a saturated calomel electrode (SCE) as a reference. All potentials are quoted vs. SCE. The electrolytes were aqueous solution of 0.1 M NaOH; a 65-W UVA lamp was used for illumination. Large area (10 × 10 cm^2) electrodes were prepared in the same way as described above and tested in a thin (1 mm) flow-through PEC reactor employing a stainless steel counter-electrode and broadband UVA irradiation using a "blacklight" lamp.

13.2.1 DEGRADATION OF METHYLENE BLUE

Methylene blue (MB, C$_{16}$H$_{18}$ClN$_3$S, AR grade, 99.9% pure, Merck) dye (200 mL, 100 ppm) was circulated through single-cell photoelectrocatalytic reactor described elsewhere.[35] Samples were collected using 1-mL plastic syringe at different time interval (0, 30, 60, 90, 120 min) until 180 min. The decolorization of methylene blue by Au:ZnO was monitored by using UV–vis spectroscopy analysis.

Degradation of MB was studied by taking FTIR spectra of samples before and after the photoelectrocatalytic process. FTIR (8400S Shimadzu, Japan) was used for investigating the changes in surface functional groups of the samples. FTIR analysis was done in the mid-IR region of 400–4000 cm⁻¹ with 16 scan speed. The pellets were prepared using spectroscopic pure KBr (5:95, w/w) and fixed in the sample holder for the analyses.

13.2.2   INACTIVATION OF E. Coli

The *E. coli* strain was cultivated in a LB medium containing Bacto-tryptone 10 g/L; Bacto-yeast extract 5 g/L; and NaCl 5 g/L at 37°C and pH 7.4 for at least 24 h so that initial cell count reached a minimum of 10^{10} CFU/mL. For photoelectrocatalytic inactivation, well-grown *E. coli* cells were suspended 200 mL in 0.8% saline water and circulating through photoelectrocatalytic single cell. The culture samples were withdrawn at regular intervals for further analysis. Both irradiated and nonirradiated cell suspensions were diluted sequentially and plated on LB-agar. Plates were then incubated at 37°C, and colony counts were taken after 24-h incubation. To study photoelectrocatalytic response, four different experiments were carried out such as bacterial culture medium in dark operated as control, only with UVA, with ZnO thin film and UVA and with Au:ZnO thin film under UVA illumination.

13.3   RESULTS AND DISCUSSIONS

The structural changes and identification of phases were studied with help of XRD technique. The diffraction angle (2θ) was varied between 20 and 80°. Figure 13.1 shows XRD pattern of the Au:ZnO thin film for various Au-doping concentrations. The XRD patterns reveal that all the films show polycrystalline in nature. The diffraction peaks are good agreement with JCPDS data card No. 80-0074. Upon increasing Au concentration, (0 0 2) plane of ZnO begins to improve up to 3 at% Au concentration and then decreases. The excess Au doping atoms can be energetically favorable to coalesce into metallic gold cluster and hence inhibit *c*-axis-preferred growth of ZnO film. Figure 13.2a–d shows the SEM micrographs at different doping concentrations of gold. With increasing Au-doping concentration in ZnO matrix, grain size increases. The grains are spherical, uniform, and compact with 3 at% Au doping and beyond morphology looks like fish scales.

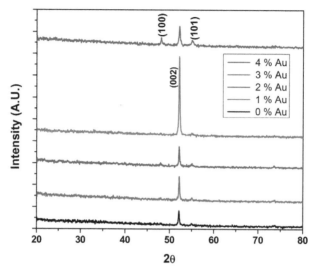

FIGURE 13.1   XRD patterns of undoped and Au:ZnO thin films prepared by spray pyrolysis technique onto glass substrates for different gold-doping concentrations.

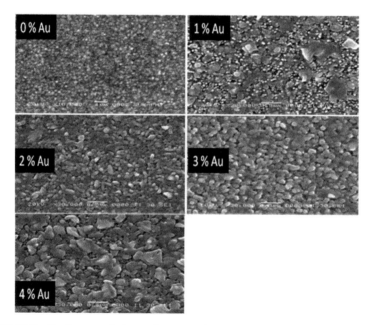

FIGURE 13.2   Surface morphology of Au:ZnO films doped at different Au concentrations: 0%, 1%, 2%, 3%, and 4%.

The transmittance spectra for Au:Zno thin films are shown in Figure 13.3. The sinusoidal nature of transmittance spectra is due to the light interference between film and glass substrate. The average transmittance is about 80% in visible region. The reduction in transmittance may be attributed to strong scattering and absorption. The strong scattering resulted from the existence of grain boundaries, the point defects, and disorders in ZnO films.

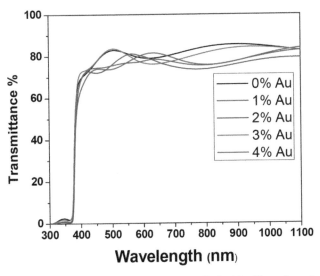

FIGURE 13.3 Optical transmittance spectra of Au:ZnO thin films deposited at various gold-doping concentrations.

The band gap energy of Au:ZnO thin films was evaluated by extending straight portion of plot of $(\alpha h v)^2$ versus hv shown in Figure 13.4. The absorption edge is observed at 389 nm. The band gap energy for typical film doped with 3 at% Au is 3.10 eV. There is slight decrease in band gap energy with increase in Au concentration, confirms the progressive improvement of the structural quality of ZnO matrix.[36]

From $I-V$ measurement, it is seen that the I_{sc} and V_{oc} values increase as the Au percentage increases. This increment in the current values is related to the surface plasmons resonance (SPR) of Au nanoparticles. The variation in I_{sc} and V_{oc} values with concentration of Au is shown in Figure 13.5. The oscillation of high density free electrons occurs under the irradiation of light, which is called SPR. Under the irradiation of incident light having wavelength larger than the particle size, the high density electrons of the noble metal nanoparticles form an electron cloud and oscillate. When Au particles

combining with ZnO, electrons accumulate at the interface between Au and
ZnO, which leads to the electron transfer from the Au particle to the ZnO
side may be the strong interaction coupling between Au and ZnO results in
electron transfer from Au to ZnO.[15]

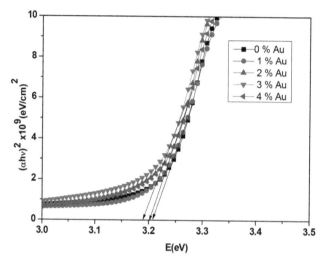

FIGURE 13.4   Plot of $(\alpha h\nu)^2$ against $h\nu$ of Au:ZnO thin films deposited at various gold-doping concentrations.

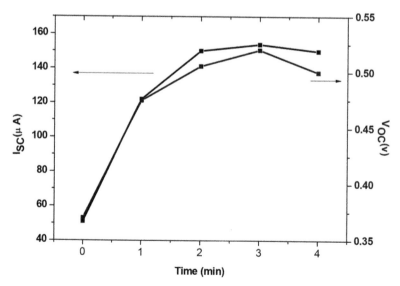

FIGURE 13.5   Variation of I_{sc} and V_{oc} for the PEC cell formed with AU:ZnO thin film vs.
Au doping.

The testing of Au:ZnO photoelectrocatalyst has been carried out by measuring *I–E* curve. An *I–E* curve of a 64-cm² Au:ZnO catalyst in 0.1-M NaOH using a steel counter electrode at a distance of 1 mm under UVA illumination is shown in Figure 13.6. The current reached to its saturation value of about 18.7 mA at bias voltage of 1.5 V. The average photocurrent during degradation of MB experiment is 18 mA for Au:ZnO electrode.

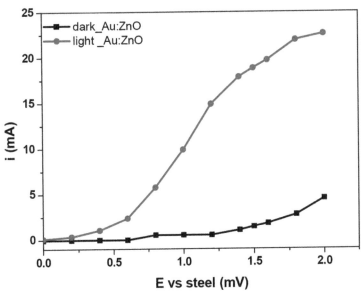

FIGURE 13.6   *I–U* curve in dark and under UVA illumination in a two electrode configuration (Au:ZnO 10 mM Na₂SO₄ steel). Active electrode area 64 cm², solution flow rate 203 mL/min. Photocurrent with respect to time for optimized Au:ZnO/FTO/glass electrode in 10 mM Na₂SO₄ at 1.5 V vs. steel.

13.3.1   DEGRADATION OF MB

Figure 13.7 shows the variation in the extinction spectra of MB collected at various intervals during its photoelectrocatalytic detoxification recorded in the wavelength range from 400 to 800 nm. During the course of the degradation experiments, the concentration of MB decreases due to its decomposition. It is found that percentage of decolorization for Au:ZnO is 94% which is comparable with earlier reported values by Baruah et al.[37] Variation of $\ln(C/C_0)$ with reaction time of MB for Au:ZnO catalyst are shown in Figure 13.8. From the kinetic parameters and degradation efficiency, it is concluded that Au: ZnO photoelectrocatalyst gives better performance in degradation of MB.

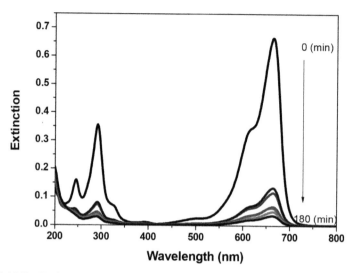

FIGURE 13.7   Extinction spectra of MB with Au:ZnO under UVA light illumination.

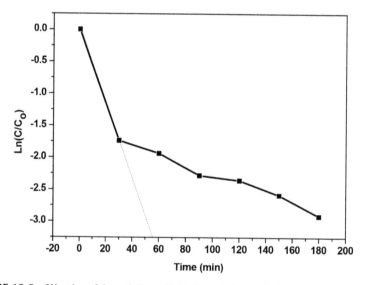

FIGURE 13.8   Kinetics of degradation of MB degradation variation with reaction time.

In the presence of Au on the surface of ZnO improves the photocatalytic activity of ZnO. The increase in photo-activity may be attributed to effect of Au in following way: Au increases absorbance in visible region, interacting in ZnO structure, electron scavenging by Au and the decrease in Fermi level to more negative side subsequently improvement in interfacial charge transfer process at ZnO interface by Au particle. Further regarding electron transfer process at the interface of ZnO and interfacial contact between Au on ZnO, so that the electron after excitation migrates to metal through conduction band driven by an external applied electric field as result the ultimate minimization of electron hole recombination and helps in improving the photo catalytic efficiency of Au:ZnO catalyst. So it combines effect of all these processes at the surface of film which results in overall increases in photo activity of Au:ZnO.

FTIR analysis of control MB showed (Fig. 13.9a) that presence of N–H stretch at 3438.20 cm⁻¹, alkanes at 2856.98–2924.24 cm⁻¹, presence of azo bond (N=N stretch) at 1632.62 cm⁻¹, aromatic ether at 1432.08 cm⁻¹, and S=O stretch at 1145.49 cm⁻¹ provide the evidence for sulfonated bond present in MB, primary alcohol (C–OH stretch) at 1012.36 cm⁻¹ confirms the MB dye with phenolic group, whereas metabolite formed after degradation of MB showed N–H stretch at 3391 cm⁻¹, presence of azo bond (N=N stretch) at 1718.66 cm⁻¹, C–H deformation at 590.80–967.84 and 1395.10 cm⁻¹, respectively, peak at 1053.04 cm⁻¹ showed presence of aliphatic ether, a peak at 2800–2929 cm⁻¹ showed presence of alkanes stretch (Fig. 13.9b). Absence of peak at 1145 cm⁻¹ for sulfonated bond in formed metabolites obtained after degradation confirms the cleavage of sulfonated bond by photoelectrocatalytic degradation.

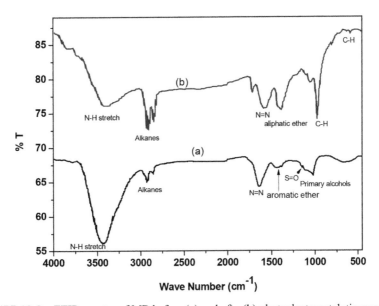

FIGURE 13.9   FTIR spectra of MB before (a) and after(b) photoelectrocatalytic reaction.

Houas et al.[38] give degradation of pathways in which MB undergo a progressive degrading oxidation of one methyl group by an attack from ·OH radical, producing an alcohol, then an aldehyde, which is spontaneously oxidized into acid, decarboxylates into CO_2. However, FTIR analysis of MB shows that azo bond present in the final intermediate formed after photoelectrocatalytic degradation.

13.3.2   INACTIVATION OF E. Coli

The initial population of *E. coli* count was around 10^{10} CFU/mL in 10 mM Na_2SO_4. In all experiments, *E. coli* suspended the solution was recirculated through the PEC cell. Photocatalytic antibacterial activity of the ZnO films against *E. coli* was investigated under various conditions (Fig. 13.10):

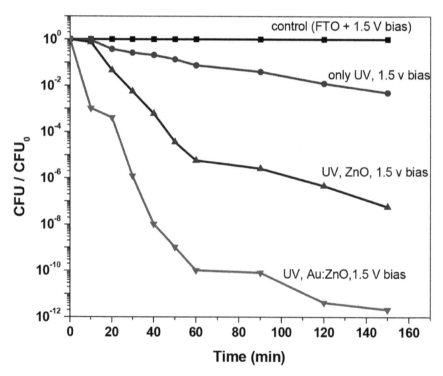

FIGURE 13.10   Antibacterial activity of an Au:ZnO thin film towards viable *E. coli* in 10-mM Na_2SO_4. Relative counts with respect to time. Total volume 0.2l, solution flow rate 203 mL/min.

i. In one of the blank experiments (no ZnO film present on the FTO substrate, application of 1.5 V to the cell), there was no noticeable decrease of bacterial count after 150 min.

ii. In an experiment with direct UVA illumination, the bacterial count was decreased to less than 1%. In this experiment, the ZnO-covered FTO/glass plate was replaced by a glass plate.

iii. A dramatic effect was noticed when a UVA illuminated ZnO electrode was used. This corresponds to the PEC mode under open circuit. Direct absorption of UV light by the bacteria was excluded as light was almost entirely absorbed by the backside illuminated ZnO layer before entering the solution. In the open circuit mode, both light induced charge carriers (e^-, h^+) are available at the surface of the semiconducting electrode but the rate of recombination is high. At the end of the experiment, the bacterial count was decreased by a factor of 10^7. Similar results on bacterial photokilling have been obtained earlier. Another study reported an *E. coli* count decrease from 10^7 to 10^3 CFU/mL due to only UV light after 4 h illumination and a decrease from 10^7 to 10 CFU/mL due to light and TiO_2 suspended particles after 1.25 h. In another report, the antibacterial activity of nanorods-based ZnO thin films synthesized by a hydrothermal process against *E. coli* bacteria was investigated under UV irradiation and showed strong antibacterial activity with an apparent first-order degradation rate constant of 0.0015 s^{-1} in a drop of 100 μL.

iv. An even larger effect was noticed when the Au:ZnO layer was electrically biased in order to suppress charge carrier recombination. In this case, conduction band electrons were not present at the interface and recombination was suppressed. As expected, the bacterial count was further reduced, killing practically all the bacteria initially present. From the initial slope, a first-order rate constant of 0.0115 s^{-1} was found (corresponding to 2.3×10^{-3} L/s if normalized to unit volume).

Such high rate constants cannot be explained alone on the basis of a single attack of a cell by a valence band hole or a surface bound ˙OH radical, based on the measured electrical currents during the reaction (around 18 mA). Colateral damage by additional reaction pathways must be envisaged. It is well known that dead bacterial cells and their contents, spilling out once the wall is damaged, is toxic for living cells. This effect and the accessibility of iron for the Fenton reaction, after wall destruction, may be responsible for the observed high rate constants. However, since the strongly oxidizing reactants (h^+, $\cdot OH$) and other potentially toxic intermediates ($\cdot O_2^-$, H_2O_2) are only available at the electrode surface, reactions involving these species are confined to the electrode surfaces.

ZnO nanoparticles exhibit strong antibacterial activities on a broad spectrum of bacteria. The antibacterial mechanism of Au:ZnO is still under investigation. The photoelectrocatalytic generation of hydrogen peroxide was suggested to be one of the primary mechanisms. In addition, penetration of the cell envelope and disorganization of the bacterial membrane upon contact with Au:ZnO nanoparticles were also indicated to inhibit bacterial growth. However, chemical interactions between Au:ZnO and the components of the cell envelope (lipid bilayer, peptidoglycan, membrane proteins, and lipopolysaccharides) and physical interactions between Au:ZnO and the cell envelope structure causes to final death of the cell.

13.4 CONCLUSIONS

Pure ZnO and Au-doped ZnO thin films have been successfully synthesized using spray pyrolysis technique and their photoelectrocatalytic activity was determined. The efficient photoelectrocatalytic degradation of methylene blue and the photoelectrocatalytic inactivation of *E. coli* were studied under UVA-light irradiation. A 94% photoelectrocatalytic degradation was observed for methylene blue in the presence of the Au:ZnO photoelectrocatalytic electrode upon UVA-light irradiation at 1120 mW/cm². Inactivation tests revealed that the photoelectrocatalytic Au:ZnO electrode is capable of completely eliminating *E. coli* within 120 min reaction time.

KEYWORDS

- Au:ZnO thin films
- photoelectrocatalysis
- degradation
- methylene blue
- *Escherichia coli*

REFERENCES

1. Nirmala, M.; Anukaliani, A. Structural and Optical Properties of an Undoped and Mn Doped ZnO Nanocrystalline Thin Film. *Photonics Lett. Poland* **2010**, *2*(4), 189–191.

2. Lee, S. H.; Goto, T.; Miyazaki, H.; Yao, T. Waveguide Lasing from V-shaped ZnO Microstructure. *Opt. Lett.* **2013**, *38*(14), 2413–2415.

3. Kong, J.; Li, L.; Yang Z.; Liu, J. Ultraviolet Light Emissions in MgZnO/ZnO Double Heterojunction Diodes by Molecular Beam Epitaxy. *J. Vac. Sci. Technol. B* **2010**, *28*(3), C3D10–C3D12.

4. Tang, Y.; Luo, L.; Chen, Z.; Jiang, Y.; Li, B.; Jia, Z.; Xu, L. Electrodeposition of ZnO Nanotube Arrays on TCO Glass Substrates. *Electrochem. Commun.* **2007**, *9*(2), 289–292.

5. Minha, V. A.; Tuanb, L. A.; Huyc, T. Q.; Hunga, V. N.; Quya, N. V. Enhanced NH_3 Gas Sensing Properties of a QCM Sensor by Increasing the Length of Vertically Orientated ZnO Nanorods. *App. Sur. Sci.* **2013**, *265*, 458–464.

6. Jongnavakit, P.; Amornpitoksuk, P.; Suwanboon, S.; Ratana, T. Surface and Photocatalytic Properties of ZnO Thin Film Prepared by Sol–gel Method. *Thin Solid Films* **2012**, *520*(17), 5561–5567.

7. Sapkal, R. T.; Shinde, S. S.; Waghmode, T. R.; Mahadik, M. A.; Mohite, V. S.; Govindwar, S. P.; Rajpure, K. Y.; Bhosale, C. H. Photoelectrocatalytic Decolorization and Degradation of Textile Effluent Using ZnO Thin Films. *J. Photochem. Photobiol. B Biol.* **2012**, *114*, 102–107.

8. Height, M. J.; Pratsinis, S. E, Mekasuwandumrong, O.; Praserthdam, P. Ag–ZnO Catalysts for UV-Photodegradation of Methylene Blue. *Appl. Catal. B* **2006**, *63*, 305–312.

9. Yuan, J.; Guang Choo, E. S.; Tang, X.; Sheng, Y.; Ding, J.; Xue, J. Synthesis of ZnO–Pt Nanoflowers and Their Photocatalytic Applications. *Nanotechnology* **2010**, *21*, 185606–185616.

10. Yuan-Yuan, N. A.; Cong, W.; Yu, L. The Influence of Au-Doping on Morphology and Visible-Light Reflectivity of TiN. Thin Films Deposited by Direct-Current Reactive Magnetron Sputtering. *Chin. Phys. Lett.* **2010**, *27*(5), 056802.

11. Shen, L.; Bao, N.; Yanagisawa, K.; Zheng Y.; Domene, K.; Gupta, A.; Grimes, C. A. Direct Growth of Comet-like Superstructures of Au–ZnO Submicron Rod Arrays by Solvothermal Soft Chemistry Process. *J. State. Chem.* **2007**, *180*, 213–220.

12. Neri, G.; Bonavita, A.; Galvagno, S.; Li, Y. X.; Galatsis, K.; Wlodarski, W. O_2 Sensing Properties of Zn- and Au-Doped Fe_2O_3 Thin Films. *IEEE Sensors J.* **2003**, *3*(2), 195–198.

13. Ramgir, N. S.; Hwang, Y. K.; Jhung, S. H.; Kim, H.-K.; Hwang, J.-S.; Mulla, I. S.; Chang, J.-S. CO Sensor Derived from Mesostructured Au-Doped SnO_2 Thin Film. *Appl. Surf. Sci.* **2006**, *252*, 4298–4305.

14. Kaneva, N. V.; Yordanov, G. G.; Dushkin, C. D. Manufacturing of Patterned ZnO Films with Application for Photoinitiated Decolorization of Malachite Green in Aqueous Solutions. *Bull. Mater. Sci.* **2010**, *33*(2), 111–117.

15. Mishra, Y. K.; Mohapatra, S.; Singhal, R.; Avasthi, D. K. Au–ZnO: A Tunable Localized Surface Plasmonic Nanocomposite. *Appl. Phys. Lett.* **2008**, *92*, 043107.

16. Kim, D. The Influence of Au Thickness on the Structural, Optical and Electrical Properties of ZnO/Au/ZnO Multilayer Films. *Optics Commun.* **2012**, *285*, 1212–1214.

17. Esmaili, M.; Habibi-Yangjeh, A. Preparation and Characterization of ZnO Nanocrystallines in the Presence of an Ionic Liquid Using Microwave Irradiation and Photocatalytic Activity. *J. Iran. Chem. Soc.* **2010**, *7*, S70–S82.

18. Mosnier, J. P.; O'Haire, R. J.; McGlynn, E.; Henry, M. O.; McDonnell, S. J.; McGuigan, B. G. ZnO Films Grown by Pulsed-Laser Deposition on Soda Lime Glass Substrates for the Ultraviolet Inactivation of *Staphylococcus epidermidis* Biofilms. *Sci. Technol. Adv. Mater.* **2009**, *10*, 045003–045012.

19. Baruah, S.; Jaisai, M.; Imani, R.; Nazhad, M. M.; Dutta, J. Photocatalytic Paper Using Zinc Oxide Nanorods. *Sci. Technol. Adv. Mater.* **2010**, *11*, 055002–055008.

20. Sapkal, R. T.; Shinde, S. S.; Babar, A. R.; Moholkar, A. V.; Rajpure, K. Y.; Bhosale, C. H. Structural, Morphological, Optical and Photoluminescence Properties of Ag-Doped Zinc Oxide Thin Films. *Mater. Express.* **2012**, *2*, 64–70.

21. Sonawane, R. S.; Dongare, M. K. Sol–gel Synthesis of Au/TiO₂ Thin Films for Photo-catalytic Degradation of Phenol in Sunlight. *J. Mol. Catal. A: Chem.* **2006**, *243*, 68–76.

22. Choi, H. W.; Kim, E. J.; Hahn, S. H. Photocatalytic Activity of Au-Buffered WO₃ Thin Films Prepared by RF Magnetron. *Chem. Eng. J.* **2010**, *161*(1–2), 285–288.

23. Gaspera, E. D.; Guglielmi, M.; Perotto, G.; Agnoli, S.; Granozzi, G.; Post, M. L.; Martucci A. CO Optical Sensing Properties of Nanocrystalline ZnO–Au Films: Effect of Doping with Transition Metal Ions. *Sens. Actuators B Chem.* **2012**, *161*, 675–683.

24. Socol, G.; Axente, E.; Ristoscu, C.; Sima, F.; Popescu, A.; Stefan, N.; Mihailescua, I. N.; Escoubas, L.; Ferreira, J.; Bakalova, S.; Szekeres, A. Enhanced Gas Sensing of Au Nanocluster-Doped or -Coated Zinc Oxide Thin Films. *J. Appl. Phys.* **2007**, *102*, 083103-1–083103-3.

25. Hongsith, N.; Viriyaworasakul, C.; Mangkorntong, P.; Mangkorntong, N.; Choopun, S. Ethanol Sensor Based on ZnO and Au-doped ZnO Nanowires. *Ceram. Int.* **2008**, *34*, 823–826.

26. Wongchoosuk, C.; Choopun, S.; Tuantranont, A.; Kerdcharoen, T. Au-doped Zinc Oxide Nanostructure Sensors for Detection and Discrimination of Volatile Organic Compounds, *Mater. Res. Innov.* **2009**, *13*(3), 185–188.

27. Hung, N. L.; Kim, H.; Hong, S. K.; Kim, D. Enhancement of CO Gas Sensing Proper-ties in ZnO Thin Films Deposited on Self-assembled Au Nanodots. *Sens. Actuators B* **2010**, *151*, 127–132.

28. Sarkar, S.; Makhal, A.; Bora, T.; Baruah, S.; Dutta, J.; Pal, S. K. Photoselective Excited State Dynamics in ZnO–Au Nanocomposites and Their Implications in Photocatalysis and Dye-Sensitized Solar Cells. *Phys. Chem. Chem. Phys.* **2011**, *13*, 12488–12496.

29. Wong, M. S.; Chu, W. C.; Sun, D. S.; Huang, H. S.; Chen, J. H.; Tsai, P. J.; Lin, N. T.; Yu, M. S.; Hsu, S. F.; Wang, S. L.; Chang, H. H. Visible-Light-Induced Bactericidal Activity of a Nitrogen-Doped Titanium Photocatalyst Against Human Pathogens. *Appl. Environ. Microbiol.* **2006**, *72*(9), 6111–6116.

30. Jalal, R.; Goharshadia, E. K.; Abareshia, M.; Moosavic, M.; Yousefid, A.; Nancarrowe, P. ZnO Nanofluids: Green Synthesis, Characterization and Antibacterial. *Mater. Chem. Phys.* **2010**, *121*, 198–201.

31. Hunge, Y. M., Mohite, V. S., Kumbhar, S.S., Rajpure, K. Y., Moholkar, A. V. (2015) Photoelectrocatalytic Degradation of Methyl Red Using Sprayed WO3 Thin Films Under Visible Light Irradiation. *J. Mater. Sci. Mater. Electron 26*(11), 8404–8412..

32. Ruffolo, S. A.; La Russa, M. F.; Malagodi, M.; Oliviero Rossi, C.; Palermo, A. M.; Crisci, G. M. *Appl. Phys. A* **2010**, *100*, 829–834.

33. Lu, W.; Liu, G.; Gao, S.; Xing, S.; Wang, J. Tyrosine-Assisted Preparation of Ag/ZnO Nanocomposites with Enhanced Photocatalytic Performance and Synergistic Antibacte-rial Activities. *Nanotechnology* **2008**, *19*, 445711–445720.

34. Babar, A. R.; Deshamukh, P. R.; Deokate, R. J.; Haranath, D.; Bhosale, C. H.; Rajpure, K. Y. Gallium Doping in Transparent Conductive ZnO Thin Films Prepared by Chemical Spray Pyrolysis. *J. Phys. D: Appl. Phys.* **2008**, *41*, 135404-1–135404-6.

35. Shinde, P. S.; Patil, P. S.; Bhosale, P. N.; Brüger, A.; Nauer, G.; Neumann-Spallart, M.; Bhosale, C. H. UVA and Solar Light Assisted Photoelectrocatalytic Degradation of AO$_7$ Dye in Water Using Spray Deposited TiO$_2$ Thin Films. *Appl. Catal. B Environ.* **2009**, *89*, 288–294.

36. Shayesteh, S. F.; Dizgah, A. A. Effect of Doping and Annealing on the Physical Properties of ZnO:Mg Nanoparticles, Pramana. *J. Phys.* **2013**, *81*(2), 319–330.

37. Baruah, S.; Jaisai, M.; Imani, R.; Nazhad, M. M.; Dutta, J. Photocatalytic Paper Using Zinc Oxide Nanorods. *Sci. Technol. Adv. Mater.* **2010**, *11*, 055002–055008.

38. Houas, A.; Lachheb, H.; Ksibi, M.; Elaloui, E.; Guillard, C.; Herrmann, J. M. Photocatalytic Degradation Pathway of Methylene Blue in Water. *Appl. Catal. B Environ.* **2001**, *31*, 145–157.

CHAPTER 14

FLAME SPRAYING OF POLYMERS: DISTINCTIVE FEATURES OF THE EQUIPMENT AND COATING APPLICATIONS

YURY KOROBOV[1] and MARAT BELOTSERKOVSKIY[2,*]

[1]UrFU, Mira St. 19, Ekaterinburg 620002, Russia

[2]JIME NSA Belarus, Academic Street 12, Minsk 220072, Republic of Belarus

[*]Corresponding author. E-mail: mbelotser@gmail.com

CONTENTS

ABSTRACT

In this chapter, new trends and developments in powder coating is reviewed in detail.

The most widely used methods for powder polymer deposition include the following: coating from the fluidized bed, electrostatic coating, and thermal spraying.

Only special chambers, baths, and ovens can realize the first two methods. Moreover, in case of large parts, their use is limited by excessive energy consumption. Therefore, thermal spraying is one of the most economical and easiest methods to implement polymer coating. It allows melting and forming a layer in a single operation.

Polymer processing should meet the following requirements:

- low heating temperature to escape a thermal destruction;
- reliability;
- low costs;
- processability.

Plasma spraying and high-velocity oxygen fuel (HVOF) allow producing polymer coatings of good quality in some regime intervals.[1-6] However, their cost is minimum 3–5 times as high comparing to flame spraying.[7] So, flame spraying is of interest to develop for the deposition of polymers because of its simplicity and low cost.

Thermal spraying of polymers can be traced back to the 1940s, when polyethylene (PE) was first produced. Early attempts were unsuccessful because flame-spraying guns, designed for spraying metals, were used. They produced a flame that was both too hot and too short to melt the PE without degradation.[8]

Since the 70s of the last century, guns were developed for flame spraying of polymer materials.[1,9-11] Common features of gun design are seen in Figure 14.1. Combustion mixtures include oxidizer (oxygen or air) and fuel gas (acetylene or propane). A powder-feed nozzle is located in the center of the flame spray gun. Two ring-shaped nozzle outlets, the inner ring being for air and the outer ring for the flame, surround this. As a result, the polymer powder is in soft heated by air blanket excluding a direct contact with the flame. Such design provides longer residence time at a lower temperature for the sprayed powder.

Heating of the powder is determined by the ratio of velocities of air blanket and the surrounding flame. Control of the ratio leads to a visible change in the shape of the torch.[1,2]

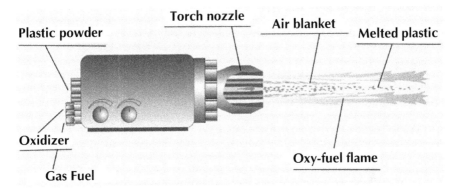

FIGURE 14.1   Scheme of polymer spraying gun.

Many plastics can now be completely melted in-flight and allow heat-sensitive components to be coated. Polymers have been sprayed by common thermal spray processes such as flame spraying, plasma spraying, and HVOF (Table 14.1).

TABLE 14.1   Thermally Sprayed Polymers and Polymer Composites.

Polymeric material	Flame	HVOF	Plasma
Bismaleimide-phenolic resin			+
Cyanate ester thermosets			+
Ethylene-methacrylic (EMAA) copolymer/Al_2O_3	+		
Epoxy or epoxy-nylon blend filled with Cu or Cu/Ni	+		
Epoxy enamels			+
Epoxy filled with TiO_2			+
Epoxy thermosets			+
Ethylene vinylene acetate (EVA)			
Ethylene-acrylic acid (EAA) copolymer	+		+
Ethylene-methyl methacrylate (EMAA) copolymer	+		
Ethyltetrafluoroethylene (ETFE)	+		
Liquid-crystalline polymers (LCP)	+		
Nylon-11/glass or Al_2O_3 or SiO_2 or carbon black	+	+	+
Polyethylene (PE)	+		+
PE modified with methacrylic acid	+		
PE (ultrahigh molecular weight)/WC–Co or $MgZrO_3$			+
PE/Al_2O_3			+

TABLE 14.1 *(Continued)*

Polymeric material	Flame	HVOF	Plasma
Phenolic thermosets			+
Polyamides (PA)—nylons	+		+
Polyarylene sulfide (PAS)		+	
Polyaryletherketone (PAEK)			+
Polycarbonate (PC) and PC/SiO$_2$		+	+
Polyester (PES)	+		+
PES or polyurethane filled with TiO$_2$, SiO$_2$, and Al$_2$O$_3$	+		
Polyester-epoxy resin	+		+
Polyether block amide copolymer	+		
Polyether-amide (PEA) copolymer			+
Polyetheretherketone (PEEK)	+	+	+
PEEK/Al$_2$O$_3$		+	
Polyethylene terephthalate (PET)	+	+	+
Polyethylene-polypropylene copolymer	+		
Polymethylmethacrylate (PMMA)			+
Polyphenylene sulfide (PPS)	+	+	+
PPS/Al$_2$O$_3$		+	
Polypropylene (PP)			+
Polysulphone			+
Polytetrafluoroethylene (PTFE) and its copolymers	+		+
Polyvinylidene fluoride (PVDF)			+
Postconsumer commingled polymers (PCCP)	+		
PVDF-hexafluoropropylene (HFP) copolymer			+
PVDF/WC-Co or MgZrO$_3$			+
Polyimide (PMR)/WC–Co		+	
Urethane	+		

Fuel mixture air–propane seems to be more suitable to spray polymers. First, combustion temperature is about 800–1000 degrees lower comparing to oxygen–fuel mixtures. Use of compressed air instead of pure oxygen decreases flame temperature due to heating of nitrogen. In addition, the actual flame temperature can be changed additionally by adjusting the oxidizer/fuel flow rate ratio (Fig.14.2).

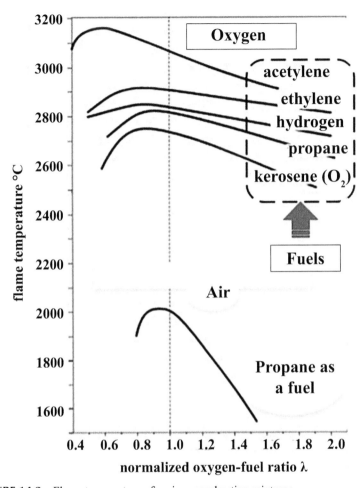

FIGURE 14.2   Flame temperature of various combustion mixtures.

In recent years, the most widespread guns operate using air as the oxidizer. Heat distribution modeling showed that specific heat input of the torch at spraying should be within a range of $(1–3) \times 10^6$ W/m² to escape its thermal destruction of polymer coatings.[13,14] In this case, short-term overheating of the material is not more than 1.5 times higher above the range of polymer melting temperatures, 360–670 K. It provides dense polymer coatings with a minimum content of low-polymeric thermal degraded substance.

As seen from Table 14.1, polymer coatings with added of inorganic fillers are also produced. It allows improving wear, corrosion, and radiation resistance. Such composite polymer coatings are intensively explored applying to various sectors of industry.

Adhesion strength of the coating to substrate is one of the main factors determining the performance of polymer coatings. An influence of air–fuel ratio of the torch, type of polymer, particle size of the polymer powder, and a share of inorganic fillers in the blend on the adhesion strength of the coatings is described below.

Typical polymer flame spraying gun Terco-P was taken to show basic coating features (Fig.14.3).

FIGURE 14.3   Polymer flame spraying gun.

Specific features of gun Terco-P are the following:

* polymer melting point—360–650 K;
* output—2.8 kg/h;
* maximum gas pressure (MPa): air—0.5; propane—0.2;
* initial polymer form—spherical powder; and
* powder diameter 50–300 μm.
* Coating properties are the following:
* adhesion strength 7.5–10.0 MPa;
* tensile strength—70 MPa;
* deposition efficiency—0.92;
* coating thickness 1–4 mm;
* thermal destruction—negligible; and
* substrates—metals, ceramics, glass, concrete.

It allows adjustment the heat exchange in the system "torch–polymer particle" by changing ratio between velocities, of the air blanket and the concurrent combustible mixture, which correspond to consumptions. Results are visible as a flame shape change (Fig.14.4).

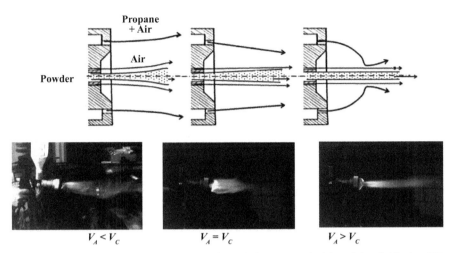

FIGURE 14.4   Torch shape depending on the ratio between velocities of the air blanket (V_A) and the combustible mixture (V_C). (a) Scheme of gas inlet and (b) view of flame.

The efficiency of flame polymer coatings during their exploitation under the influence of external factors is determined, above all, by the strength of the coating adhesion to the substrate. The efficiency of the application process, the adhesive strength, and properties of the resulting flame composite polymer coatings obtained with flame devices depends on the composition of the material being deposited and the technological features of spraying. The value of the adhesive strength of the polymer coating composition onto the prepared substrate surface depends on the structure of sprayed powder particles, and the type and percentage of fillers.

To improve the adhesion of flame sprayed coatings volume modification of applied polymer with inorganic fillers was used. It was successfully applied in different methods of formation of polymer coatings.[1,10] Fillers are used to modify the mechanical, structural, chemical, and other properties of polymers, including strength of adhesive polymer–metal compounds. Depending on the type and nature of fillers, they can increase or decrease the adhesion strength of polymer coatings to the substrate.

Therefore, the influence of the type and the percentage of filler of the polymer on adhesion strength of the polymer coating was investigated.[15] The following polymer powders were tested: polyamide, high-density PE (HDPE), and PE terephthalate (PET). The following components were added to polymers: 200–300 μm glass–ceramic (GC), 25–50 μm aluminum powder (Al). The thickness of the coatings was 0.5–1 mm.

Adhesion strength of the coatings was evaluated by the method of separation of a conical pin.[16] According to a schematic view of the test (Fig.14.5), the washer 1 serves as the basis; pin 2 is inserted into its hole so that its end face surface is flush with the external plane of the washer. The total surface of the pin and the washer after preparation is coated with the coating 3. The pin and the washer were made from low-carbon steel like C1020 ASTM.

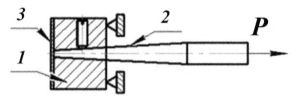

FIGURE 14.5   Scheme of conical pin separation.

The test consists of pulling the pin by applying a force (P). The adhesion strength (σ) is calculated by an equation:

$$\sigma = P/S,$$

where S is the square of the end face surface. The results were averaged for five samples.

The adhesion strength of the coating HDPE + GC/Al changes due to filler share (5–30 vol%). Peak value is indicated at 15–20% of fillers (Fig.14.6).

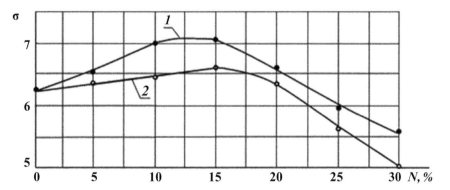

FIGURE 14.6   The change of the adhesion strength σ of polymer coating due to the share of filler content (N, vol%): 1—glass–ceramic and 2—aluminum powder.

The adhesion strength of the coating PA/HDPE/PET + 10 vol% Al changes due to polymer particle size. Min/max ratio of diameters is 1.5. Peak value is indicated at mean particle diameter 200–250 μm (Fig.14.7).

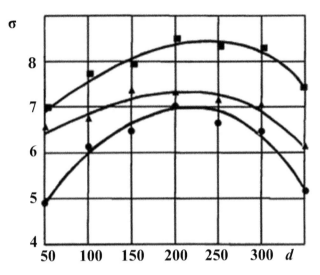

FIGURE 14.7 The change of the adhesion strength σ of polymer coating due to particle size of polymers: PET (■), HDPE (▲), and PA (●).

These results can be explained by the following:

- Particles less than 50 μm are subjected to burning when spraying. Such burning products may be placed on a surface to be coated which leads to reduction of the adhesion strength;
- Particle has no time to melt by the heat of the torch if its size exceeds 300 μm. Presence of incompletely melted particles in coating-substrate interface also leads to reduction of the adhesion strength.

The results of experimental studies showed that the maximum adhesive strength is observed when ratio (max particle diameter, d_{max})/(min particle diameter, d_{min}) is less than 1.8–2.0 (Fig.14.8). Polymer powders sizes range from 100 to 300 μm. So, use powders of uniform particle size leads to improving adhesion strength.

In addition, studies have shown (Figs.14.7 and 14.8) that coating based on PET have the highest adhesion strength comparing to PA and HDPE. As noted, there is a certain correlation with the adhesion energy of the molecules bond in the polymer (cohesive energy).[8] The higher the energy of cohesion of the functional groups of the polymer, the higher the adhesiveness. PET, due to presence of high-energy functional groups, has the highest adhesion.

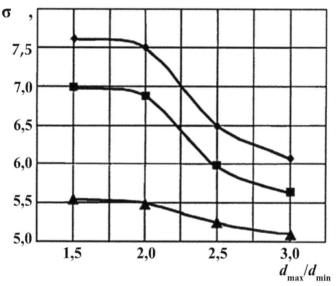

FIGURE 14.8   The change of the adhesion strength σ of polymer coating due to the d_{max}/d_{min} ratio (filler content—10 vol% Al):♦ PET; ■ HDPE; and ▲ PA.

As known, the adhesion strength of polymer coatings with surface of mild steel is drastically increased when oxygen-containing groups appear in polymers (–OH, –COOH, etc.).[8,17]

Flame, depending on the share of the combustible gas in the gas mixture, obtains "oxidizing," "normal," or "reducing" mode.[18] Normal flame is formed by combustion of the stoichiometric mixture of air–fuel, when all of the hydrocarbon molecules are reacted with oxygen molecules. Oxidizing flame is formed by combustion with excess oxygen in the mixture. An excess of fuel gas forms reducing flame. Oxidizing flame has a limit of oxidant concentration above which the combustion process is terminated.

For flame spray guns, the lower limit of the reducing propane–air flame is 16 volumes of air per 1 volume of propane (oxidizer-to-fuel ratio $\beta = 16$). With a further decrease in air content, there is a large amount of unreacted carbon. The upper limit of the oxidizer-to-fuel ratio is $\beta = 32$. The subsequent increase of oxidant leads to the disruption of the flame.

The adhesion strength of the coating changes due to oxidizer-to-fuel ratio and particle size. The tests showed that its highest values are obtained at oxidizing flame (β = 22–32) and with particles of smaller size (Fig.14.9). It results from the largest specific surface area of particles–air interaction, which promotes the formation of a significant number of oxygen-containing groups.

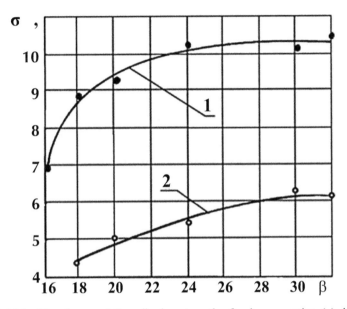

FIGURE 14.9 The change of the adhesion strength of polymer coating (σ) due to the oxidizer-to-fuel ratio (β). Polymer powder was PA–HDPE 50/50, particle size: 50–60 μm—1 and 100–200 μm—2.

So, various stages of the process require various modes. Spraying the underlayer should be performed with the powder of particle size less than 60 m, β = 24–32. Spraying the base coating should be performed with the powder of particle size 100–300 μm, β = 20–24. In this case, a significant oxidation of the coating material, which may cause some reduction in polymer properties and adversely affect the adhesion, is eliminated. Melting the coating should be carried out at β = 16–20 to decrease oxidation that can be realized by reducing flame treatment.

Polymer coatings are sprayed on metals, ceramics, glass, wood, and construction materials (concrete, brick, and slate). With regard to these materials, the polymer coatings are applied to a variety of purposes:

- protection of structural elements from aggressive environments;
- wear protection, application of antifriction layer;

- electrical connections of electrical power fittings;
- isolation of the contact of dissimilar metals to avoid electrochemical processes;
- restoration of defective polymer coating on-site;

The addition of fillers to the starting polymer (see Fig.14.10) changes the properties of the coatings as follows: rubber crumb improves anti-friction properties; alumina, metals improve durability; and elements for thermal neutron absorption (boron and its compounds) increase the radiation resistance.

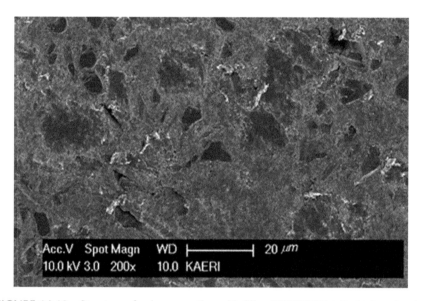

FIGURE 14.10 Structure of polymer coating with filler: UHMW PE (ultrahigh molecular weight polyethylene) + 20% B4C.

Some examples are shown below.

This process has been successfully used to protect concrete structures (foundations, walls), which can be destroyed by the corrosion of steel reinforcement (Fig.14.11).

Polymeric coatings are characterized by high performance in a dry friction conditions. Bench tests have shown that coefficient of dry sliding friction is 0.15 at specific load of 5 MPa and 0.08 at specific load of 10 MPa. It allows decreasing of cranking shall effort, contact temperature, and wear of mating parts (see Table 14.2).

FIGURE 14.11   Polymer coating is sprayed on concrete wall of reservoir.

TABLE 14.2   Results of Bench Tests.

Mated parts	Section moment to the cranking shall (N m)	Temperature in the friction zone (°C)	Linear wear (μm)
Steel–steel	110–130	55–65	60–85
Steel–polymer coating	60–80	30–35	10–20

Such coatings have been used on the surface of a cereal overload sleeve of combine harvester (Fig. 14.12). Spraying process is shown in Figure 14.13.

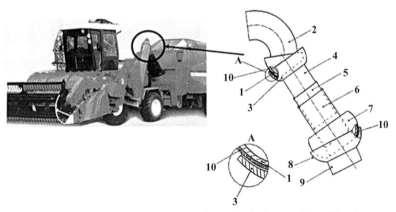

FIGURE 14.12   Polymer coating on a cereal overload sleeve of combine harvester. 1, 7—truncated spherical nozzles; 2—feed conveyor; 3—hollow truncated spherical support; 4, 5, and 6—upper, middle, and lower tube adapters of telescopic extension; 8—truncated spherical tank; 9—box-shaped receptacle; 10—friction reducing polymer coating.

FIGURE 14.13   Polymer spraying of the truncated spherical support.

During the run tests, the sleeve with a polymer coating showed minimum grain loss and excellent durability of spherical joints at overload of the cereal. In truncated spherical nozzles without coating failure was observed after 80 h, the damaged span was 45–70% of the contact area. Coated joints worked without failure 220 h in the absence of wear.

Polymer coating is applied at processing galvanic baths that operate at aggressive chemical environment (Fig.14.14).

FIGURE 14.14   Polymer coating is sprayed on a galvanic bath.

Polymer coating are used as road markings and concrete block paving (Fig.14.15). Concrete block of which the surface is coated by powder polymer satisfied all the standards of the traffic paint in Japan in terms of whiteness, reflectance ratio, weathering resistance, and abrasion resistance.[19]

FIGURE 14.15   Polymer coating at road markings and concrete block paving.

14.1   CONCLUSIONS

1. Flame sprayed coatings can be produced from various polymers of 360–670 K without destruction.
2. Applying to various polymeric coatings, maximum adhesion strength is achieved by spraying the powders with a particle size 150–300 μm, the ratio between the minimum and maximum diameter of the particles should be less than 2.0.
3. Inorganic fillers in the polymer powder allow to increase the adhesion strength by 15–20%, and peak value of the filler content is 15 vol%.
4. To get the highest coating properties, the process should be divided to stages of spraying intermediate layer, base layer, and melting which differ by size of used powder and oxidizer-to-fuel ratio.

5. Polymer coatings are applied to protect structural elements against aggressive environments, wear, and to provide antifriction properties and electro-insulation.

KEYWORDS

- **low heating**
- **thermal destruction**
- **reliability**
- **low costs**
- **processability**

REFERENCES

1. Petrovicova, E.; Schadler, L. S. Thermal Spraying of Polymers. *Int. Mater. Revue.* **2002,** *47*(4), 169–190.
2. Gupta, V. Thermal Spraying of Polymer–Ceramic Composite Coatings with Multiple Size Scales of Reinforcements. Master Thesis, Drexel University, Germany, 2006.
3. Rodchenko, D. A.; Kovalkov, A. N.; Barkan, A. I. Heating of the Polymer Particles by Spraying a Jet of Plasma. *Proc. Univ. Mech. Eng.* [In Russian]. **1985,** *9*, 108–113.
4. Bao, Y. Plasma Spray Deposition of Polymer Coatings. PhD Thesis, Brunel University, Uxbridge, West London, 1995.
5. *HVOF Sprayed Multi-Scale Polymer/Ceramic Composite Coatings;* Gupta, V.; Niezgoda, S.; Knight, R. In *Thermal Spray 2006: Science, Innovation, and Application*, Proceedings of International Thermal Spray Conference, Materials Park, OH, 2006; Marple, B. R. Moreau, C. Eds.; ASM International: Materials Park, 2006.
6. Vuoristo, P. M. J. Functional and Protective Coatings by Novel High-Kinetic Spray Processes. In *Kokkola Material Week—2014*, Proceedings of International Conference, Kokkola, Finland, 2014. TUT, Tampere, 2014.
7. Brogan, J. A. Thermal-Spraying of Polymers and Polymer Blends. *MRS Bull.* **2000,** *25*(7), 48–53.
8. Wicks, Z. W.; Jones, F. N.; Pappas, S. P. *Organic Coatings. Science and Technology*; John Wiley & Sons Ltd.: Chichester, 1994; Vol. 2, p 438.
9. Thermal Spray Coatings. *ASM Handbook; Welding, Brazing, and Soldering*; ASM International, 1993; Vol. 6, pp 1004–1009.
10. Chen, H.; Qu, J.; Shao, H. Erosion-Corrosion of Thermal-Sprayed Nylon Coatings. *Acta Materialia.* **1999,** *57*, 980–992.
11. Belotserkovsky, M. A.; Chekylaev, A. V.; Korobov, Y. S. Flame Spraying of Polymer Coatings. In *Technologies of Repair, Restoration and Strengthening of Machine Parts,*

Machinery, Equipment, Tools and Tooling, Proceedings of 9th International Conference. In 2 Parts. St. Petersburg, Russia, 2007, Polytechnic. University Publishing House; St. Petersburg [In Russian], 2007; Part 1, p 26–33.

12. Kreye, H.; Gartner, F.; Kirsten, A.; Schwertzke, R.In *High Velocity Oxy-Fuel Flame Spraying. State of Art, Prospects and Alternatives.* 5th Colloquium "High Velocity Oxy-Fuel Spraying"; GTS e.V.: Erding, Germany, 2000.

13. Belotserkovsky, M. *Development of Up-to-Date Coating and Hardening Technologies in Belarus*; Reports of the Korea—Eurasia Technology Cooperation Workshop: Seoul, Korea, 2007.

14. Rekhlitsky, O.; Solovei, N.; Belotserkovsky, M.; Korotkevitch, S. In *Anti-Friction and Mechanical Properties of Flame Sprayed Polymer Coatings.* Proceedings of "Balt-Trib—2013," Lithuania, Kaunas, 2013.

15. Korobov, Y. S.;Belotserkovsky, M. A.; Timofeev, K. M.; Thomas, S. In *Adhesive Strength of Flame-Sprayed Polymer Coatings*, AIP Conference Proceedings, 1785, 030011, 2016; 16–20 May 2016, Ekaterinburg, Russia, 2016. http://dx.doi.org/10.1063/1.4967032.

16. Tyshinski, L. I.; Plokhov, A. V.; Tokarev, O. A.; Sindeev, V. I. *Methods of Study the Materials*; Mir: Moscow [in Russian], 2004, p 384.

17. Rama, K.; Layek, A.; Nandi, K. A Review on Synthesis and Properties of Polymer Functionalized Grapheme. *Polymer.* **2013,** *54*(19), 5087–5103.

18. *Welder's Handbook for Gas Shielding Arc Welding, Oxy Fuel Cutting & Plasma Cutting,* 3rd ed.; Air Products PLC: NY, 1999; p 274.

19. Sano, Y.; Fukushima, N.; Niki, T. Consideration on Powder Flame Sprayed Concrete Blocks [online]. https://docs.google.com/viewerng/viewer?url=http://www.sept.org/tech-papers/133.pdf.

CHAPTER 15

INFLUENCE OF V DOPING ON PHYSICOCHEMICAL PROPERTIES OF MESOPOROUS TITANIA WITH POTENTIAL APPLICATION IN SOLAR PHOTOCATALYSIS

RANJITH G. NAIR[1,*], ABINASH DAS[1], KARIM BOUCHMELLA[2], and BRUNO BOURY[2,*]

[1]*Department of Physics, National Institute of Technology Silchar, Silchar 788010, Assam, India*

[2]*CMOS/Institut Charles Gerhardt Montpellier, Université de Montpellier, 34095 Montpellier, France*

Corresponding author. E-mail: bruno.boury@univ-montp2.fr; rgnair2007@gmail.com

CONTENTS

ABSTRACT

The physicochemical properties of semiconductor metal-oxides play a vital role in determining the photocatalytic performance. The present chapter scrutinizes the physicochemical characterization of mesoporous V-doped titania nanomaterials synthesized through nonhydrolytic sol–gel technique with different V doping. Structural, surface, elemental, and optical properties of the samples have been evaluated using different characterization techniques such as X-ray Diffraction (XRD), Field Emission Scanning Electron Microscopy (FESEM), Energy Dispersive X-ray analysis (EDX), Brunauer–Emmett–Teller (BET) analysis, UV-vis Diffuse Reflectance Spectroscopy (UV-DRS), and Photoluminescence (PL) Spectroscopy. XRD spectra reveal that samples are highly crystalline along with change in strain and phase transformation at different V concentrations and calcination temperatures. Surface analyses of the samples confirm that the agglomerates are spherical with an average grain size distribution 10–20 nm. Nitrogen adsorption desorption isotherms of the samples calcined at 500°C shows that samples are mesoporous in nature with high surface area and pore size distribution ranging from 5.8 to 9.1 nm at different V loading. Crystallite size, particle size, and grain sizes of the samples calculated using XRD, BET, and FESEM are in good agreement with each other. The elevated absorption of the samples observed in UV-DRS spectra confirms the V doping. The detailed UV-DRS analysis of the samples corroborates the band transformation from indirect to direct type at high V concentration, and it is in agreement with the XRD data. The PL analysis shows the type of electronic transitions and confirms that the carrier recombination rate decreases with the increase in V doping. All the above analyses confirm that V-loaded titania samples can be used as an effective solar photocatalyst.

15.1 INTRODUCTION

Semiconductor metal oxide is one of the most emerging areas of research, due to their ability to address a wide range of environmental and energy applications. Titania is one of the most popular inorganic semiconductor oxide materials suitable for various applications such as photocatalysis, photochromism, Li-ion batteries, and chemical and gas sensors.[1-4] Titania can also be used as an effective photoanode in dye sensitized solar cells.[4] Among the different metal oxide semiconductor materials, titania-based photocatalysis is the most emerging technology to address energy and environmental issues, simultaneously.[5,6] The advantages of titania over others are structural firmness, low cost, inertness and high stability, long durability, and nontoxicity.[4,7–10] Titania can exist in three different polymorphs such as rutile, anatase, and brookite with bandgap energy 3.0 eV, 3.2 eV, and 3.2 eV, respectively.[10,11] Literature reveals that anatase is the most photoactive phase.[10] However, recent researches showed that mixed phase titania can perform better, compared to its pristine phases.[12,13] Photocatalytic performance of the pristine titania is limited due to its wide band gap, low surface, area, and high charge carrier recombination.[8–10] Recombination of photo-generated carriers (electron and hole) reduces the overall quantum efficiency of the catalyst.[14]

Different strategies have been reported to address the limitations of titania. Among them doping is a well-accepted method; various dopants, like cationic and anionic dopants, sensitization, metal oxide complexes, and so on, are well effective.[10,15,16] Doping is one of the best known techniques to extent the spectral response of pristine TiO_2 into the visible region and reduces the charge carrier recombination.[9,15] V-doped TiO_2 shows higher photocatalytic performance compared other cationic dopants.[17] The red shift of the absorption depends on the type of metal ion implanted which follows the sequence V > Cr > Mn > Fe > Ni.[7,9]

The performances of TiO_2 photocatalyst is also depend on their texture and much research has been focused on obtaining high specific surface area mesoporous materials. Indeed, mesopores facilitate the rapid diffusion of molecules, thus improving the kinetics of photocatalytic reactions. [1,18,19] Among the methods proposed for the synthesis of mesoporousTiO_2, sol–gel technique is the most versatile. Recent research showed that nonhydrolytic sol–gel technique is a powerful process to obtain mesoporous oxides and mixed oxides by avoiding the use of expensive reactants and templates.[20–24] For instance, V-doped TiO_2 catalysts were prepared by reaction of $TiCl_4$ and $VOCl_3$ precursors with diisopropyl ether. The proposed nonhydrolytic reaction involves the formation of intermediate isopropyl-oxide groups that further condense with the remaining chloride groups with production of isopropyl chlorides as shown below.

$$M–Cl + iPr–O–iPr \rightarrow M–O–iPr + iPr–Cl$$

$$M–Cl + M–OiPr \rightarrow M–O–M + iPr–Cl$$

The present chapter reports the synthesis and characterization of high surface area metal oxide materials with mesoporosity through nonhydrolytic sol–gel technique. To the best of our knowledge, such detailed analyses of these oxides and an attempt to correlate the characteristic of these materials with the V-content and calcination temperature have never been reported. Such attempt has been made in order to achieve a possible extend of the absorption of pristine mesoporous titania toward visible region. With this aim, the effect of V doping on structural, surface, and optical performance has been thoroughly investigated at different calcination temperature.

15.2   EXPERIMENTAL

15.2.1   MATERIALS AND SYNTHESIS

Titanium tetrachloride (99.9%), diisopropyl ether (99%), and anhydrous dichloromethane (100%) were obtained from Across (France). Vanadium (V) oxychloride (99%) was obtained from Aldrich. Diisopropyl ether and dichloromethane were dried by distillation over sodium wire and phosphorus pentoxide, respectively. All the experiments were carried out under argon atmosphere in a glove box using 80-mL glass tubes (heavy-wall Pyrex tube) (H_2O < 10 ppm, O_2 < 20 ppm).

Syntheses were performed using nonhydrolytic sol–gel technique as reported earlier.[25,26] The reagents $TiCl_4$, $VOCl_3$, and CH_2Cl_2 were added to the glass tube in different Ti:V weight ratio. Finally, the desired quantity of diisopropyl ether was mixed; then, the glass tube was frozen in liquid nitrogen and sealed under vacuum. The $TiCl_4/VOCl_3$ to diisopropyl ether ratios was kept 1:3 during the synthesis. Precise amount of the precursors used for the synthesis is given in Table 15.1. All the sealed glass tubes were kept in an oven at a reaction temperature of 110°C for 3 days under autogeneous pressure, leading to dark brown gels. The gel was washed with dichloro methane to remove the by-products and then dried under vacuum at room temperature at 120°C for 5 h. Finally, the xero-gel obtained was crushed in a mortar and a part of the sample was calcined for 2 h at 500°C and the rest for 2 h at 800°C with a ramping rate of 4°C min⁻¹. The pristine and V-doped samples calcined at 500°C were labeled as Ti-500, TiV(1)-500, TiV(2.5)-500, and TiV(5)-500 where V% is the atomic percentage of V doping.

The crystallite size of the samples were calculated using Scherer's formula[27]

$$D = (k\lambda/\beta\cos\theta) \tag{15.1}$$

where D is the crystallite size, λ is the wavelength of the X-ray radiation (CuKα =0.15406 nm), k is the shape factor (0.94), β is the full width at half maxima of the most intense peak, and θ is the diffraction angle.

The limitation of this method is that, it is not considering structural parameters like lattice strain which contributes in the broadening of the diffraction peak and which plays a major role in calculating crystallite size. To overcome this limitation, the crystallite sizes of the samples were also calculated using Williamson Hall method.[27,28] In this technique, it is considered that the crystallite size and strain contributes in the line broadening. It is also assumed that the line broadening β of a Bragg reflection originating from the small crystallite size follows Scherer's equation, the strain induced broadening β_ε, which is given by the Williamson formula as $\beta_\varepsilon = 4\varepsilon \tan \theta$, where ε is the root mean square value of the micro strain. Therefore, the observed line breadth is given by

$$\beta = [K\lambda/D\cos\theta] + [4\varepsilon\tan\theta]. \tag{15.2}$$

By plotting, the value of $\beta\cos\theta$ as a function of $4\sin\theta$ the micro strain ε has been estimated from the slope of the line and the crystallite size from the intersection with the vertical axis.

The anatase-to-rutile phase ratio of the samples was calculated using Spur's formula[29]

$$f(a) = (1 + 1.26\ I(r)/I(a))^{-1} \tag{15.3}$$

$$f(r) = (1\ f(a)) \tag{15.4}$$

$$A/R \text{ ratio} = f(a)/f(r) \tag{15.5}$$

where $f(a)$ is the intensity function of anatase phase, $f(r)$ is the intensity function of rutile phase, $I(r)$ is the intensity of rutile reflection, and $I(a)$ is the intensity of anatase reflection.

15.2.2 CHARACTERIZATION

The X-ray diffraction (XRD) patterns of the samples were recorded using a diffractometer (Philips X'Pert Pro, Netherlands). The FESEM micrographs were recorded using a Hitachi S-4800 scanning electron microscope (Japan); energy dispersive X-ray analysis (EDAX) was performed using the EDAX attachment (Oxford Instruments, UK). N$_2$-physisorption isotherms were obtained at 77K using Micrometrics Tristar, USA. The samples were out gassed for 12 h at 150°C under vacuum (2 Pa). UV–Visible diffuse reflectance spectra were recorded using Jasco V-670 UV–Visible–NIR Spectrophotometer (UK) with diffuse reflectance attachment. The photoluminescence (PL) measurements were carried out using a LS 55 Perkin Elmer spectrophotometer.

15.3 RESULTS AND DISCUSSION

Elemental compositions of the samples calcined at 500°C were evaluated using EDAX analysis and the values are given in Table 15.1. It confirms the presence of V in titania, a rise in the vanadium atomic percentage is observed with increase of V content in the starting mixture, and we note a good correlation between the experimental doping concentration and the starting mixture.

TABLE 15.1 Details of the Reactants Quantities Used for the Synthesis along with EDAX Data on Prepared Materials.

Sample	Experimental						EDAX		
	$TiCl_4$ (g)	$VOCl_3$ (g)	iPr_2O (g)	CH_2Cl_2 (mL)	Ti (%)	V (%)	Elements (at%)		
							Ti	V	O
Ti-500	11.89	0	19.21	20	100	0	21.05	0	78.95
TiV(1)-500	11.77	0.10	19.2	20	99	1	24.64	0.25	75.11
TiV(2.5)-500	11.6	0.25	19.18	20	97.5	2.5	23.91	0.61	75.48
TiV(5)-500	11.29	0.51	19.15	20	95	5	25.22	1.37	73.41

XRD patterns of the pristine and V-doped titania samples calcined at different temperatures are depicted in Figure 15.1. Diffraction patterns of the samples calcined at 500°C shows that the samples are in anatase phase (JCPDS data PDF86-1156).

A peak shift is observed for the samples TiV(1)-500, TiV(2.5)-500, and TiV(5)-500 as shown in Figure 15.1 (inset) and which can be ascribed to the imperfections developed in titania matrix due to the foreign atom V.[30–32] In the reported hydrolytic sol–gel technique, V doping lead to the phase transformation at low calcinations (~600°C) temperature and the nonhydrolytic sol–gel technique improved the thermal stability resulting the phase transformation at higher calcination temperature. The diffraction patterns of the samples calcined at 800°C show mixed phase with anatase and rutile

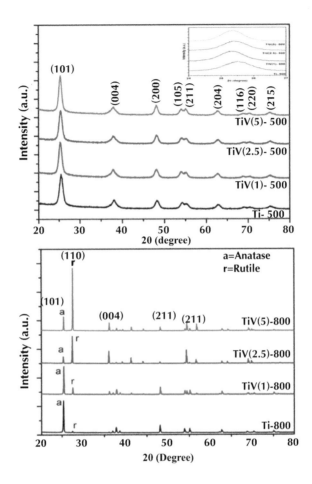

FIGURE 15.1   X-ray diffraction spectra of the samples calcined at 500°C (top) and 800°C (bottom).

phases (JCPDS data PDF86-1156 & PDF89-4920). The spectra reveal that the temperature required for the phase transformation of titania from anatase to rutile is increased compared to the reported values.[10,33] This high thermal stability of anatase titania may be distinctive feature of the mesoporous nature of titania along with synthesis conditions of nonhydrolytic technique. Interestingly, XRD patterns show that with increase in V concentration the phase transformation also increases. The above results substantiate that along with calcination temperature, doping and synthesis procedure also plays a role in anatase to rutile phase transformation. The crystallite sizes of the samples were calculated using both Scherrer and WH methods and depicted in Table 15.2. The breadth of the diffraction pattern can be attributed to the size of the crystals and to their strains, but in Scherrer method, strain was not considered. So from WH plot, we can find the strain developed in the crystal and its consequences on crystallite size. For the samples calcined at 500°C, the crystallite size calculated from Scherrer formula is close to 10 nm and increases with the V doping. The crystallite sizes obtained from W–H method are similar, except for the sample with the highest V content and which may be attributed due to negative slope. Negative slope is due to compressive strain and positive slope is due to the tensile strain.[34,35] It is further reported that strain can modify the band structure of semiconductors by customizing interatomic distances and relative positions of atoms.[36] As expected, calcination at 800°C leads to a significant increase in the particle size. From the Scherrer formula, the crystallite size increases monotonically with the proportion of V.[37] The phase transformation of the samples from anatase to rutile at 800°C is calculated from the intensity of rutile and anatase reflections using Spur's equation.[29] The *A/R* values displayed in Table 15.2 show that the relative abundance of rutile and anatase phases strongly depends on the synthesis conditions, doping, and thermal treatment, and it is in agreement with earlier reports.[32, 38] Samples calcined at 800°C showed mixed phase with anatase and rutile phase and the rutile content increases with the increase in V content.

TABLE 15.2   Crystallite Size, Strain, $F(a)$, and A/R Ratio of the Samples Calcined at 500°C and 800°C.

Samples	Ti-500	TiV(1)-500	TiV(2.5)-500	TiV(5)-500	Ti-800	TiV(1)-800	TiV(2.5)-800	TiV(5)-800
Crystallite size (nm) (Scherrer)	8.5	9.4	9.7	10.2	52	60.3	68.2	73.6
Crystallite size (nm) (WH plot)	9.2	9.6	10.2	8.1	76.1	41.3	39.1	46.7
Strain ($\times 10^{-3}$)	2	2	1	−0.3	1	−0.7	−1	−0.9
$F(a)$ (Spurr) (%)	100.0	100.0	100.0	100.0	93.9	85.0	24.0	22.0
A/R ratio	∞	∞	∞	∞	15.39	5.7	0.32	0.29

The surface morphology of the samples calcined at 500°C has been evaluated using FESEM at different magnifications (Fig. 15.2) Micrographs of the samples showed spherical primary nanoparticles, aggregating to form larger secondary particles, up to a few μm in size. The presence of V favors the formation of spherical secondary particles. The shapes of the agglomerates

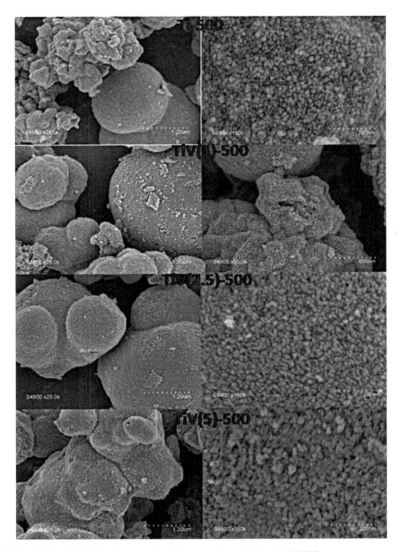

FIGURE 15.2 FESEM micrographs of the samples calcined at 500°C.

are comparable with earlier reported work on nonhydrolytic Ti–V mixed oxide catalysts.[25,26] At the highest magnification, it may be seen that the grains are of spherical shape and the average sizes of the grains are in the 12–16-nm range (Table 15.3) depending on the V-content, and the observed crystallite sizes are in good agreement with the XRD data. However, it is observed that at higher concentration, most of the agglomerates are converted into spherical agglomerates. The measured primary particle size for the different samples is given in Table 15.3. The micrographs also show the presence of pores between the primary particles. The details of the texture of this material were evaluated through N_2 adsorption–desorption measurements.

The specific surface area has obviously an influence on the specific photocatalytic activity. N_2 adsorption–desorption isotherms of the samples calcined at 500°C (Fig. 15.3) showed type IV isotherms with type H2 hysteresis loops, typical of mesoporous materials.[18,39] All together, the increase of the V concentration leads to larger particle size and consequently to lower specific surface area, pore width and pore volume. The Barrett–Joyner–Halenda (BJH) pore size distributions, shown in Figure 15.3 (inset), confirm the presence of well-defined mesopores in the 3–10-nm range. BET specific surface area, pore volume, and average pore size of the calcined samples are displayed in Table 15.3. From the BET, specific surface area A, density ρ, and average particle size D can be calculated.[40]

$$D = 6/(A \times \rho) \tag{15.6}$$

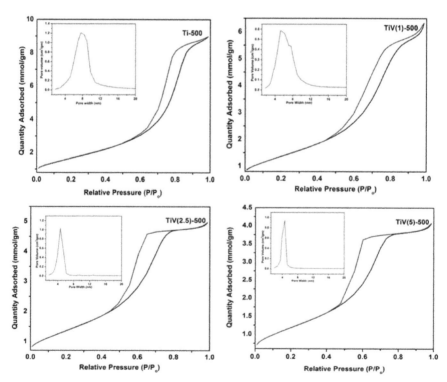

FIGURE 15.3 N_2 adsorption–desorption isotherms of the samples along with pore width distribution.

TABLE 15.3   Surface Area, Pore Width, Pore Volume, Particle Size, Grain Size, and Data of Samples Calcined at 500°C.

Sample	BET				FESEM
	Surface area (m²/g)	Pore width (nm)	Pore volume (cm³/g)	Particle size (nm)	Primary particle size (nm)
Ti-500	136	9.1	0.31	11.6	12
TiV(1)-500	113	7.8	0.22	14.1	12.1
TiV(2.5)-500	113	6.1	0.17	14.0	12.5
TiV(5)-500	97	5.8	0.14	16.3	13.5

UV–Visible DRS Kubelka–Munk absorbance spectra of the samples calcined at 500°C and 800°C are shown in Figure 15.4. The spectra reveal that there is a bathochromic shift in Kubelka–Munk absorbance, and it can be the sign of V doping in comparison with pristine titania. This shift is increasing with the increase in V concentration. This red shift can be ascribed a narrowing of the band gap of pristine TiO_2 and/or to the formation of color centers.[41,42] Interestingly at the spectra of the samples calcined at 800°C showed an increased red shift toward visible region (Fig. 15.4). These data evidenced that for the samples calcined at 800°C, crystallinity, and proportion of rutile increases compared to the samples calcined at 500°C, but none of them should enhance the adsorption in the visible. Possible hypothesis for the reported redshift of the samples calcined at higher temperature may be the compressive strain leads to the modification of the band transition from indirect to direct type and which is in good agreement with the results.[36]

The band gap of the samples were calculated using the modified Tauc's relation[13]

$$(\alpha h\upsilon)^n = A(h\upsilon - Eg) \tag{15.7}$$

where Eg is the absorption band gap, α is the absorption coefficient, $h\upsilon$ is the photon energy, A is the absorbance, and n is an integer either 2 for a direct band gap material or 1/2 for indirect band gap material. In this transition, if the electron momentum is conserved, then the transition is direct, but if the momentum does not conserve, then the transition is an indirect transition. The band gap of the samples were determined by plotting $[F(R)hu]^n$ vs. $h\upsilon^8$ as shown in Figure 15.5, and the values are depicted in Table 15.4.

The analyzed optical properties of the samples show both indirect and direct band gap properties. The band gap of the samples calcined at 500°C calculated using direct type transition shows unrealistic band gap values of

3.45 eV, which are not expected for anatase phase. The band gap values of the same samples calculated through indirect type transition showed the band gap values of between 2.9 and 3.2 eV which is in agreement with the anatase phase.[43]

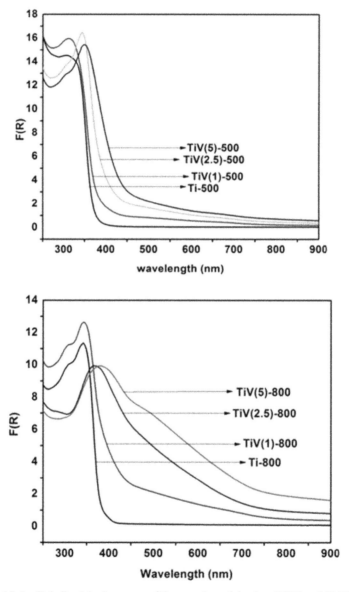

FIGURE 15.4   Kubelka–Munk spectra of the samples calcined at 500°C and 800°C.

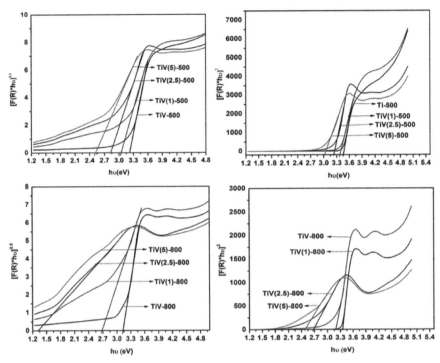

FIGURE 15.5 Tauc's plot of the samples calcined at 500°C and 800°C for direct and indirect transitions.

TABLE 15.4 Direct and Indirect Band Gap of the Samples Calculated Using Tauc's Relation.

Sample	Ti-500	Ti-800	TiV(1) -500	TiV(1) -800	TiV(2.5) -500	TiV(2.5) -800	TiV(5) -500	TiV(5) -800
Band gap (eV) (indirect)	3.2	3.15	3.0	2.7	2.85	1.35	2.55	1.0
Band gap (eV) (direct)	3.45	3.3	3.35	3.15	3.2	2.7	3.0	2.5

The band gap calculated using the indirect type transition is below 1.6 eV for the samples calcined at 800°C with high V concentration (2.5% and 5%) which is also impractical and the band gap calculated using direct type transition are in order. It is in good agreement with the earlier report.[43] With the help of the above results, it can be concluded that high V concentration/calcination temperature can change the type of transition of titania from indirect to direct. However, more detailed analysis is needed to confirm this. From literature, it is found that rutile phase direct band gap value is 3.06 eV and an indirect band gap is 3.10 eV, and anatase has only band gap at 3.23 eV, which is in good agreement with our result.[44] The elevated Kubelka–Munk absorbance of the doped samples compared to pristine will lead to the enhancement of the photocatalytic activity.

For better understanding of the samples, PL analysis was carried out to study the trap levels and recombination rate which play a major role in photocatalytic performance.[45] PL spectroscopy gives information on opto-electronic properties and charge transfer efficiencies of the samples.[46] PL emission is the result of the recombination of excited electrons and holes, and a lower PL intensity indicates a lower recombination rate under irradiation.[47] The PL spectra of the samples recorded with an excitation wavelength of 280 and 340 nm in emission mode are given in Figure 15.6. The spectra show distinct peaks at 418, 444, 482, and 527 nm. The trap levels identified at 418 nm can be assigned to the band-to-band transition of self-trapped excitons localized on TiO_6 octahedral.[48] The remaining trap levels ranging from 444 to 527 nm can be assigned to the oxygen vacancy levels.[48] These trap levels may play a positive role in reducing the recombination.[8] The low PL intensity observed in the V-doped samples may be ascribed to the reduction in electron–hole recombination compared to pristine TiO_2, which plays a positive role in the enhancement in the photocatalytic activity.

In addition, the decreased PL intensity with increase in doping concentration can be attributed to the reduction of surface area of the sample[49] which is in good agreement with the reported BET result. Usually, the smaller the particle size, larger the oxygen vacancy content, higher the probability of exciton production and stronger the PL intensity.[43] But, when there are too many surface defects and oxygen vacancies existing in TiO_2 crystal, the oxygen vacancies and defects become the electron–hole trap center which result in the lower recombination rate of electron–hole pairs and lower PL intensity. Reduced PL intensity of the doped samples can be assigned to the same.

15.4   CONCLUSION

Vandium-doped titania nanoparticles were successfully synthesized using nonhydrolytic sol–gel technique. This technique has a good control in homogeneous incorporation of vanadium in TiO_2. The detailed analyses of the samples calcined at a particular temperature corroborate the fact that V doping has a direct impact on porosity, crystallinity, anatase/rutile ratio, and bandgap. Additionally, a band transition from indirect to direct is observed

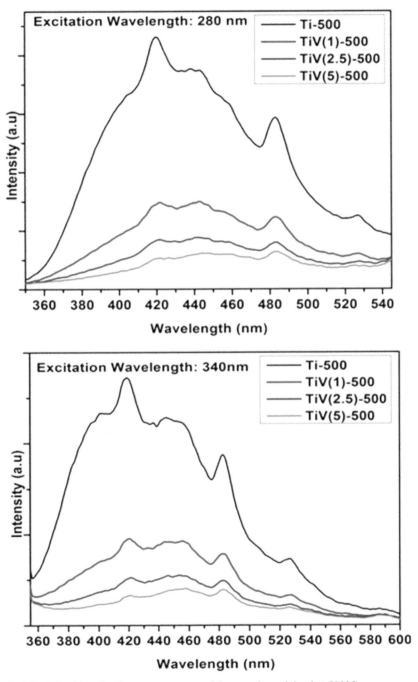

FIGURE 15.6   Photoluminescence spectra of the samples calcined at 500°C.

for V-doped samples in which V concentration plays a crucial role. The low carrier recombination of the V-doped samples has been validated through PL analysis. The overall analysis of this work indicates that the V-doped titania samples synthesized through nonhydrolytic technique could be effective photocatalyst for solar energy applications.

ACKNOWLEDGMENTS

Financial support from the Ministère de l'EnseignementSupérieuret de la Recherche and from the Sandwich Ph.D. Fellowship Programme of the Science and Technology Department of the French Embassy in India and TEQIP-II is gratefully acknowledged.

KEYWORDS

- **mesoporous titania**
- **band transition**
- **solar energy**
- **photocatalyst**
- **nonhydrolytic sol–gel**

REFERENCES

1. Li, W.; Wang, F.; Liu, Y.; Wang, J.; Yang, J.; Zhang, L.; Elzatahry, A. A.; Al-Dahyan, D.; Xia, Y.; Zhao, D. General Strategy to Synthesize Uniform Mesoporous TiO2/Graphene/ Mesoporous TiO2 Sandwich-Like Nanosheets for Highly Reversible Lithium Storage. *Nano Lett.* **2015,** *15*(3), 2186–2193.

2. Sauvage, F.; Chen, D.; Comte, P.; Huang, F.; Heiniger, L.-P.; Cheng, Y.-B.; Caruso, R. A.; Graetzel, M. Dye-Sensitized Solar Cells Employing a Single Film of Mesoporous TiO2 Beads Achieve Power Conversion Efficiencies Over 10%. *ACS Nano.* **2010,** *4*(8), 4420–4425.

3. Jung, H.-G.; Yoon, C. S.; Prakash, J.; Sun, Y.-K. Mesoporous Anatase TiO_2 with High Surface Area and Controllable Pore Size by F^--Ion Doping: Applications for High-Power Li-Ion Battery Anode. *J. Phys. Chem. C.* **2009,** *113*, 21258–21263.

4. Nakataa, K.; Fujishimaa, A. TiO2 Photocatalysis: Design and Applications. *J. Photochem. Photobiol. C: Photochem. Rev.* **2012,** *13*, 169–189.

5. Li, L.; Wang, L.; Hu, T.; Zhang, W.; Zhang, X.; Chen, X. Preparation of Highly Photocatalytic Active CdS/TiO$_2$ Nanocomposites by Combining Chemical Bath Deposition and Microwave-Assisted Hydrothermal Synthesis. *J. Solid State Chem.* **2014**, *218*, 81–89.

6. Zhang, Z.; Shao, C.; Zhang, L.; Li, X.; Liu, Y. Electrospun Nanofibers of V-Doped TiO$_2$ with High Photocatalytic Activity. *J. Colloid Interface Sci.* **2010**, *351*, 57–62.

7. Zhou, W.; Liu, Q.; Zhu, Z.; Zhang, J. Preparation and Properties of Vanadium-Doped TiO$_2$ Photocatalysts. *J. Phys. D: Appl. Phys.* **2010**, *43*, 035301 (6pp).

8. Wang, S.; Yi, L.; Halpert, J. E.; Lai, X.; Liu, Y.; Cao, H.; Yu, R.; Wang, D. Li, Y. A Novel and Highly Efficient Photocatalyst Based on P25–Graphdiyne Nanocomposite. *Small* **2012**, *8*, 265–271.

9. Carp, O.; Huisman, C. L.; Reller, A.; Carp, O.; Huisman, C. L.; Reller, A. Photoinduced Reactivity of Titanium Dioxide. *Prog. Solid State Chem.* **2004**, *32*, 33–177.

10. Pelaez, M.; Nolan, N.; Pillai, S.; Seery, M. K.; Falaras, P. A Review on the Visible Light Active Titanium Dioxide Photocatalysts for Environmental Applications. *Appl. Catal. B.* **2012**, *125*, 331–349.

11. Suwarnkar, M. B.; Dhabbe, R. S.; Kadam, A. N.; Garadkar, K. M. Enhanced Photocatalytic Activity of Ag Doped TiO$_2$ Nanoparticles Synthesized by a Microwave Assisted Method. *Ceram. Int.* **2014**, *40*, 5489–5496.

12. Ji, H.; Qu, Z.; Jia, Q.; Gao, P.; Li, X.; Sun, X. The Phase Transformation, Morphology Evolution and Visible-Light Photocatalytic Activity of V-Doped TiO$_2$ Thin Films. *Integr. Ferroelectr.* **2012**, *138*(1), 105–110.

13. Nair, R. G.; Paul, S.; Samdarshi, S. K. High UV/Visible Light Activity of Mixed Phase Titania: A Generic Mechanism. *Sol. Energy Mater. Sol. Cells* **2011**, *95*(7), 1901–1907.

14. Haber, J.; Nowak, P.; Zurek, P. Charge Transfer in Photocatalytic Systems: V and Mo Doped TiO$_2$/Ti Electrodes. *Catal. Lett.* **2008**, *126*, 43–48.

15. Chen, J.; Yao, M.; Wang, X. Investigation of Transition Metal Ion Doping Behaviors on TiO$_2$ Nanoparticles. *Nanopart. Res.* **2008**, *10*, 163–171.

16. Lachène, D.; Ayral, A.; Boury, B.; Laine, R. M. Surface Modification of Titania Powder P25 with Phosphate and Phosphonic Acids—Effect on Thermal Stability and Photocatalytic Activity. *J. Colloid Interface Sci.* **2013**, *393*, 335–339.

17. Songara, S.; Patra, M. K.; Manoth, M.; Saini, L.; Gupta, V.; Gowd, G. S.; Vadera, S. R.; Kumar, N. Synthesis and Studies on Photochromic Properties of Vanadium Doped TiO$_2$ Nanoparticles. *J. Photochem. Photobiol. A* **2010**, *209*, 68–73.

18. Sing, K. S. W.; Everett, D. H.; Haul, R. A. W.; Moscou, L.; Pierotti, R. A.; Rouquerol, J.; Siemieniewska, T. Reporting Physisorption Data for Gas/Solid Systems with Special Reference to the Determination of Surface Area and Porosity. *Pure Appl. Chem.* **1985**, *57*(4), 603–619.

19. Na, L.; Liu, G.; Zhen, C.; Li, F.; Zhang, L.; Cheng, H.-M. Battery Performance and Photocatalytic Activity of Mesoporous Anatase TiO$_2$ Nanospheres/Graphene Composites by Template-Free Self-Assembly. *Adv. Funct. Mater.* **2011**, *21*, 1717–1722.

20. Mutin, P. H.; Vioux, A. Nonhydrolytic Processing of Oxide-Based Materials: Simple Routes to Control Homogeneity, Morphology, and Nanostructure. *Chem. Mater.* **2009**, *21*(4), 582–596.

21. Debecker, D. P.; Hulea, V.; Mutin, P. H. Mesoporous Mixed Oxide Catalysts via Non-Hydrolytic Sol–Gel: A Review. *Appl. Catal. A.* **2013**, *451*, 192–206.

22. Vivero-Escoto, J. L.; Chiang, Y.-D.; Wu, K. C.-W.; Yamauchi, Y. Recent Progress in Mesoporous Titania Materials: Adjusting Morphology for Innovative Applications. *Sci. Technol. Adv. Mater.* **2012**, *13*, 013003.

23. Zhou, W.; Li, W.; Wang, J.-Q.; Qu, Y.; Yang, Y.; Xie, Y.; Zhang, K.; Wang, L.; Fu, H.; Zhao, D. Ordered Mesoporous Black TiO_2 as Highly Efficient Hydrogen Evolution Photocatalyst. *Am. Chem. Soc.* **2014**, *136*, 9280–9283.

24. Linares, N.; Silvestre-Albero, A. M.; Serrano, E.; Silvestre-Albero, J.; Garcia-Martinez, J. Mesoporous Materials for Clean Energy Technologies. *Royal Soc. Chem.* **2014**, *43*, 7681–7717.

25. Debecker, D. P.; Bertrand, P.; Bouchmella, K.; Gaigneaux, E. M.; Delaigle, R.; Mutin, P. H.; Eloy, P.; Poleunis, C. One-Step Non-Hydrolytic Sol–Gel Preparation of Efficient V_2O 5-TiO_2 Catalysts for VOC Total Oxidation. *Appl. Catal. B: Environ.* **2010**, *94*, 38–45.

26. Mutin, P. H.; Popa, A. F.; Vioux, A.; Delahay, G.; Coq, B. Nonhydrolytic Vanadia-Titania Xerogels: Synthesis, Characterization, and Behavior in the Selective Catalytic Reduction of NO by NH_3B. *Appl. Catal. B* **2006**, *69*, 49–57.

27. Mote, V. D.; Purushothamand, Y.; Dole, B. N. Williamson-Hall Analysis in Estimation of Lattice Strain in Nanometer-Sized ZnO Particles. *J. Theor. Appl. Phys.* 2012, *6*(6), 1–8.

28. Choudhury, B.; Choudhury, A. Ce3+ and Oxygen Vacancy Mediated Tuning of Structural and Optical Properties of CeO_2 Nanoparticles. *Mater. Chem. Phys.* **2012**, *131*, 666–671.

29. Spurr, R. A.; Myers, H. Quantitative Analysis of Anatase-Rutile Mixtures with an X-Ray Diffractometer. *Anal. Chem.* **1957**, *29*(5), 760–762.

30. Tian, B.; Li, C.; Gu, F.; Jiang, H.; Hu, Y.; Zhang, J. Flame Sprayed V-Doped TiO_2 Nanoparticles with Enhanced Photocatalytic Activity Under Visible Light Irradiation. *Chem. Eng. J.* **2009**, *151*, 220–227.

31. Nair, R. G.; Tripathi, A. M.; Samdarshi, S. K. Impact of Ti–V Ratio on the Crystalline Phase/Visible Light Activity of TiV-Oxide Photocatalyst. *Environ. Prog. Sustainable Energy* **2012**, *31*, 107–113.

32. Nair, R. G.; Roy, J. K.; Samdarshi, S. K.; Mukherjee, A. K. Mixed Phase V Doped Titania Shows High Photoactivity for Disinfection of *Escherichia coli* and Detoxification of Phenol. *Sol. Energy Mater. Sol. Cells* **2012**, *105*, 103–108.

33. Peng, T.; Zhao, D.; Dai, K.; Shi, W.; Hirao, K. Synthesis of Titanium Dioxide Nanoparticles with Mesoporous Anatase Wall and High Photocatalytic Activity. *J. Phys. Chem.* **2005**, *109*, 4947–4952.

34. Rahmani, N.; Dariani, R. S. Strain-Related Phenomena in TiO_2 Nanostructures Spin-Coated on Porous Silicon Substrate. *Superlattices Microstruct.* **2015**, *85*, 504–509.

35. Singh, M. K.; Mathpal, M. C. Study of Structural Transformation in TiO_2 Nanoparticles and its Optical Properties. *J. Alloys Compd.* **2013**, *549*(5), 114–120.

36. Zhou, W.; Liu, Y.; Yang, Y.; Wu, P. Band Gap Engineering of SnO_2 by Epitaxial Strain: Experimental and Theoretical Investigations. *J. Phys. Chem. C.* **2014**, *118*, 6448–6453.

37. Samet, L.; Ben Nasseur, J.; Chtourou, R.; March, K.; Stephan, O. Heat Treatment Effect on the Physical Properties of Cobalt Doped TiO_2 Sol–Gel Materials. *Mater. Charact.* **2013**, *85*, 1–12.

38. Yang, X.; Fengyan, M.; Li, K.; Guo, Y.; Hu, J.; Li, W.; Huo, M.; Guo, Y. Mixed Phase Titania Nanocomposite Co doped with Metallic Silver and Vanadium Oxide: New Efficient Photocatalyst for Dye Degradation. *J. Hazard. Mater.* **2010**, *175*, 429–438.

39. Sing, K. The Use of Nitrogen Adsorption for the Characterisation of Porous Materials. *Colloids Surf.* **2001**, *187–188*, 3–9.

40. Raj, K. J. A.; Viswanathan, B. Effect of Surface Area, Pore Volume and Particle Size of P25 Titania on the Phase Transformation of Anatase to Rutile. *Indian J. Chem.* **2009**, *48A*, 1378–1382.

41. Serpone, N. Is the Band Gap of Pristine TiO_2 Narrowed by Anion- and Cation-Doping of Titanium Dioxide in Second-Generation Photocatalysts? *J. Phys. Chem. B* **2006**, *110*, 24287–24293.

42. Chen, K.; Li, J.; Wang, W.; Zhang, Y.; Wang, X.; Su, H. The Preparation of Vanadium-Doped TiO_2—Montmorillonite Nanocomposites and the Photodegradation Under Visible Light Irradiation of Sulforhodamine. *Appl. Surf. Sci.* **2011**, *257*(9), 7276–7285.

43. Valencia, S.; Marín, J. M.; Restrepo, G. Study of the Bandgap of Synthesized Titanium Dioxide Nanoparticles Using the Sol–Gel Method and a Hydrothermal Treatment. *Open Mater. Sci. J.* **2010**, *4*, 9–14.

44. Welte, A.; Waldauf, C.; Brabec, C.; Wellmann, P. J. Application of Optical Absorbance for the Investigation of Electronic and Structural Properties of Sol–Gel Processed TiO_2. *Thin Solid Films* **2008**, *516*, 7256–7259.

45. Wu, Y.; Liu, H.; Zhang, J.; Chen, F. Enhanced Photocatalytic Activity of Nitrogen-Doped Titania by Deposited with Gold. *J. Phys.* **2009**, *113*, 14689–14695.

46. Liang, Y. T.; Vijayan, B. K.; Lyandres, O.; Gray, K. A.; Hersam, M. C. Effect of Dimensionality on the Photocatalytic Behavior of Carbon–Titania Nanosheet Composites: Charge Transfer at Nanomaterial Interfaces. *J. Phys. Chem. Lett.* **2012**, *3*, 1760−1765.

47. Li, X. Z.; Li, F. B.; Yang, C. L.; Ge, W. K. Photocatalytic Activity of WO_x–TiO_2 Under Visible Light Irradiation. *J. Photochem. Photobiol. A* **2001**, *141*, 209–217.

48. Sun, Z.; Kim, D. H.; Wolkenhauer, M.; Bumbu, G. G.; Knoll, W.; Gutmann, J. S. Synthesis and Photoluminescence of Titania Nanoparticle Arrays Templated by Block-Copolymer Thin Films. *Chem. Phys. Chem.* **2006**, *7*, 370–378.

49. Chen, Y.; Cao, X.; Lin, B.; Gao, B. Origin of the Visible-Light Photoactivity of NH_3-Treated TiO_2: Effect of Nitrogen Doping and Oxygen Vacancies. *Appl. Surf. Sci.* **2013**, *264*, 845–852.

PART IV
Special Topics

CHAPTER 16

RESEARCH OF HYDRODYNAMIC PARAMETERS OF A TURBULENT FLOW IN THE INERTIA APPARATUSES WITH ACTIVE HYDRODYNAMICS

R. R. USMANOVA[1] and G. E. ZAIKOV[1,2,*]

[1]*Ufa State Technical University of Aviation, Ufa 450000, Bashkortostan, Russia*

[2]*Institute of Biochemical Physics, Russian Academy of Sciences, Moscow 119991, Russia*

[*]*Corresponding author. E-mail: chembio@chph.ras.ru*

CONTENTS

ABSTRACT

In this chapter, new research and developments on application of hydrodynamic parameters of a turbulent flow in the inertia apparatuses with active hydrodynamics are discussed in detail.

16.1 INTRODUCTION

Working out of new mathematical model approaches to calculation of the turbulent twirled currents is the important step to creation of adequate methods of calculation of the inertia apparatuses for the purpose of their optimization technological and design data and exclusion of expensive experimental researches.[1] Now, there were considerable changes in the areas of mathematical modeling connected with application of computing production engineering and software packages that gives the chance to predict integrated characteristics of apparatuses already on a design stage, it is possible to provide such constructive solutions of separate knots of the apparatus which will allow to increase efficiency of a gas cleaning considerably.

Mathematical models of a current of multiphase medium should predict as much as possible precisely, on the one hand, gas cleaning parameters at inoculation of any parameter, and on the other to show possible ways of an intensification of process of separation. For this purpose, the model should provide characteristics of all prominent aspects of a current (boundary conditions, physical parameters of multiphase medium, turbulence, and geometrical characteristics) with possibility of the solution of such equations.

Modeling of a current of a dispersoid in the inertia apparatuses becomes complicated stochastic character traffic of corpuscles in the turbulent twirled stream which becomes complicated interacting of corpuscles with each other and with apparatus walls, complexity of the task of entrance conditions, modification of corpuscles as a result of crushing and concretion.

Calculation of the turbulent twirled currents at creation of adequate methods of calculation of the inertia apparatuses should yield exact enough results in a wide range of variables and combine with simple and inexpensive laboratory researches of characteristics of the dust which results can be used in the capacity of an input information.

16.2 SAMPLING OF PARAMETERS FOR THE DESCRIPTION OF A MULTIPHASE STREAM

The analytical approach of the majority of researchers[2] to the description of hydrodynamic characteristics of the inertia apparatuses is based on system of the equations of the Nave-Stokes added with continuity equations of the installed axisymmetric twirled gas stream.

$$\frac{\partial}{\partial t}(\rho v_i) + \frac{\partial}{\partial q_j}(\rho v_i v_j) = -\frac{\partial}{\partial q_i} + \frac{\partial}{\partial q_j}\left[\mu\left(\frac{\partial v_i}{\partial q} + \frac{\partial v_j}{\partial q_j}\right)\right]\frac{\partial \rho}{\partial t} + \frac{\partial}{\partial q_j}(\rho v_i) = 0 \quad (16.1)$$

where v_i is a component of speed on an axis; P, ρ, μ is the pressure, density, and turbulent viscosity of a stream; t is the a time; and q is the a direction of a coordinate axis.

The necessity of adoption of variety not absolutely correct assumptions that reduces adequacy of offered analytical descriptions to a real flow pattern in the inertia apparatuses and, finally, leads to essential divergences of results of scalings with empirical data.

The great interest represents research of effective numerical methods of the solution of the multidimensional equations of hyperbolic type or the parabolic equations containing a hyperbolic part. Such mathematical models present many nonstationary problems of mechanics of multiphase currents. Construction of a computational algorithm for the specified sort of problems represents a challenge which usually dares stage by stage. However, now principles of rational numerical modeling allow to promote essentially in the field of construction of the systems simulating such phenomena that gives a basis for progress in creation of simulars and calculation of currents of multiphase medium.

16.3 DEFINITION OF BOUNDARY CONDITIONS

At the task of boundary conditions statements of a problem on an exit and an entry in the setting of the counted volume are specified. In the capacity of boundary conditions, it was set: a sticking condition on walls (speed and a temperature gradient on walls are equal to null); distribution of dispersion particles in entrance cross-section was accepted the uniform; distribution of all a component of speed in entrance cross-section was set; on an entrance surface the total charge on weight was set, and on a target surface the condition on pressure[3,4] was laid down. On target boundary line, the most reliable way of the task of boundary conditions is full definiteness of values ψ, ω, v and statement of "soft" boundary conditions of Neumann is applied:

$$\frac{d\psi}{dz} = 0; \quad \frac{d\omega}{dz} = 0; \quad \psi_{\bar{\lambda}} = \psi_{\bar{\lambda}-1}; \quad \omega_{\bar{\lambda}} \omega_{\bar{\lambda}-1} \qquad (16.2)$$

$$\frac{\partial \omega}{\partial z} = \frac{\partial v_r}{\partial z} = 0 \quad \frac{\partial^2 \psi}{\partial z^2} = 0$$

These conditions have the second order of accuracy.

The numerical solution of model was carried out for one midflight pass from entrance cross-section of a swept volume to the day off by means of integration in a neighborhood of each knot of an is final-difference grid by which all space of a gas bottle (Fig. 16.1), unknown values of speed and pressure are found in knots of this grid.

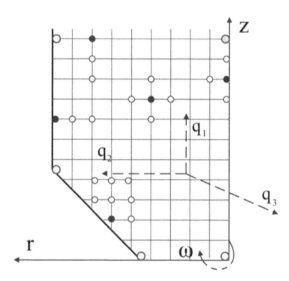

FIGURE 16.1   The settlement grid.

Each knot of a grid is defined by values of projections of speed of a stream: the radial v_r, tangential v_φ, and axial v_z. Transitions between knots are carried out in steps by replacement of one value of speed with another or a finding of intermediate values between knots by means of interpolation. At such statement of a regional problem, the sticking condition was realized on each time step, is analogous to conditions for functions ψ and ω, and was put on different boundary lines. This results from the fact that use of conditions of sticking on the same boundary line changes a modeling problem, and at its numerical solution, there can be an accuracy decrease.

On boundary lines of settlement area for each knot of a grid, it is possible to write down:

$$\frac{d}{dr}\left(r\frac{d\varphi}{dr}\right)+\frac{d}{dr}\left(r\frac{d\varphi}{dz}\right)=0 \qquad (16.3)$$

For the function addressing in a zero on boundary line of a grid, we will compute a scalar product and norms:

$$(y,v)=\sum_{i,j=1}^{N}h_r h_z y_{i,j}\cdot v_{i,j}, \quad \|y\|=\sqrt{(y,y)}$$

Let's execute approximation of eq 16.3 on a grid step h, having made replacement of derivatives with the following function:

$$\frac{i+(1/2)}{h_r}\cdot\phi_{i+1_j}+\frac{i\cdot h_r}{h_z^2}\cdot\phi_{i,j+1}-\left(\frac{2\cdot i}{h_r}+\frac{2\cdot i\cdot h_r}{h_z^2}\right)\cdot\phi_{i,j}+\frac{i-(1/2)}{h_r}\cdot\phi_{-1_j}+\frac{i\cdot h_r}{h_z^2}\cdot\phi_{i,j-1}=0$$

We will inject a designation:

$$(Ay)_{ij}=-\frac{i+(1/2)}{h_r}y_{i+1,j}-\frac{i\cdot h_r}{h_z^2}y_{i,j+1}+\left(\frac{2i}{h_r}+\frac{2i\cdot h_r}{h_z^2}\right)\cdot y_{i,j}-\frac{i-(1/2)}{h_r}y_{i-1j}-\frac{i\cdot h_r}{h_z^2}y_{i,j-1}$$

Then, eq 16.3 will register as

$$Ay=f;(Ay,y)\geq0;(Ay,v)=(y,Av)$$

Over the range sizes $0\leq r_{min}\leq r\leq r_{max}$ of parameter will be in limits $\gamma_1\leq A\leq\gamma_2$, forming system of the linear equations for each knot of a grid

$$\gamma_1=2\cdot r_{min}\left(\frac{4}{h_z^2}\sin^2\frac{\pi\cdot h_r}{2l_r}+\frac{4}{h_z^2}\sin^2\frac{\pi\cdot h_r}{2l_z}\right),$$

$$\gamma_2=2\cdot r_{max}\left(\frac{4}{h_z^2}\cos^2\frac{\pi\cdot h_r}{2l_r}+\frac{4}{h_z^2}\cos^2\frac{\pi\cdot h_z}{2l_z}\right)$$

(16.4)

At $r_{min}=h_r, r_{max}=l_r$, where l is the length of settlement area, and h is a grid step. The system of the linear is final-difference equations which solution allows to define value of potential in grid knots is gained.[5] By results of scalings, pictures of stream-lines and profiles of speed in various cross-sections of a stream were under construction.

16.4   VISUALIZATION OF RESULTS OF CALCULATION

By results of scalings, pictures of stream-lines and profiles of speed in various cross-sections of a stream were under construction. The analysis of hydrodynamics and distribution of precipitated corpuscles in a dynamic gas bottle has shown that because of presence of a turbulent diffusion, a corpuscle concentrate at an apparatus wall not a dense bed and in the form of the loosened concentrated gas layer. At dust passage through shovels, there is a concentration of corpuscles on a peripheral zone of shovels. We will note that the peripheral velocity profile v_φ essentially changes on radius of a gas bottle and on an axis x that means presence of differential twirl owing to what whirl lines begin to twist on a spiral, as shown in-process.[2] In Figure 16.2, peripheral velocity lines are presented, it is visible that near to walls whirlwind zones are formed.

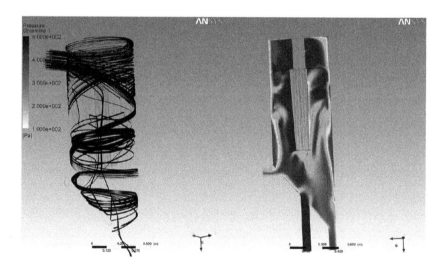

FIGURE 16.2   Projections of a peripheral velocity of a gas stream.

It is installed that at increase in a Reynolds number, the current structure changes from layered to complicated by the developed secondary whirl-winds. Three types of a current are qualitatively discriminated: a layered current, a current with an axial whirl, a current with axial, and peripheral whirlwinds. At considerable intensity of process, $Re = 6 \times 104$, the forming has big tangential speeds, it leads to the considerable pressure gradients calling a reverse flow along an axis, reducing efficiency of separation. It is installed that a condition necessary for origination of return currents, the twisting, and twisting falling is not. Character and intensity of return currents depends on intensity and character of falling of a twisting. This leading-out is necessary for considering in practice and in appropriate way to organize hydrodynamics of streams in the apparatus.

16.5 CONCLUSIONS

The algorithm of modeling of process of separation of a dispersoid in a gas stream has been developed. The carried out calculations allow to define potential possibilities of a dynamic gas bottle at its use in the capacity of the apparatus for clearing of gas emissions. Verification of the data gained by calculation was spent by modeling of process of a current of a gas stream in a package of computing hydrodynamics *ANSYS CFX*.

The calculations of currents defined by a regional problem were spent for values of a Reynolds number from 1×10^2 to 6×10^4. By results of scalings, pictures of stream-lines and profiles of speed in various cross-sections of a gas bottle were under construction. The analysis of the gained profiles of speed allows to reveal three characteristic areas on an apparatus axis: area of formation of a gas stream, area of a stable stream, and damping area.

The changing twisting of a stream can call emersion near to the walls of a gas bottle of zones of a return current reducing efficiency of separation of fine fractions of a dust.

KEYWORDS

- vortical zone
- solid wall
- velocity profiles
- ANSYS CFX
- boundary conditions
- stream function

REFERENCES

1. Varkasin, A. J. Turbulent Flows of Gas with Firm Corpuscles (in Russian). Moscow: Engineering, 2013.
2. Kochevskiy, A. N.; Nenja V. G. Modem the Approach to Modelling and Calculation of Currents of a Liquid in Bladed Hydromachines (in Russian). *Vestnik SumGU* **2003**, *13*(2), 195–210.
3. Usmanova, R. R.; Zaikov G. E. Choice of Boundary Conditions to Calculating Parameters Movement of Gas-dispersion Streams (in Russian). *Enciklopedia ingenera-chimika.* **2015**, *3*(37), 36–42.
4. Usmanova, R. R.; Zaikov, V. S. Simulation and Research of Factors Affecting Aerodynamic Indices of the Gas Purification Process (in Russian). *Vestnik SGAU* **2014**, *1*(43), 173–180.
5. Usmanova, R. R.; Zaikov G. E. The Modern Approach to Modeling and Calculation of Efficiency of Process of a Gas Cleaning (in Canada). *Chem. Eng. Chemoinf.* **2016**, *3*, 36–42.

CHAPTER 17

PROBING TERPENOIDS: TMAP AND TMPMP FROM *Acorus Calamus* FOR ITS POSSIBLE ANTIFUNGAL MECHANISM

SHARANYA M.[*]

Assistant Professor, Department of Bioinformatics,
Vels Institute of Science, Technology & Advanced Studies (VISTAS),
Chennai 621117, India

[*]*Corresponding author. E-mail: sharanya.bioinfo@gmail.com*

CONTENTS

ABSTRACT

Fungal infections are the cause of high mortality and morbidity rate especially among the immunocompromised patients. Few classes of antifungal agents exists, perhaps, its concurrency for number of side effects persists high. Though Food and Drug Administration in 2015 approves a new azole drug, cresemba, the goal in achieving a reliable medical therapy is indeed the real destiny in the field. The current pace is probing for an ideal one from the medicinal plants which highly subsists among the researchers. The terpenoid molecules 1-2,4,5-tri methoxy phenyl-1'-methoxy propionaldehyde (TMPMP) and 1-2,4,5-tri methoxy acetophenone (TMAP) isolated from the plant *Acorus calamus*, known as Vasambu in Tamil, were reported for its significant effect against *Candida albicans*, *Candida tropicalis*, *Trichophyton rubrum*, *Trichophyton mentagrophytes*, *Trichophyton tonsurans*, and *Trichophyton ajelloi*. The present study is designed in an intention to ascertain its probable means of mechanism through in silico techniques. Fungi being eukaryote shares most of its biochemical and cell biological process with human, the fungal cell wall targets alone considered prominently unique to cause lethality. Thenceforward, the favorable binding efficiency of TMPMP and TMAP with the enzymes chitinase A, exo-β-1,3-glucanase, chitin synthase 2, and glucan synthase were analyzed using molecular docking studies.

## 17.1	INTRODUCTION

### 17.1.1	BIOINFORMATICS IN DRUG DESIGNING

Drug discovery includes high-throughput screening and preclinical studies to evaluate the potentiality of healing as well as drug-like property and the final phase includes the clinical development.[7] Drug discovery is a time-consuming and also expensive process which approximately requires 14 years to launch a new drug into the market.[57] Despite the time involved, only 10% of the drugs would successfully pass through the latter phases.[99] Drug discovery has achieved tremendous modifications when compared to the late 20th and early 21st centuries.

In silico strategies like ADME-Tox property analysis, lead designing, and optimization would greatly assist the researcher to predict whether a compound is likely to be successful and enable the identification of molecules as early as possible.[5] Basically, drug discovery is of two types ligand based and target based where the former depends on the knowledge of molecules that has the ability to bind the biological target of interest and the latter begins with the thorough understanding of the disease mechanisms and the role of enzymes, receptors or protein which are involved in the disease pathology. Target-based approach is a preferred method to identify the drug molecule specific for the disease as it could provide better understanding on the mechanisms of disease, structural and active site details of the targets. Both the approaches involve the screening of small molecules and the two main sources of these molecules are commercial libraries and in-house libraries. ChemBank, PubChem, and ChemDB databases store thousands of existing drugs, constituents from natural products and the bioactive compounds of known or unknown action, whereas the in-house small molecule libraries are generated based on the objectives of diversity (compounds that are diverse in structure) or target-oriented (analogs of specific structures called scaffold).[13] The present day modern technology focuses on the quantity and quality of these small molecules in the process of their translation from hit-to-lead and lead-to-drug.[63]

In addition, bioinformatics softwares and tools contribute a key role in drug designing by providing a clear perceptive knowledge on the interaction of small molecules with the molecular targets. A novel triazole, 1-(1*H*-1,2,4-triazol-1-yl)-2-(2,4-di fluorophenyl)-3-[(4 substituted trifluoromethyl phenyl)-piperazin-1-yl]-propan-2-ols targeting CYP51, an enzyme that catalyzes the oxidative removal of 14α-methyl group of (C-32) lanosterol in the biosynthesis of ergosterol, was synthesized and investigated for its antifungal potency by both in vitro and molecular docking study.[20] This was achieved based on the available knowledge about the structure of the enzyme and functional regions such as the heme group, the hydrophilic H-bonding region, the narrow hydrophobic cleft, substrate access channel, and the active site residues.[21] The study assisted in developing a lead molecule in which the piperazinyl side chain interacts with the difluorophenyl group is located in the hydrophobic binding cleft lined with Phe126, Ile304, Met306, Gly307, and Gly308 residues.[19]

A novel compound, 5-(2,4-dimethylbenzyl)pyrrolidin-2-one, extracted from *Streptomyces* VITSVK5 spp. showed strong binding with 14, α-sterol demethylase (cyp51) where the binding energy was −6.66 kcal/mol.[78] A synthesized 3-azolyl-4-chromanone phenylhydrazone was evaluated for its biological activity against the pathogenic fungus *Candida albicans*, *Saccharomyces cerevisiae*, *Aspergillus niger*, and *Microsporum gypseum*. In silico analysis like toxicity risks, drug-likeness was predicted and in vitro antifungal analysis revealed that the compounds to possess activity comparable to the standard drug fluconazole.[10] Screening a series of triazolopyrimidine-sulfonamide compounds revealed the broad-spectrum of antifungal activity.[74]

The chemogenomic profiling suggested that these compounds possess high similarity with the mechanism of acetolactate synthase (enzyme that catalyzes the first common step in branched-chain amino acid biosynthesis) inhibitors. Richie et al.[74] cocrystallized chlorimuron-ethyl with acetolactate synthase and from the 3D structure of the complex, ligand was separated and docked in silico which bond to the same binding pocket of acetolactate synthase which was observed in the crystallized complex.

Synthesis and screening of azole derivatives is increased, in order to improve the efficacy and overcome the drug resistance of the organisms. Benzimidazole, benzotriazole, aminothiazole derivatives, novel derivatives of phenyl(2H-tetrazol-5-yl) methanamine, and pyrazino[2,1-a]-isoquinolin exhibited ergosterol biosynthesis inhibition.[47,77,94,103] Apart from azoles, a series of N-1, C-3, and C-5 substituted bis-indoles showed strong interaction in the active sites of lanosterol demethylase, dihydrofolate reductase, and topoisomerase II enzymes.[85] Investigation of small molecules from natural sources, synthesized compounds, and their derivatives using in silico methods, quietly supports the screening of antifungal potency as well as in predicting the drug-like properties. Certainly, applications of bioinformatics immensely reduces the time and cost investment in drug designing.

The drug-like properties of a molecule are said to have functional groups and/or physical properties in consistent with the majority of known drugs. They might also show therapeutic potentiality like the drugs.[98] According to Lipinski et al.,[52] the drug-like compounds have sufficiently acceptable ADME/T (absorption, distribution, metabolism, excretion, and toxicity) properties to survive through the Phase I clinical trials. Likewise, the properties of synthetic ease, stability, oral availability, good pharmacokinetic properties and lack of toxicity are equally important.[45] It is an attractive approach of evaluation since it helps in reducing the time consumption by sorting the candidate molecule at the early stage of drug development. Additionally, this could guide chemists to structure–activity relationship-based modifications to optimize "drug-like properties."[44] Several methods have been used to predict the drug-like properties, which include simple counting, knowledge-based, functional-group filters, and chemistry space methods.[45] Lipinski's rule,[51] which comes under the simple counting method states that, in general, a drug has no more than one violation of the following criteria:

- Not more than five hydrogen bond donors (the total number of nitrogen–hydrogen and oxygen–hydrogen bonds)
- Not more than 10 hydrogen bond acceptors (all nitrogen or oxygen atoms)

- A molecular mass less than 500 Da
- An octanol–water partition coefficient log P not greater than 5.

Probing the natural products is essential, since the available antifungals are restricted to target only limited mechanisms of the fungal cell. The treatment using medicinal plants has historical basis of therapeutic health care. Therefore, identifying new and effective drugs from plant source would be economically accessible and affordable in developing countries.[9] Terpenes, terpenoids, saponins, alkaloids, phenolic compounds, flavonoids, coumarins, xanthones, tannins, lignans, and other secondary metabolites of plants are reported for promising antifungal activity.[61] An oxygenated xanthone, 1,2-dihydroxyxanthone showed activity on clinical strains of *Candida*, *Cryptococcus*, *Aspergillus*, and *Trichophyton Mentagrophytes*, and the effect of xanthone on sterol biosynthesis was validated by a simple and efficient High performance liquid chromatography (HPLC) method along with UV detection.[70] Similarly, terpenes including methyl chavicol and linalool of *Ocimum sanctum* affected the synthesis of ergosterol and caused cell membrane damage in *Candida* species.[8] Furthermore, the fungicidal effect of carvacrol and thymol are originated from the inhibition of ergosterol biosynthesis which correspondingly disrupts the membrane integrity.[4]

Ajoene [(*E,Z*)-4,5,9-trithiadodeca-1,6,11-triene-9-oxide)] derived from allicin of garlic juice had a strong antifungal activity against the itraconazole resistant *Fusarium* spp.[61] Phytolaccosides B, E, and F, mono desmosidic triterpenoid saponins obtained from the butanolic extract of *Phytolacca tetramera* (berries) showed the broadest spectrum of antifungal action. Specifically, phytolaccoside B was active against *T. mentagrophytes*.[33] Steroid saponins isolated from *Tribulus terrestris* were effective against *Candida* sp. and *C. neoformans*. The inhibitory effect of steroid saponins on hyphal formation and cell membrane destruction were observed using the phase contrast microscopy.[104] Styraxjaponoside C, a glycoside derivative of lignans from the stem bark of *Styrax japonica*, inhibited the mycelial growth of *C. albicans* with low cytotoxicity to human erythrocytes.[65] In this regard, plant sources are considered to be an inexhaustible resource by being a solution for a number of health hazards.

17.1.2 NEED FOR EFFECTIVE ANTIFUNGAL AGENTS

The prevailing environment adds more number of factors for acquiring fungal infections. The population undergoing invasive surgery, immunosuppressive therapy during organ transplantation, treatment with broad-spectrum antibiotics and glucocorticoids, receipt of peritoneal dialysis or hemodialysis and patients infected with immunodeficiency disorders, AIDS, and cancer are critically susceptible to fungal infections.[55] In such cases, treatment to eliminate the fungal cells from the host becomes a challenge. Antifungal drug development has to focus on rapid reduction in pathogen number, reduced drug toxicity for the host and to show low relapse rates after therapeutic course is completed. Therefore, the prime need is to develop a novel and unique antifungal agent that comprises all the facts mentioned previously.

The antimycotic drugs are extremely limited. The discovery of penicillin in 1928 by Alexander Fleming is the driving force for both pharmaceutical and academic researchers to seek an equivalent agent to combat fungal infections.[55] The drugs for fungal infections started emerging in the 1950s. The mechanisms of action exerted by the available antifungal agents on the infectious fungi include:

1. alteration of membrane function,
2. inhibition of DNA or RNA synthesis,
3. inhibition of ergosterol biosynthesis, and
4. inhibition of glucan synthesis.[68]

Even though the antifungal agents act on the pathogens in their unique mode of action, interrelationship between the host and pathogen, both being eukaryotes turn out to be rationale in the development of antifungals a difficult task. The major difference between fungi and mammals is the presence of ergosterol as the primary sterol in the cell membrane in fungi instead of cholesterol which forms a part in the composition of mammalian membrane. Hence, ergosterol and its biosynthetic pathway serve as unique targets. Amphotericin B (AmB) is one of the efficacious fungicidal agents that binds to ergosterol in the membrane, subsequently forms pores in channels which leads to the loss of transmembrane potential, cell lysis, and impaired cellular function.[102] AmB remained as the "gold standard" for treating fungal infections over 40 years.[32] Later the treatment with AmB becomes limited which may be due to the following liabilities:

1. Pharmacokinetics and distribution are poor which allows some of the fungi to hide in niches,
2. AmB induces idiosyncratic systemic reactions involving fever and tremors, and
3. Frequency of toxic reactions like hypokalemia and nephrotoxicity due to the decreased filtration.[53,67,83]

Fungal resistance is one of the major emerging issues in the world. However, the evolution of antimicrobial drug resistance is inevitable and ubiquitous and also the frequency of drug resistance has been rising day by day.[100] In general, drug resistance is said to be the failure of drugs to eliminate the particular infection, where the resistance can be categorized as primary or intrinsic, acquired, and clinical. Intrinsic or primary resistance occurs without exposure to antifungal drugs, whereas acquired or secondary resistance develops during treatment. Clinical resistance is multifactorial which includes immune status of the host, pharmacokinetics of the antifungal agent and the species of infecting fungus.[69] In other words, resistance is a complex phenomenon involving multiple mechanisms such as increased efflux of the drug, phenotypic alteration in the drug target site, genomic recombinations that minimize toxic effect of the drug and biofilm formation.[86] However, the resistance mechanism differs for each class of antifungals. In the case of drug resistance, AmB treatment is the only option left to deal with the resistant developed organisms.[97] In the present scenario, overcoming the antifungal resistance is the foremost requirement. The development and identification of novel antimycotic drugs are of higher priority, because of the limited treatment options and the resistant mechanisms exhibited by the pathogen to the available drugs.

Several molecules that have been approved by the Food and Drug Administration (FDA) with significant antifungal activity had been rapidly identified using high throughput screening methodologies.[93] However, these techniques are time consuming and expensive. In silico methodologies contribute efficiently to screen millions of chemical compounds in libraries at reduced cost.[14] In addition, the accumulation of genomic data has created more possibilities for drug development.[12] The identification of the molecular targets of antifungal drugs and elucidating their mechanisms of action in the fungus is an important strategy for the development of new pharmacological probes.[2] Hence, the advancement in the field of structural biology, medicinal chemistry and in silico technologies have allowed the development of promising research and proposed antifungals based on the targets, to increase the specificity and to reduce side effects.[29]

17.1.3 ANTIFUNGAL TARGETS

To develop a potent antifungal agent, the target should be unique and specific, leading to the death of the pathogenic organism without affecting the host.[56] The following are considered to be the basic and fundamental categories of targets: (1) alteration of membrane function, (2) inhibition of DNA or RNA synthesis, (3) inhibition of ergosterol biosynthesis, and (4) inhibition of glucan synthesis. Apart from the four means of mechanism, Kathiravan et al.[46] suggested eight different targets, namely, ergosterol synthesis, chitin synthesis, ergosterol disruptors, glucan synthesis, squalene epoxidase (SE), nucleic acid synthesis, protein synthesis, and microtubules synthesis, where all these come under the four main categories.

The present study focuses on targets involved in the mechanisms of membrane function, ergosterol synthesis, and inhibition of glucan synthesis.

17.1.3.1 FUNGAL CELL STRUCTURE AND TARGETS

Fungal cell wall is an attractive antifungal target because of its dynamic nature and its significance for cell viability and pathogenicity. The cell wall is a polysaccharide-based three-dimensional network which apart from playing an important role in both protective and aggressive functions, acts as an initial barrier in a hostile environment, therefore the fungi with weakened cell wall cannot survive. Further, its rigid structure keeps the insoluble substrates from penetrating into the cell.[49]. The cell wall retains adequate plasticity during the cell growth and division. The cell wall mediates the adhesion of cells to one another and the substratum, also serves as a signaling center to activate signal transduction pathways within the cell. The structure and biosynthesis of a fungal cell wall is unique to the fungi. Cell wall disruption has profound effect on the growth and morphology of the fungal cell and is therefore an excellent target for the development of antifungal drugs.[15]

Cell wall is a complex structure mainly composed of chitin, 1,3-β- and 1,6-β-glucan, mannan, and proteins with additional minor components which vary amongst fungal species. The glycoproteins are extensively modified with both N- and O-linked carbohydrates and, in many instances, contain a glycosylphosphatidylinositol anchor. Glucans have alternate linkages of 1,3-β- and 1,6-β-glucans, whereas chitin is linear polysaccharides synthesized as chains of β-1,4-linked N-acetylglucosamine residues and is typically less abundant than either of glycoprotein or glucan portions of the wall. The composition of the cell wall undergoes changes and may vary within a single fungal isolate depending upon the conditions and stages of growth.[15] Due to this nature, the knowledge on fungal cell wall is insufficient and their biosynthesis is not completely understood.[50]

17.1.3.2   *ERGOSTEROL SYNTHESIS*

Sterols in the eukaryotic cells are involved in structural organization and signaling functions and their fundamental contribution are fluidity, permeability, microdomain formation, protein functionality, and membrane activities.[30]. The three predominant forms of sterols are cholesterol in vertebrates, phytosterols in plants, and ergosterol in fungi. All the three kingdoms have a common biosynthetic pathway up to SE. Ergosterol is obtained from the consecutive five steps after SE of ergosterol biosynthesis pathway (EBP). Allylamines and azoles are the drugs specifically targets SE and lanosterol-14-α-demethylase of EBP, respectively. SE is a key flavin adenine dinucleotide-dependent enzyme that catalyzes the stereospecific epoxidation of squalene to lanosterol. Since the crystal structure of SE is not available, the domains responsible for enzymatic activity are not well understood.[62] *ERG*1 which encodes SE, on disruption have deleterious effects in yeast cells.[95] Conditional *ERG*1 mutant reduced the drug mediated efflux by ABC transporter, increases its susceptibility to drugs and forms defected hyphae formation.[66]

Azoles are the first class of antifungals that target the demethylation of sterol precursors, namely lanosterol at position 14. The enzyme sterol 14-α demethylase belongs to the superfamily mono-oxygenases called cytochrome P450.[59]

The fungi-specific cytochrome P450 enzyme 14α-sterol demethylase catalyzes the oxidative removal of 14α-methyl group from the sterol precursors.[71] Targeting the ergosterol biosynthesis would affect the above-said functionalities and also lead to the accumulation of metabolic intermediates that are toxic to the fungal cells.[60] Apart from targeting the SE and lanosterol-14-α-demethylase, Rong-mei et al.[75] focused on inhibiting the sterol C-14 reductase with a 2-aminotetralin derivatives, called 10b, which had strongest binding efficiency to it. The residues Val59, Phe60, Tyr61, Trp62, Ala69, Tyr72, Gly73, Tyr77, Phe79, Phe80, Tyr87, Phe90, Leu95, Leu96, Arg98, Tyr126, Phe236, Pro238, Phe241, Gly310, His381, Ser382, Leu383, His389, Ile471, Phe506, Ser508, and Met509 were represented as actively participating in inhibiting the protein, lanosterol-14-α-demethylase.[59] The treatment with 2-amino-nonyl-6-methoxyl-tetralin muriate (10b) had ergosta-8,14,22-trienol accumulation along with ignosterol, Rong-mei et al.[75] deducted that 2-amino-nonyl-6-methoxyl-tetralin to target the sterol C-5 desaturase which catalyzes the conversion of ergosta-8,14,22-trienol to ergosta-5,8,14,22-tetraenol.

17.1.3.3   CELL WALL BIOSYNTHETIC PATHWAY

Enzymes involved in the cell wall biosynthesis are considered to be ideal drug targets as the fungal cell wall plays an essential role during fungal growth.[34] Chitin is a white, hard, inelastic polysaccharide that accounts for 7–15% of dry weight in the exoskeleton of fungi. Accumulation of chitin by organisms is modulated by chitin synthase (CS)-mediated biosynthesis and by chitinase-mediated hydrolytic degradation.[38] Both the β-1,3-glucanase and chitinase are the key enzymes for the fungal cell, sclerotial wall lysis, and degradation.[31] CS 1, 2, and 3 (CS1, CS2, CS3) are found to be involved in the biosynthesis of chitin, where the CS2 is an essential enzyme for primary septum formation and cell division, CS3 is responsible for chitin in the ring at bud emergence and in the lateral cell wall, also for the formation of glucan-chitin linkages.[40] The chitin degradation takes place in two steps; the first is the cleavage of chitin polymer into chitin oligosaccharides followed by the cleavage of *N*-acetylglucosamine and monosaccharides by chitobiases.[92]

Chitinases are glycosyl hydrolases and are divided into two main groups endochitinases and exochitinases, based on their activity. Endochitinases break the chitin molecules at internal sites, whereas the exochitinases breaks at the terminals.[37] In general, the chitinases are subdivided into two subfamilies, bacterial-type and plant-type family 18 chitinases. Fungi express both subfamilies, where the bacterial-type chitinases are involved in processing chitin as a carbohydrate source in both fungi and bacteria. However, the fungal plant-type chitinases are having function in remodeling cell wall and maintenance.[18,43] Degradation of chitin leads to the cell separation, morphogenesis, and sporulation of fungi.[6] This activity may be counter-balanced by the "repair" enzyme CS3.[16]

Natural inhibitors for chitinase are allosamidin, styloguanidines, Cl-4 (cyclo-L-Arg-D-Pro) which were isolated from *Bombyx mori*, *Stylotella aurantium*, *Pseudomonas* sp., *Aplysinella rhax*, respectively.[38] Though allosamidin had potential inhibition on plant-type and bacterial-type chitinases, it had poor drug-like properties (high molecular weight, glycosidic bonds and cLog*P* of −5.2). Hirose et al.[38] also evaluated argadin and argifin, cyclic pentapeptides, produced by soil microorganisms *Clonostachys* sp. and *Gliocladium* sp. for antifungal activity and found that those two compounds were potentially targeting chitinase. Rush et al.[76] investigated the efficient binding of dimethylguanylurea fragment obtained from argifin, with the fungal plant-type chitinase A of *A. fumigatus* (*Af*ChiA) and also described the residues Tyr23, Gln37, Phe60, Asn76, Ala124, Tyr125, Asp170, Asp172,

Glu174, Gln207, Gln230, and Trp312 are actively involved in the binding. The same research group compared the binding efficiency of allosamidin and acetazolamide, where acetazolamide being small drug had efficient binding than allosamidin.[80]

Noncellulosic β-glucans consist of a backbone of glucose residues linked by β-(1,3)-glycosidic bonds, often attached with side-chain glucose residues joined by β-(1,6) linkages. The frequency of the branching may differ from one species to another. Similar to chitinases, glucanases have a role during cell separation and exhibit transglycosylase activity and also found to be involved in extending and rearranging 1,3-β-glucan chains and crosslinking the polymers to other wall components.[3]. Exo-β-(1,3)-glucanase involved in the metabolism of cell wall glucan by catalyzing the hydrolytic removal of a glucose residue at the nonreducing end of β-(1,3) and (1,6)-glucan.[24] The investigation by Jiang et al. (1995) illustrated the importance of exo-glucanase in developing the sensitivity and resistance to the K1 killer toxin on disrupting and overexpression of *EXG*1 gene of *S. cerevisiae*. According to Jiang et al. (1995), overexpression of this gene also showed slight increase in the β-1,6-glucan component in the cell wall, and he proposed that exo-β-glucanase may have functional role in cell wall glucan metabolism.

1,3-β-Glucan synthases (GS) are glycosyltransferase enzyme, also known as callose synthase that catalyzes the formation of β-1,3-glucan polymer. It is a membrane enzyme activated by Guanosine-5'-triphosphate (GTP) and has been fractionated as soluble (GTP-binding) and membrane-bound (catalytic) components.[58] The studies using resistant mutants, gene disruption experiments, and biochemical characterization lead to design a model for GS, which was found to be encoded by two genes, *FKS*1 and *FKS*2 and functions as the catalytic subunit for which a small GTP-binding subunit, Rho1p is necessary for activity.[28] It is found that *FKS*1 expression is cell cycle regulated which is abundant during vegetative growth. Alike, *ERG*1 of exo-glucanase, the disruption of *FKS*1 gene rendered resistance to lipopeptides (semisynthetic echinocandin B) inhibition.[27,58] Targeting these enzymes would greatly affect the growth of fungal cells.

17.1.4 EFFECT OF ACORUS CALAMUS ON FUNGUS

Acorus calamus is commonly called sweet flag (Family: Araceae) and is anaromatic herb, with creeping rhizomes, sword-shaped leaves of spadix inflorescence, hasperennial semiaquatic habitat. In India, the rhizome of *A. calamus* has been utilized as Ayurvedic medication to cure diseases like fever, asthma, and bronchitis and even as a sedative.[11] He, who reviewed the plant *A. calamus*, has mentioned the important pharmacological activities such as antimicrobial activity, antioxidant and insecticidal activity of the plant.

In the year 2009a, Subha and Gnanamani[88] had identified the active fraction of methanolic extracts from *A. calamus* to exhibit the antidermatophytic effect against *Trichophyton rubrum*, *T. mentagrophytes*, *Trichophyton tonsurans*, and *Trichophyton ajelloi*, in specific, the reduction in secreted aspartyl proteases, keratinases, alkaline, and acid proteases. The active fractions also demonstrated its efficacy in controlling the biofilm development of *C. albicans* and *Candida tropicalis*, compared to the standard drugs ketoconazole and amphotericin B, the fractions are capable to perfuse through biofilm as well as lead to cell death, therefore would greatly reduce the recurrence of infections.[89] It was identified that the fractions had impact on reduction in ergosterol biosynthesis and preventing germ tube formation, thus causing the cell death of *C. albicans*.[87] The terpenoid components extracted from *A. calamus* are 1-2,4,5-tri methoxy phenyl-1'-methoxy propionaldehyde (TMPMP) and 1-2,4,5-tri methoxy acetophenone (TMAP) has effectively reduced the major virulence factors namely, acidic, alkaline and neutral proteases, secreted aspartyl protease, lipase, phospholipase, elastase, and keratinase activities.[90,91] Hence, in this study, the possible mechanism of both TMAP and TMPMP has been assessed using the in silico approach.

17.1.5 ROLE OF TERPENOIDS AS ANTIFUNGALS

Recently, the research focuses on reverting the usage of plant and plant-based products for treating many diseases, the scientific inventions also revealed the effective plant compounds responsible for different pharmacological activities. Terpenoids are the large class plant compound, which have major contribution as scent in eucalyptus, flavors in cinnamon, cloves, ginger, and also the reason for colors, yellow in sunflowers and red in tomatoes. The terpenoid molecules linalool, nerol, isopulegol, menthol, carvone, α-thujone, and farnesol were observed to exhibit biofilm-specific activity for *C. albicans* and has MIC at the concentration of <2 mg mL^{-1},[73] which indicated the promising activity of the terpenoids of plant origin. Moreover, several researchers has proven the effect of terpenoids on fluconazole sensitive *C. albicans* through in vitro and in vivo studies.[17,25,105] Fernandez et al.[35] have described that lindenanolides, a terpenoid from *Hyalis argentea* var. *latisquama* to possess strong antifungal activity against *Cryptococcus neofromans* and *C. albicans*. The sesquiterpenes and sesquiterpene lactones are a group of terpenoid compounds that have been reported to a greater extent for its antifungal efficiency, where few has been reviewed by Abad et al.[1] In consequence, an in silico study conducted by our research team, also revealed that sesquiterpenes to exhibit significant hydrogen bonding interactions with the antifungal targets (not published). Moreover, geraniol, a terpene compound, was found to exhibit synergistic in combination with fluconazole with greater inhibition of *Candida* strains growth than the fluconazole alone.[42] It is reported that carvacrol, the components of oregano and plant essential oils involved in upregulating the genes of alternate metabolic and energy pathways, stress response, autophagy and drug efflux, which indicating its similarity to the effects of rapamycin in the inhibition of TOR pathway.[72]

17.2 MATERIALS AND METHODS

The 3D structures of proteins chitinase (2XVP) and β-(1,3)-glucanase (1EQP) were retrieved from Protein Data Bank (PDB), and the sequences for CS2 (P14180) and β-(1,3)-glucan synthase (B8XH77) were retrieved from Uniprot database, since the latter proteins do not have 3D structure in PDB, the structures were modeled using online server, I-Tasser which is available at http://zhanglab.ccmb.med.umich.edu/I-TASSER. The modeled structures were validated in Structural Analysis and Verification Server (http://nihserver.mbi.ucla.edu/SAVES/). The binding sites for the modeled proteins were identified using the online server LIGSITE (http://projects.biotec.tu-dresden.de/pocket/). Further, the 3D structure of terpenoid molecules reported in the plant *A. calamus*, that is, TMPMP and TMAP were retrieved from the PubChem database of ID CID_104221941 and CID_104221942, respectively. The synthetic drugs acetozolamide, azaserine, benzofuran, 5-flurocytosine, ketoconazole, manumycin A, polyoxin B, and terbinafine were also retrieved from PubChem database (Table 17.1).

The docking study was conducted in the Glide module of Schrodinger software which comprises four main steps, protein preparation, ligand preparation, grid generation, and XP ligand docking. The mode of binding between the ligand and protein was visualized using XP visualize and the interactions were observed in PyMol software.

TABLE 17.1　The Structure of TMPMP and TMAP Compounds, Synthetic Drugs, and Inhibitor Molecules from PubChem Database.

2D structure of PubChem compounds

TMPMP (CID 104221941)

TMAP (CID 104221942)

Acetozolamide (CID 1986)

Ketoconazole (CID 456201)

Azaserine (CID 5284344)

Polyoxin B (CID 181352)

17.3 RESULTS AND DISCUSSION

17.3.1 TARGET PROTEINS

The 3D structures of the proteins were represented in Table 17.2. The 3D structure of GS and CHS2 proteins were modeled from their sequences of length 240 and 489, respectively, and since the BLAST hit against PDB structure database indicated that only 22% and 29% of identity, which is less than 30%, the structures were modeled using online server I-Tasser. The predicted structure models were selected based on the confidence scores (C.scores), which were −3.875 and −1.10 for GS and Chs2, respectively. Energy minimization for the predicted 3D structure was carried out using SwissPDB, which was −6283.78 and −7982.428 kcal/mol for GS and Chs2, respectively.

TABLE 17.2 3D Structure of Selected Antifungal Protein Targets.

S. no.	Protein name	ID	3D structure
1	Chitinase A	PDB ID 2XVP	
2	Exo-β-(1,3)-glucanase	PDB ID 1EQP	

TABLE 17.2   *(Continued)*

S. no.	Protein name	ID	3D structure
3	Chitin synthase 2 (Chs2)	UniProt ID P14180	
4	β-1,3-Glucan synthase (GS)	UniProt ID B8XH77	

The structure validation for both GS and Chs2 indicated the presence of 97.3% of amino acids within the allowed region of Ramachandran plot, therefore confirmed the good quality of the predicted structures (Fig. 17.1). The Ramachandran plot has been divided into most favorable region, additionally allowed, generously allowed and disallowed regions which are indicated in red, yellow, light yellow, and white color fields, respectively. Accordingly, the proteins modeled in the present study, that is, Chs2 had 77.9, 14.5, and 4.8% of amino acids in the three differently partitioned allowed regions of Ramachandran plot and only 2.7% of amino acids in the disallowed region. While in GS protein, 84.1, 10.1, and 3.9% of amino acids located in the most favorable, additionally allowed and generously allowed regions of the plot and 2.2% in the disallowed region. Therefore, it is observed that the total of 97.2 and 98.1% of amino acids respective for Chs2 and GS were located in the allowed region, indicating the good quality of the modeled protein structures.

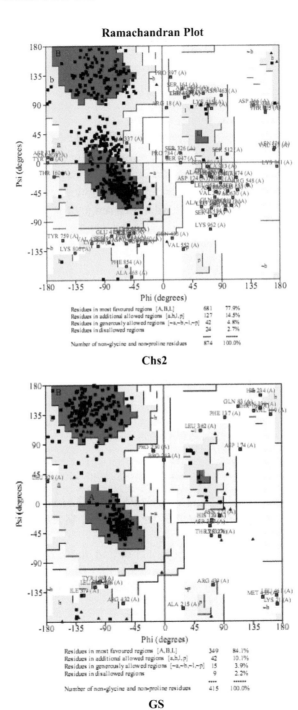

FIGURE 17.1   Ramachandran plot for modeled protein structures.

The active sites for the modeled proteins were predicted using the bioinformatics online tool, LIGSITE, which has been indicated in Table 17.3, along with the amino acid residues of chitinase A and exo-β (1,3)-glucanase. Determining the active site pocket for the modeled protein structures are considered important, since, there is no experimentally verified available, the online tools and software could be used for the prediction purpose. Schneider and Fechner[79] have mentioned methods like HSITE, HIPPO, GRID, LigBuilder, MCSS, serving the purpose. However, the present study utilized LigSite online tool for prediction. In general, these tools and software identify the hydrogen bond acceptor and donor regions which would be ideal center allowing the certain tolerance range of bond length and bong angle occur around the centers.

TABLE 17.3	The Active Site Residues for Selected Proteins.

S. no.	Protein name	Active site residues
1.	Chs2	Ile310, His311, Met314, Lys315, and Ala318
2.	GS	Tyr88, Leu89, Trp91, Phe92, Leu93, Ser118, Thr119, Met122, Trp129, Trp130, Ile133, Val134, Lly140, Ile141, Val142, Leu143, Gly144, Leu145, Met146, Tyr147, Leu156, Met160, Ile163, Ile164, and Trp185
3	Chitinase A	Gln34, Phe60, Asn76, Ala124, Tyr125, Asp170, Asp172, Glu174, Gln207, Gln230, Tyr232, and Trp312
4	Exo-β (1,3)-glucanase	Glu27, Arg92, His135, Phe144, Asn191, Glu192, His253, Tyr255, Phe258, Glu292, Trp363, Ser364, Trp373, and Pro387

17.3.2 INTERACTIONS OF TMPMP AND TMAP WITH ANTIFUNGAL TARGETS INVOLVE IN CELL WALL SYNTHESIS

Focusing the fungal cell wall components and the enzymes involved in synthesis as antifungal targets is an outstanding idea to be considered as novel therapy, since the host cell lacks these specific components. In this study, the proteins chitinase A, exo-β-1,3-glucanase, CS 2, and β-1,3-glucan synthase were selected as antifungal targets, to which the binding affinity of the terpenoids TMPMP and TMAP was analyzed and compared with the synthetic drugs (Table 17.4 and Fig. 17.2). The enzymes chitinase and β-1,3-glucanase have important role in fungal cell.[31] The structure of chitinase A1 has been explained by Rush et al.[76], who crystallized the structure with and without the inhibitor molecule. He explained the structure has $(\beta/\alpha)_8$

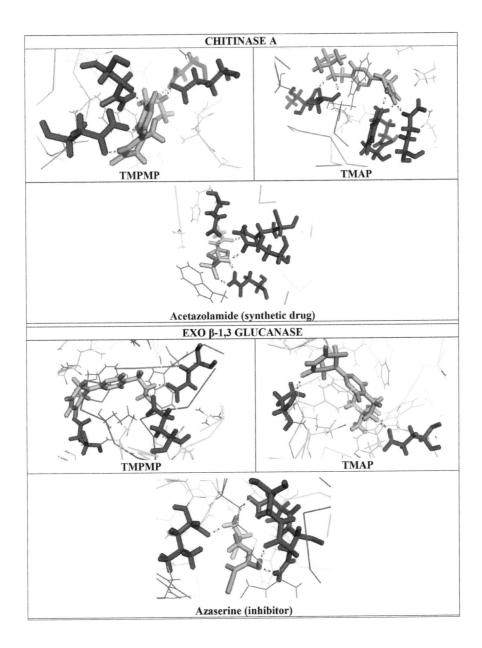

CHITINASE A

TMPMP

TMAP

Acetazolamide (synthetic drug)

EXO β-1,3 GLUCANASE

TMPMP

TMAP

Azaserine (inhibitor)

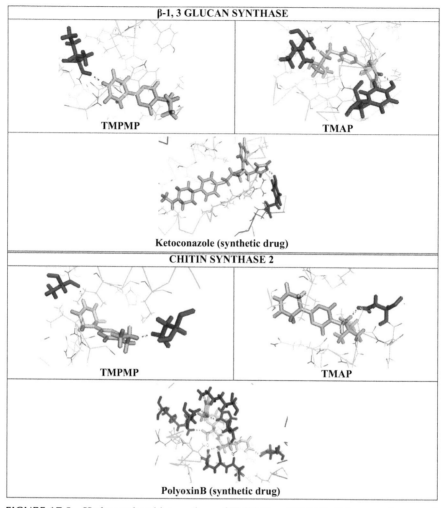

FIGURE 17.2   Hydrogen bond interactions of TMPMP and TMAP of *Acorus calamus* and respective synthetic drugs with the selected antifungal targets. Green color represents the ligand molecule; pink color represents the interacting amino acid residues, blue dots represent the hydrogen bond.

barrel fold connected by short loops. Naturally occurring components like allosamidin, styloguanidines, cyclo-L-Arg-D-Pro, psammaplin, argadin, and argifin were reported as inhibitors for family-18 chitinases (exochitinases).[38,80] To chitinase A1, the natural compound argifin had interaction with the residues Glu174 (the catalytic acid), Asp172 and weak hydrogen bond with Tyr232 and Trp312. It is reported that the residues Trp312 and Gln37 are lining the groove of protein, whereas Ala124 and Tyr125 were the backbone atoms, and the floor was contributed by the phenyl moiety of Phe60. Perhaps the hydrophobic amino acid residue Phe60 is located in the β-barrelas not like other residues represented before which located in the loops connecting the β-barrel.

The G.score for TMPMP and TMAP along with the number of hydrogen bonds with the chitinase, exo-β-1,3-glucanase, β-1,3-glucan synthase, and CS 2 protein residues and their bond length were tabulated (Table 17.4). The docking of TMPMP and TMAP of *A. calamus* with chitinase A was compared with the inhibitor acetozolamide. The G.score of acetozolamide was −3.48 kcal/mol, and five bonds interactions were observed with the polar amino acids Gln and Asn. The residue Asn at 233rd position had two hydrogen bonds with acetozolamide where the bond length was 2.5 and 2.2 Å. The interaction was also found with the active site residues Gln207, Gln230, and Glu174 of bond length 1.9, 2.0, and 2.0 Å, respectively. Terpenoid, TMPMP scored −6.64 kcal/mol of G.score and had five interactions as like acetozolamide; however, the single bond interaction was observed with the residues Asp172, Glu174, Asn233, whereas two bond formations with Asn281. Compound TMAP also had G.score of −6.62 kcal/mol and interactions with Gln230, Tyr232, Asn233, Ala279, and Thr278. Only Asn233 was involved in hydrogen bond formation with TMPMP, TMAP and also with acetazolamide. In all the three compounds, the residue Asn233 showed hydrogen bond length above 2.0 Å. The residue Glu174 had interaction with both TMPMP and acetazolamide.

The structure of exo-β-1,3-glucanase was described to have an irregular $(\beta/\alpha)_8$ barrel with its active site pocket formed by the extended loop regions, where this region function as close the substrate binding site as a shallow groove.[24] The residue Glu at the position 192 and 292 were the glucose binding site of the protein, to which 2-fluoroglycosylpyranoside and a transition state analogue, castanospermine were found to interact with these residues; however, they are also represented as the catalytic residues.[22,54] The residues Arg92 (β2 strand), His135 (β3), Asn191 (between β4 and α4), His253, Tyr255 (end of β6), and Trp363 (end of β8) are recognized as conserved to form a distinctive pocket. The negatively charged Glu27 is referred to serve as a dominant hydrogen-bonding in the exo-glucanase inhibition. Additionally, the residues located in the loop 1, 3, and 7 region act as the sides for the active site pocket, where Cutfield et al.[24] reported that the residues located in and around the pocket are highly conserved regions in the fungal exo-β-(1,3)-glucanases.

TABLE 17.4 Interaction Profile of TMPMP, TMAP, and Respective Synthetic Drug and Inhibitor with Selected Antifungal Targets.

Compounds	G.score (kcal/mol)	No. of hydrogen bonds	Interacting residues	Bond length (Å)
		Chitinase		
TMPMP	−6.64	5	ASP172 (O–H)	2.0
			GLU174 (O–H)	1.6
			ASN233 (H–N)	2.8
			ASN281 (O–H)	2.0
			ASN281 (O–H)	1.7
TMAP	−6.62	5	GLN230 (O–H)	1.8
			TYR232 (H–O)	2.2
			ASN233 (H–O)	2.1
			ALA279 (O–H)	2.0
			THR278 (O–H)	2.2
Acetazolamide	−3.48	5	GLN207 (H–O)	1.9
			GLN230 (H–O)	2.0
			ASN233 (H–O)	2.5
			ASN233 (H–N)	2.2
			GLU174 (H–O)	2.0
		Exo β-1,3-glucanase		
TMPMP	−7.62	3	ASP145 (H–O)	2.0
			GLU192 (O–H)	2.1
			GLU292 (O–H)	2.0
TMAP	−5.48	2	ASP145 (H–O)	2.1
			GLU292 (O–H)	1.9
Azaserine	−7.73	4	ASN146 (H–O)	2.0
			GLU27 (O–H)	1.9
			ASP145 (O–H)	2.3
			LEU304 (O–H)	2.0
		β-1,3-Glucan synthase		
TMPMP	−4.01	1	ILE141 (O–H)	2.0
TMAP	−4.77	2	ILE141 (O–H)	1.9
			TYR88 (H–O)	2.2
Ketoconazole	−3.27	1	TYR76 (H–N)	2.0
		Chitin synthase 2		
TMPMP	−2.42	2	ALA307 (O–H)	2.2
			ASP393 (O–H)	2.1
TMAP	−2.30	1	ASP393 (O–H)	1.6

TABLE 17.4 *(Continued)*

Compounds	G.score (kcal/mol)	No. of hydrogen bonds	Interacting residues	Bond length (Å)
PolyoxinB	−8.54	10	ARG222 (H–O)	1.7
			LYS315 (H–O)	2.1
			ASP272 (O–H)	1.9
			HIS311 (H–O)	2.0
			HIS311 (H–O)	1.9
			GLN387 (O–H)	2.3
			ALA307 (O–H)	1.6
			ARG308 (O–H)	2.1
			ARG308 (O–H)	2.6
			ARG308 (O–H)	2.1

Terpenoid, TMPMP, and synthetic drugs had G.score of −7.62 and −7.73 kcal/mol, while TMAP scored −5.48 kcal/mol. The number of interactions is 3, 2, and 4 for TMPMP, TMAP, and azaserine, respectively, where the residue Asp145 was forming hydrogen bond with all the three of bond length 2.0, 2.1, and 2.3 Å. Both TMPMP and TMAP had interaction with Glu292, where the compound TMPMP in addition interacted with Glu192. Therefore, it is understood that TMPMP do interaction as like 2-fluoroglycosylpyranoside in the glucose binding site of the protein. None of the residues located in the loop region had interaction with both TMPMP and TMAP, except azaserine which had interacted with Asn146 and Leu304 of loop 7 and 3, respectively. One more interaction with Glu27 indicated that azaserine is capable to inhibit exo-glucanase by forming a dominant hydrogen bond with this residue according to Cutfield et al.[24] Besides, not only azaserine, TMPMP is equally capable to inhibit the exo-glucanase, since it blocks the glucose binding site of the protein.

The other antifungal targets focused in the present study are CS and glucan synthase, which takes an important role in fungal cell morphogenesis. The enzymes CS and chitinase interplay in the chitin synthesis, assembly and chitin hydrolysis.[81]. Addition to it, these enzymes involve in the regulation of spore germination, budding, hyphal growth, hyphal branching, and septum formation.[36] Though, in general, chitinases are reported to function as lytic enzyme to degrade the cell wall, apart from that, maintaining plasticity and insertion of chitin fibrils are also been its role to play during the expansion of cell surface.[26] CS 1 (Chs1) and 2 (Chs2) are the existing type of enzyme to play an essential role in fungal cell wall, where Chs2 is recognized to involve in both septum formation and cell division.[82,84] Moreover, mutation in class II CS encoding gene *ChsA* reduced the conidiation efficiency,[23] whereas in the case of *ChsC* (endocing class I CS), mutation in both *ChsC* and *ChsA* had reduced the hyphal density.[39] Addition to that, Ichinomiya et al.[41] observed the defects like formation of abnormal metulae and phialides, larger septal pores with aberrant distribution and abnormal nuclear distribution. However, the Chs2 has been concentrated in the present study. Hwang et al.[40] reported that aneolignan compound, obovatol isolated from *Magnolia obovata* and tetrahydroobovatol, derivative of obovatol had potent inhibitory effect on CS 2. Here, in the present study, the drug molecule polyoxin B had 10 hydrogen bonds with 7 residues predicted as active site residues.

Targeting fungal cell wall glucan synthesis has been validated as an effective method of treating fungal infections.[96] The existing antifungal lipo-peptides (caspofungin, echinocandins, papulacandins, and penumocandins), though act as an alternative to the ergosterol-directed antimycotic agents, too have limitations because of the poor oral absorption and limited potency on animal models.[101] Therefore, screening for novel chemical entities to inhibit GS with improved pharmacokinetic properties is going on at many laboratories.[64] A set of acidic terpenoids (ascosteroside, arundifungin, enfumafungin, ergokonin A) were found effectively inhibiting the *Candida* sp. by affecting the morphology of filamentous fungi and yeast and completely prevented the normal polarized growth and the effect was as that of other GS inhibitors.[48] The fungicidal effect of GS inhibitors takes place by reversing the osmotic support in the media, which disturb the maintenance of structural integrity and isotonic environment from preserving the cells.[64]

KEYWORDS

- **terpenoids**
- **TMAP**
- **TMPMP**
- ***Acorus calamus***
- **antifungal mechanism**
- **docking study**

REFERENCES

1. Abad, M. J.; Ansuategui, M.; Bermejo, P. Active Antifungal Substances from Natural Sources. *ARKIVOC.* **2007**, *2007*, 116–145.

2. Abadio, A. K. R.; Kioshima, E. S.; Teixeira, M. M.; Martins, N. F.; Maigret, B.; Felipe, M. S. S. Comparative Genomics Allowed the Identification of Drug Targets Against Human Fungal Pathogens. *BMC Genomics* **2011**, *12*(75), 1–10.

3. Adams, D. J. Fungal Cell Wall Chitinases and Glucanases. *Microbiology* **2004**, *150*, 2029–2035.

4. Ahmad, A.; Khan, A.; Akhtar, F.; Yousuf, S.; Xess, I.; Khan, L. A.; Manzoor, N. Fungicidal Activity of Thymol and Carvacrol by Disrupting Ergosterol Biosynthesis and Membrane Integrity Against Candida. *Eur. J. Clin. Microbiol. Infect. Dis.* **2011**, *30*, 41–50.

5. Ahmet, S.; Sean, E.; Sandhya, K. Applications and limitations of *in silico* models in drug discovery. *Bioinform. Drug Discov.* **2012**, *910*, 87–124.

6. Alcazar-Fuoli, L.; Clavaud, C.; Lamarre, C.; Aimanianda, V.; Seidl-Seiboth, V.; Mellado, E.; Latge. J. P. Functional Analysis of the Fungal/Plant Class Chitinase Family in *Aspergillus fumigatus*. *Fungal Genet. Biol.* **2011**, *48*, 418–429.

7. Ali, G.; Khan, A.; Samiullah New Drug Development Process-Today: A Review. *Pharmacol. Online* **2013**, *1*, 1–10.

8. Amber, K.; Aijaz, A.; Immaculata, X.; Luqman, K. A.; Nikhat, M. Anticandidal Effect of *Ocimum sanctum* Essential Oil and its Synergy with Fluconazole and Ketoconazole. *Phytomedicine* **2010**, *17*, 921–925.

9. Ashcroft, D. M.; Po, A. L. W. Herbal Remedies. *Pharmacoeconomics* **1999**, *16*, 321–328.

10. Ayati, A.; Falahati, M.; Irannejad, H.; Emami, S. Synthesis, In Vitro Antifungal Evaluation and In Silico Study of 3-Azolyl-4-Chromanone Phenylhydrazones. *DARU J. Pharm. Sci.* **2012**, *20*(46), 1–7.

11. Balakumbahan, R.; Rajamani, K.; Kumanan, K. *Acorus calamus*: An Overview. *J. Med. Plants Res.* **2010**, *4*(25), 2740–2745.

12. Basak, S. C. Chemobioinformatics: The Advancing Frontier of Computer-Aided Drug Design in the Post-Genomic Era. *Curr. Comput. Aided Drug Des.* **2012**, *8*, 1–2.

13. Bevan, P.; Ryder, H.; Shaw, I. Identifying Small-Molecule Lead Compounds: The Screening Approach to Drug Discovery. *Science* **1995**, *287*, 1964–1969.

14. Bleicher, K. H.; Böhm, H. J.; Müller, K.; Alanine, A. I. Hit and Lead Generation: Beyond High-Throughput Screening. *Nat. Rev. Drug Discov.* **2003**, *2*, 369–378.

15. Bowman, S. M.; Free, S. J The Structure and Synthesis of the Fungal Cell Wall. *BioEssays* **2006**, *28*, 799–808.

16. Cabib, E.; Dong-Hyun, R.; Schmidt, M.; Crotti, L. B.; Varma, A. The Yeast Cell Wall and Septum as Paradigms of Cell Growth and Morphogenesis. *J. Biol. Chem.* **2001**, *276*, 19679–19682.

17. Campbell B. C.; Chan, K. L.; Kim, J. H. Chemosensitization Asa Means to Augment Commercial Antifungal Agents. *Front Microbiol.* **2012**, *3*, 1–20.

18. Cantarel, B. L.; Coutinho, P. M.; Rancurel, C.; Bernard, T.; Lombard, V.; Henrissat, B. The Carbohydrate-Active EnZymes database (CAZy): An Expert Resource for Glycogenomics. *Nucleic Acids Res.* **2009**, *37*, D233–D238.

19. Chai, X.; Yang, G.; Zhang, J.; Yu, S.; Zou, Y.; Wu, Q.; Zhang, D.; Jiang, Y.; Cao. Y.; Sun, Q. Synthesis and Biological Evaluation of Triazole Derivatives as Potential Antifungal Agent. *Chem. Biol. Drug Des.* **2012**, *80*, 382–387.

20. Chai, X.; Zhang, J.; Cao, Y.; Zou, Y.; Wu, Q.; Zhang, D.; Jiang, Y.; Sun, Q. New Azoles with Antifungal Activity: Design, Synthesis, and Molecular Docking. *Bioorg. Med. Chem. Lett.* **2011**, *21*, 686–689.

21. Chai, X.; Zhang, J.; Cao, Y.; Zou, Y.; Wu, Q.; Zhang, D.; Jiang, Y.; Sun, Q. Design, Synthesis and Molecular Docking Studies of Novel Triazole as Antifungal Agent. *Eur. J. Med. Chem.* **2011**, *46*, 3167–3176.

22. Chambers, R. S.; Walden, A. R.; Brooke, G. S.; Cutfield, J. F. Sullivan, P. A. Identification of a Putative Active Site Residue in the Exo-β-(1,3)-Glucanase of *Candida albicans*. *FEBS Lett.* **1993**, *327*, 366–369.

23. Culp, D. W.; Dodge, C. L.; Miao, Y.; Li, L.; Saq-Ozkal, D.; Borgia, P. T. The *chsA* Gene from *Aspergillus nidulans* in Necessary for Maximal Conidiation. *FEMS Microbiol. Lett.* **2000**, *182*, 349–353.

24. Cutfield, S. M.; Davies, G. J.; Murshudov, G.; Anderson, B. F.; Moody, P. C. E.; Sullivan, P. A.; Cutfield, J. F. The Structure of the Exo-β-(1,3)-Glucanase from *Candida albicans* in Native and Bound Forms: Relationship Between a Pocket and Groove in Family 5 Glycosyl Hydrolases. *J. Mol. Biol.* **1999**, *294*, 771–783.

25. Devkatte, A.; Zore G. B.; Karuppayil, S. M. Potential of Plantoils as Inhibitors of *Candida albicans* Growth. *FEMS Yeast Res.* **2005**, *5*, 867–873.

26. Dickinson, K.; Keer, V.; Hitchcock, C. A.; Adams, D. J. Microsomal Chitinase Activity from *Candida albicans*. *Biochim. Biophys. Acta* **1991**, *1073*, 177–182.

27. Douglas, C. M.; Foor, F.; Marrinan, J. A.; Morin, N.; Nielsen, J. B.; Dahl, A. M.; Mazur, P.; Baginsky, W.; Li, W.; El-Sherbeini, M.; Clemas, J. A.; Mandala, S. M.; Frommer B. R.; Kurtz, M. B. The *Saccharomyces cerevisiae* FKS1 (ETG1) Gene Encodes an Integral Membrane Protein Which is a Subunit of 1,3-β-D-glucan synthase. *Proc. Natl. Acad. Sci.* **1994**, *91*, 12907–12911.

28. Douglas, C. M.; D'Ippolito, J. A.; Shei, G. J.; Meinz, M.; Onishi, J.; Marrinan, J. A.; Li, W.; Abruzzo, G. K.; Flattery, A.; Bartizal, K.; Mitchell, A.; Kurtz, M. B. Identification of the *FKS*1 Gene of *Candida albicans* as the Essential Target of 1,3-β-D-Glucan Synthase Inhibitors. *Antimicrob. Agents Chemother.* **1997**, *41*(11), 2471–2479.

29. Drews, J. Drug Discovery: A Historical Perspective. *Science* **2000**, *287*, 1960–1964.

30. Dupont, S.; Lemetais, G.; Ferreira, T.; Cayot, P.; Gervais, P.; Beney, L. Ergosterol Biosynthesis: A Fungal Pathway for Life on Land? *Evolution* **2011**, *66*(9), 2961–2968.

31. El-Katatny, M. H.; Somitsch, W.; Robra, K. H.; El-Katatny, G. M.; Gubitz, G. M. Production of Chitinase and β-1,3-Glucanase by *Trichodermaharzianum* for Control of the Phytopathogenic Fungus *Sclerotium rolfsii*. *Food Technol. Biotechnol.* **2000**, *38*(3), 173–180.

32. Ellis, D. Amphotericin B: Spectrum and Resistance. *J. Antimicrobial. Chemother.* **2002**, *49*(S1), 7–10.

33. Escalante, A. M.; Santecchia, C. B.; López, S. N.; Gattuso, M. A.; Gutiérrez Ravelo, A.; DelleMonache, F.; Gonzalez Sierra, M.; Zacchino, S. A. Isolation of Antifungal Saponins from *Phytolaccate tramera*, An Argentinean Species in Critic Risk. *J. Ethnopharmacology* **2002**, *82*, 29–34.

34. Fang, W.; Robinson, D. A.; Raimi, O. G.; Blair, D. E.; Harrison, J. R.; Lockhart, D. E. A.; Torrie, L. S.; Ruda, G. F.; Wyatt, P. G.; Gilbert, I. H.; van Aalten, D. M. F.

N-Myristoyltransferase is a Cell Wall Target in *Aspergillus fumigatus*. *ACS Chem. Biol.* **2015**, *10*, 1425–1434.

35. Fernandez, L. R.; Butassi, E.; Svetaz, L.; Zacchino, S. A.; Palermo, J. A.; Sanchez, M. Antifungal Terpenoids from *Hyalisargentea* var. *latisquama*. *J. Nat. Prod.* **2014**, *77*, 1579–1585.

36. Gooday, G. W.; Zhu, W. Y.; O'Donnell, R. W. What are the Roles of Chitinases in the Growing Fungus? *FEMS Microbiol. Lett.* **1992**, *100*, 387–392.

37. Hamid, R.; Khan, M. A.; Ahmad, M.; Ahmad, M. M.; Abdin, M. Z.; Musarrat, J.; Javed, S. Chitinases: An Update. *J. Pharm. Bioallied Sci.* **2013**, *5*(1), 21–29.

38. Hirose, T.; Sunazuka, T.; Omura, S. Recent Development of Two Chitinase Inhibitors, Argifin and Argadin Produced by Soil Microorganisms. *Proc. Jpn. Acad. Ser.* **2010**, *B86*(2), 85–101.

39. Horiuchi, H. Functional Diversity of Chitin Synthases of *Aspergillus nidulans* in Hyphal Growth, Conidiophores Development and Septum Formation. *Med. Mycol.* **2009**, *47*(1), S47–S52.

40. Hwang, E. I.; Kwon, B. M.; Lee, S. H.; Kim, N. R.; Kang, T. H.; Kim, Y. T.; Park, B. K.; Kim, S. U. Obovatols, New Chitin Synthase 2 Inhibitors of *Saccharomyces cerevisiae* from *Magnolia obovata*. *J. Antimicrobial. Chemother.* **2002**, *49*, 95–101.

41. Ichinomiya, M.; Yamada, E.; Yamashita, S.; Ohta, A.; Horiuchi, H. Class I and Class II Chitin Synthases Are Involved in Septum Formation in the Filamentous Fungus *Aspergillus nidulans*. *Eukaryot Cell* **2005**, *4*, 1125–1136.

42. Ismail, H.; Kamara, J. Novel Antifungal Activity of Geraniol and Its Synergistic Effect in Combination with Fluconazole Against Resistant *Candida albicans*. B.Sc., Project. 2012.

43. Jaques, A. K.; Fukamizo, T.; Hall, D.; Barton, R. C.; Escott, G. M.; Parkinson, T.; Hitchcock, C. A.; Adams, D. J. Disruption of the Gene Encoding the ChiB1 Chitinase of *Aspergillus fumigatus* and Characterization of a Recombinant Gene Product. *Microbiology* **2003**, *149*(10), 2931–2939.

44. Jianling, W.; Laszlo, U. The Impact of Early ADME Profiling on Drug Discovery and Development Strategy. *Drug Discov. World Fall* **2004**, *5*, 73–86.

45. Kadam, R. U.; Roy, N. Recent Trends in Drug-Likeness Predictions: A Comprehensive Review of *In Silico* Methods. *Indian J. Pharam. Sci.* **2007**, *69*(5), 609–615.

46. Kathiravan, M. K.; Salake, A. B.; Chothe, A. S.; Dudhe, P. B.; Watode, R. P.; Mukta, M. S.; Gadhwe, S. The Biology and Chemistry of Antifungal Agents: A Review. *Bioorg. Med. Chem.* **2012**, *20*, 5678–5698.

47. Khabnadideh, S.; Rezaei, Z.; Pakshir, K.; Zomorodian, K.; Ghafari, N. Synthesis and Antifungal Activity of Benzimidazole, Benzotriazole and Aminothiazole Derivatives. *Res. Pharm. Sci.* **2012**, *7*(2), 65–72.

48. Kurtz, M. B.; Abruzzo, G.; Glattery, A.; Bartizal, K.; Marrinan, J. A.; Li, W.; Milligan, J.; Nollstadt, K.; Douglas, C. M. Characterization of Echinocandin-Resistant Mutants of *Candida albicans*: Genetic, Biochemical and Virulence Studies. *Infect. Immun.* **1996**, *64*, 3244–3251.

49. Latge, J. P. The Cell Wall: A Carbohydrate Armour for the Fungal Cell. *Mol. Microbiol.* **2007**, *66*(2), 279–290.

50. Lesage, G.; Bussey, H. Cell Wall Assembly in *Saccharomyces cerevisiae*. *Microbiol. Mol. Biol. Rev.* **2006**, *70*, 317–343.

51. Lipinski, C. A. Lead- and Drug-Like Compounds: The Rule-of-Five Revolution. *Drug Discov. Today Technol.* **2004**, *1*(4), 337–341.

52. Lipinski, C. A.; Lombardo, F.; Dominy, B. W.; Feeney, P. J. Experimental and Computational Approaches to Estimate Solubility and Permeability in Drug Discovery and Development Settings. *Adv. Drug Deliv. Rev.* **1997**, *23*, 3–25.

53. Loeffler, J.; Stevens, D. A. Antifungal Drug Resistance. *Infect. Diseases Soc. Am.* **2003**, *36*(1), S31–S41.

54. Mackenzie, L. F.; Brooke, G. S.; Cutfield, J. F.; Sullivan, P. A.; Withers, S. G. Identification of Glu-330 as the Catalytic Nuclephile of Candida Albicansexo-β-(1,3)-Glucanase. *J. Biol. Chem.* **1997**, *272*, 361–3167.

55. Maertens, J.; Vrebos, M.; Boogaerts, M. Assessing Risk Factors for Systemic Fungal Infections. *Eur. J. Cancer Care* **2001**, *10*(1), 56–62.

56. Martinez, L.; Falson, P. Multidrug Resistance ATP-Binding Cassette Membrane Transporters as Targets for Improving Oropharyngeal Candidiasis Treatment. *Adv. Cell. Mol. Otoaryngol.* **2014**, *2*, 1–8.

57. Matthew, H. The Cost of Creating a New Drug now $5 Billion, Pushing Big Pharma to Change. *Pharma Healthcare* **2013**, *11*, 183.

58. Mazur, P.; Morin, N.; Baginsky, W.; El-Sherbeini, M.; Clemas, J. A.; Nielson, J. B.; Foor, F. Differential Expression and Function of Two Homologous Subunits of Yeast 1,3-β-D-Glucan Synthase. *Mol. Cell Biol.* **1995**, *15*, 5671–5681.

59. Monk, B. C.; Tomasiak, T. M.; Keniya, M. V.; Huschmann, F. U.; Tyndall, J. D. A.; Connell, J. D.; Cannon, R. D.; McDonald, J. G.; Rodriguez, A.; Finer-Moore, J. S.; Stroud, R. M. Architecture of a Single Membrane Spanning Cytochrome P450 Suggests Constraints that Orient the Catalytic Domain Relative to a Bilayer. *PNAS* **2014**, *111*(10), 3865–3870.

60. Muller, C.; Staudacher, V.; Krauss, J.; Giera, M.; Bracher, F. A Convenient Cellular Assay for the Identification of the Molecular Target of Ergosterol Biosynthesis Inhibitors and Quantification of Their Effects on Total Ergosterol Biosynthesis. *Steroids*, **2013**, *78*, 483–493.

61. Negri, M.; Salci, T. P.; Shinobu-Mesquita, C. S.; Capoci, I. R. G.; Svidzinski, T. I. E.; Kioshima, E. S. Early State Research on Antifungal Natural Products. *Molecules* **2014**, *19*, 2925–2956.

62. Nowosielski, M.; Hoffmann, M.; Wyrwicz, L. S.; Stepniak, P.; Plewczynski, D. M.; Lazniewski, M.; Ginalski, K.; Rychlewski, L. Detailed Mechanism of Squalene Epoxidase Inhibition by Terbinafine. *J. Chem. Inf. Model.* **2011**, *51*, 455–462.

63. Ohlmeyer, M.; Zhou, M. M. Integration of Small-Molecule Discovery in Academic Biomedical Research. *Mt. Sinai J. Med.* **2010**, *77*, 350–357.

64. Onishi, J.; Meinz, M.; Thompson, J.; Curotto, J.; Dreikorn, S.; Rosenbach, M.; Douglas, C.; Abruzzo, G.; Flattery, A.; Kong, L.; Cabello, A.; Vicente, F.; Pelaez, F.; Diez, M. T.; Martin, I.; Bills, G.; Giacobbe, R.; Dombrowski, A.; Schwartz, R.; Morris, S.; Harris, G.; Tsipouras, A.; Wilson, K.; Kurtz, M. B. Discovery of Novel Antifungal (1,3)-β-D-Glucan Synthase Inhibitors. *Antimicrob. Agents Chemother.* **2000**, *44*(2), 368–377.

65. Park, C.; Woo, E. R.; Lee, D. G. Anti-*candida* Property of a Lignan Glycoside Derived from *Styrax japonica* s. Et z. via Membrane-Active Mechanisms. *Mol. Cells* **2010**, *29*, 581–584.

66. Pasrija, R.; Krishnamurthy, S.; Prasad, T.; Ernst, J. F.; Prasad, R. Squaleneepoxidase Encoded by *ERG*1 Affects Morphogenesis and Drug Susceptibilities of *Candida albicans*. *J. Antimicrobial Chemother.* **2005**, *55*, 905–913.

67. Paterson P. J.; Seaton, S.; Prentice, H. G.; Kibbler, C. C. Treatment failure in invasive Aspergillosis: susceptibility of Deep Tissue Isolates Following Treatment with Amphotericin B. *J. Antimicrob. Chemother.* **2003**, *52*, 873–876.

68. Perea, S.; Patterson, T. F. Antifungal Resistance in Pathogenic Fungi. *Antimicrob. Resist.* **2002**, *35*(9), 1073–1080.

69. Pfaller, M. A. Antifungal Drug Resistance: Mechanisms, Epidemiology, and Consequences for Treatment. *Am. J. Med.* **2012**, *125*(1A), S3–S13.

70. Pinto, E.; Afonso, C.; Duarte, S.; Silva, L.; Costa, E.; Sousa, E.; Pinto, M. Antifungal Activity of Xanthones: Evaluation of Their Effect on Ergosterol Biosynthesis by High-Performance Liquid Chromatography. *Chem. Biol. Drug Des.* **2011**, *77*(3), 212–222.

71. Podust, L. M.; Poulos, T. L.; Waterman, M. R. Crystal Structure of Cytochrome P450 14α-Sterol Demethylase (CYP51) from *Mycobacterium tuberculosis* in Complex with Azole Inhibitors. *Proc. Natl. Acad. Sci.* **2001**, *98*(6), 3068–3073.

72. Rao, A.; Zhang, Y.; Muend, S.; Rajini, R. Mechanism of Antifungal Activity of Terpenoid Phenols Resembles Calcium Stress and Inhibition of the TOR Pathway. *Antimicrobial. Agents Chemother.* **2010**, *54*(12), 5062–5069.

73. Raut, J. S.; Shinde, R. B.; Chauhan, N. M.; Karuppayil, S. M. Terpenoids of Plant Origin Inhibit Morphogenesis, Adhesion, and Biofilm Formation by *Candida albicans*. *Biofouling.* **2013**, *29*(1), 87–96.

74. Richie, D. L.; Thompson, K. V.; Studer, C.; Prindle, V. C.; Aust, T.; Riedl, R.; Estoppey, D.; Tao, J.; Sexton, J. A.; Zabawa, T.; Drumm, H.; Cotesta, S.; Eichenberger, J.; Schuierer, S.; Hartmann, N.; Movva, N. R.; Tallarico, J. A.; Ryder, N. S.; Hoepfner, D. Identification and Evaluation of Novel Acetolactate Synthase Inhibitors as Antifungal Agents. *Antimicrobial. Agents Chemother.* **2013**, *57*(5), 2272–2280.

75. Rong-mei, L.; Yong-bing, C.; Kai-hua, F.; Yi, X.; Ping-hui, G.; You-Jun, Z.; Bao-di, D.; Yong-hong, T.; Shi-hua, W.; Hui, T.; Honh-tao, L.; Yuan-ying, J. 2-Amino-Nonyl-6-Methoxyl-Tetralin Muriate Inhibits Sterol C-14 Reductase in the Ergosterol Biosynthetic Pathway. *Acta Pharmacol. Sin.* **2009**, *30*, 1709–1716.

76. Rush, C. L.; Schuttelkopf, A. W.; Hurtado-Guerrero, R.; Blair, D. E.; Ibrahim, A. F. M.; Desvergnes, S.; Eggleston, I. M.; van Aalten, D. M. F. Natural Product-Guided Discovery of a Fungal Chitinase Inhibitor. *Chem. Biol.* **2010**, *17*, 1275–1281.

77. Salake, A. B.; Chothe, A. S.; Nilewar, S. S.; Khilare, M.; Meshram, R. S.; Pandey, A. A.; Kathiravan, M. K. Design, Synthesis, and Evaluations of Antifungal Activity of Novel Phenyl(2*H*-Tetrazol-5-yl)Methanamide Derivatives. *J. Chem. Biol.* **2014**, *7*, 29–35.

78. Saurav, K.; Kannabiran, K. Interaction of 5-(2,4-Dimethylbenzyl) Pyrrolidin-2-One with Selected Antifungal Drug Target Enzymes by *In Silico* Molecular Docking Studies. *Interdiscip. Sci. Comput. Life Sci.* **2011**, *3*, 198–203.

79. Schneider, G.; Fechner, U. Computer-Based De Novo Design of Drug-Like Molecules. *Nat. Rev. Drug Discov.* **2005**, *4*, 649–663.

80. Schuttelkopf, A. W.; Gros, L.; Blair, D. E.; Frearson, J. A.; van Aalten, D. M. F.; Gilbert, I. H. Acetazolamide-Based Fungal Chitinase Inhibitors. *Bioorg. Med. Chem.* **2010**, *18*(23), 8334–8340.

81. Selvaggini, S.; Munro, C. A.; Paschoud, S.; Sanglard, D.; Gow, N. A. R. Independent Regulation of Chitin Synthase and Chitinase Activity in *Candida albicans* and *Saccharomyces cerevisiae*. *Microbiology* **2004**, *150*, 921–928.

82. Shaw, J. A.; Mol, P. C.; Bowers, B.; Silverman, S. J.; Valdivieso, M. H.; Duran, A. Cabib, E. The Function of Chitin Synthases 2 and 3 in the *Saccharomyces cerevisiae* Cell Cycle. *J. Cell Biol.* **1991**, *114*, 111–123.

83. Shigemi, A.; Matsumoto, K.; Ikawa, K.; Yaji, K.; Shimodozono, Y.; Morikawa, N.; Takeda, Y.; Yamada, K. Safety Analysis of Liposomal Amphotericin B in Adult Patients: Anaemia, Thrombocytopenia, Nephrotoxicity, Hepatotoxicity and Hypokalaemia. *Int. J. Antimicrob. Agents* **2011**, *38*, 417–420.

84. Silverman, S. J.; Sburlati, A.; Slater, M. L.; Cabib, E. Chitin Synthase 2 is Essential for Septum Formation and Cell Division in *Saccharomyces cerevisiae*. *Proc. Natl. Acad. Sci.* **1988**, *85*, 4735–4739.

85. Singh, P.; Verma, P.; Yadav, B.; Komath, S. S. Synthesis and Evaluation of Indole-Based New Scaffolds for Antimicrobial Activities-Identification of Promising Candidates. *Bioorg. Med. Lett.* **2011**, *21*, 3367–3372.

86. Srinivasan, A.; Lopez-Ribot, J. L.; Ramasubramanian, A. K. Overcoming Antifungal Resistance. *Drug Discov. Today Technol.* **2014**, *11*, 65–71.

87. Subha, T. S.; Gnanamani, A. Effect of Active Fraction of Methanolic Extract of *Acorus calamus* on Sterol Metabolism of *Candida albicans*. *J. Appl. Biosci.* **2008**, *8*(1), 243–250.

88. Subha, T. S.; Gnanamani, A. *In Vitro* Assessment of Anti-Dermatophytic Effect of Active Fraction of Methanolic Extracts of *Acorus calamus*. *J. Anim. Plant Sci.* **2009**; *5*(1), 450–455.

89. Subha, T. S.; Gnanamani, A. Candida Biofilm Perfusion Using Active Fractions of *Acorus calamus*. *J. Anim. Plant Sci.* **2009**, *4*(2): 363–371.

90. Subha, T. S.; Gnanamani, A.; Mandal, A. B. Role of TMPMP and TMAP on Virulence Factors of *C. albicans*. *Global J. Biochem.* **2011**, *2*(1), 74–80.

91. Subha, T. S.; Gnanamani, A.; Mandal, A. B. Pharmacognostic Evaluation of *Acorus calamus* L. *Phcog. J.* **2011**, *3*(23), 24–27.

92. Suginta, W.; Robertson, P. A.; Austin, B.; Fry, S. C.; Fothergill-Gillmore, L. A. Chitinases from Vibrio: Activity Screening and Purification of ChiA from *V. carchariae*. *J. Appl. Microbiol.* **2000**, *89*, 76-84.

93. Sun, W.; Park, Y. D.; Sugui, J. A.; Fothergill, A.; Southall, N.; Shinn, P.; McKew, J. C.; Kwon-Chung, K. J.; Zheng, W.; Williamson, P. R. Rapid Identification of Antifungal Compounds Against *Exserohilum rostratum* using High Throughput Drug Repurposing Screens. *PLoS One* **2013**, *8*, e70506.

94. Tang, H.; Zheng, C.; Lv, J.; Wu, J.; Li, Y.; Yang, H.; Fu, B.; Li, C.; Zhou, Y.; Zhu, J. Synthesis and Antifungal Activities In Vitro of Novel Pyrazino [2,1-*a*] Isoquinolin Derivatives. *Bioorg. Med. Chem. Lett.* **2010**, *20*, 979–982.

95. Tsai, H. F.; Bard, M.; Izumikawa, K.; Krol, A. A.; Sturm, A. M.; Culbertson, N. T.; Pierson, C. A.; Bennett, J. E. *Candida glabrata* Mutant with Increased Sensitivity to Azoles and to Low Oxygen Tension. *Antimicrob. Agents Chemother.* **2004**, *48*, 2483–2489.

96. Turner, W. W.; Current, W. Echinocanding Antifungal Agents. In *Biotechnology of Antibiotics*; Strohl, W. R., Eds.; Marcel-Dekker, Inc.: New York, 1997, pp 315–334.

97. Vincent, B. M.; Lancaster, A. K.; Scherz-Shouval, R.; Whitesell, L.; Lindquist, S. Fitness Trade-Offs Restrict the Evolution of Resistance to Amphotericin B. *PLoS Biol.* **2013**, *11*(10), 1–17.

98. Walters, W. P.; Murcko, A.; Murcko, M. A. Recognizing Molecules with Drug-Like Properties. *Curr. Opin. Chem. Biol.* **1999**, *3*, 384–387.

99. Wong, S. S. W.; Samaranayake, L. P.; Seneviratne, C. J. In Pursuit of the Ideal Antifungal Agent for *Candida* infections: High-Throughput Screening of Small Molecules. *Drug Discov. Today* **2014**, *19*(11), 1721–1730.

100. Xie, J. L.; Polvi, E. J.; Shekhar-Guturaj, T.; Cowen, L. E. Elucidating Drug Resistance in Human Fungal Pathogens. *Future Microbiol.* **2014**, *9*(4), 523–542.

101. Yeung, C. M.; Klein, L.; Lartey, P. A. Preparation and Antifungal Activity of Fusacandin Analogs: C-6′ Sidechain Esters. *Bioorg. Med. Chem. Lett.* **1996**, *6*, 819–822.

102. Young, L. Y.; Hull, C. M.; Heitman, J. Disruption of Ergosterol Biosynthesis Confers Resistance to Amphotericin B in *Candida lusitaniae. Antimicrobial. Agents Chemother.* **2003**, *47*(9), 2717–2724.

103. Yu, S.; Chai, X.; Wang, Y.; Cao, Y.; Zhang, J.; Wu, Q.; Zhang, D.; Jiang, Y.; Yan, T.; Sun, Q. Triazole Derivatives with Improved In Vitro Antifungal Activity Over Azole Drugs. *Drug Des. Dev. Ther.* **2014**, *8*, 383–390.

104. Zhang, J. D.; Xu, Z.; Cao, Y. B.; Chen, H. S.; Yan, L.; An, M. M.; Gao, P. H.; Wang, Y.; Jia, X. M.; Jiang, Y. Y. Antifungal Activities and Action Mechanisms of Compounds from *Tribulus terrestris* L. *J. Ethnopharmacol.* **2006**, *103*, 76–84.

105. Zore, G. B.; Thakre, A. D.; Jadhav. S.; Karuppayil, S. M. Terpenoids Inhibit *Candida albicans* Growth by Affecting Membrane Integrity and Arrest of Cell Cycle. *Phytomedicine* **2011**, *18*, 1181–1190.

CHAPTER 18

SUPPORT VECTOR MACHINE AS A CLASSIFIER FOR FEATURE-BASED CLASSIFICATION: A TECHNICAL NOTE

SURENDRA KUMAR*

Department of Bioengineering, Birla Institute of Technology, Mesra, Ranchi, India

Corresponding author. E-mail: Sikuranchi@gmail.com

CONTENTS

ABSTRACT

The classification task is quite common in making decisions on the information available at a given time. The classification of the signals involves investigation of different attributes to decide that on which group or class the signal belongs. The resulting segmentation can be mapped back into the physical world to extract information about the physical phenomena and interaction that created the signal. The algorithm that performs the classification task is commonly referred as a classifier. Support vector machine (SVM) is a supervised learning model based on statistical learning theory. While classifying by SVM, a set of hyperplanes are constructed in a high-dimensional space. The hyperplanes are constructed by mapping of n-dimensional feature vectors in to a k-dimensional space via a nonlinear function $\Phi(x)$ with the aim of minimizing the margin between two classes of data. For the present classification, kernel functions have been used for mapping the feature space to a high-dimensional space in which the class is linearly separable.

18.1 INTRODUCTION

Today, any kind of experimental research requires a lot of data analysis to find a conclusion. Without application of soft-computing tools like dimension reduction and classification, it is not possible to handle a big feature data. The classification task is quite common in making decisions on the information available at a given time. Some of the common classification tasks include the approaches undertaken for categorization of letters and consignments based on automated detected postcodes, denoting individuals to credit status on the basis of information related with the initial diagnosis of a patient's disease, so as to determine the first level treatment, when test result are not available.[1] The classification of the signals involves investigation of different attributes to decide that on which group or class the signal belongs. The resulting segmentation can be mapped back into the physical world to extract information about the physical phenomena and interaction that created the signal. The classification aims at assigning class labels to the features obtained from the analysis of dataset in each problem. The algorithm that performs the classification task is commonly referred as a classifier. The term classifier, sometimes also related to the mathematical function, is mapped by a given classification algorithm that maps input data to a particular class. Classifiers are trained to detect the class of a feature vector and assign them accordingly. These train datasets are composed of feature vectors with information about their class of belonging.

18.2 MATERIALS AND METHODS

Support vector machine (SVM) is a supervised learning-based algorithm for solving binary (two class) classification problems. It was proposed by Vapnik in 1995. Since its inception, it has been widely applied in different fields of machine learning and pattern recognition. Classification of linearly separable feature data obtained from research are quite simple and can be easily handled, but it became difficult when feature vectors are nonlinear like EEG data.[3,10] Therefore, in the present work, the SVM classifier has been applied to separate extracted feature of EEG signals into the two categories of alcoholic and nonalcoholic data.

This study considers a total of 40 right handed male subjects between 32 and 38 years of age and were equally divided in two groups: (1) alcoholic and (2) control. After recoding the data, the artifacts were removed and conditioned by applying suitable digital filters. The two-second epochs were selected, and Hilbert Huang Transform has been used for feature extraction of the brain signals, and a combined framework linear discriminate analysis (LDA) and SVM has been used for classification. The feature vector can be used as a distinguishing feature for gathering EEG into alcoholic and control groups. The intrinsic mode functions (IMFs) obtained from the empirical mode decomposition algorithm was used to obtain the feature vector. The IMFs itself can be used as discriminating feature for classifying alcoholics and control subjects. Instantaneous amplitude and instantaneous frequency was also obtained for differentiating alcoholics and control subject. The different number of samples of EEG data was classified using the combined framework of LDA and SVM.

Ideally, an SVM can identify a hyperplane that separates alcoholic and control feature vector of EEG signal data in its high-dimensional feature space. SVM performs the classification task by constructing a set of hyperplanes in a high-dimensional space.[11] This is achieved by mapping n-dimensional feature vector to a higher dimensional space using a function $\varphi(.)$. The function aims at maximizing the margin of separability between the classes. For the simplest case, where the classification can be performed using a linear function, the hyperplane equation is given by

$$y = \langle w.x \rangle + b \qquad (18.1)$$

$$\varphi(y) = sign(y) \qquad (18.2)$$

where $\langle w.x \rangle$ represents the dot product of the weight vector w and feature vector x, and b represents the bias weight. However, for many feature dataset, the classes may not be separable using a linear hyperplane because the feature versus class relationship may be nonlinear. For such cases, the SVM solves the nonlinear classification problem using kernel functions. For classification, each sample in a set of training samples consisting of feature vector is matched into two categories before being reflected into a high-dimensional space using the appropriate kernel function. Further, the SVM aims at creating a model and uses it to group the samples to a definite class. The nonlinear kernel used for mapping the input patterns into a higher dimensional feature space is chosen a priori. Based on the mapping, the model constructs a linear separating hyperplane in the high-dimensional space. The hyperplane on both sides is covered by parallel hyperplanes, which groups the data points into two categories between the two hyperplanes (Fig. 18.1). A greater distance or difference between parallel hyperplanes indicates a smaller total SVM error rate. Thus, SVM is a linear classifier in the parameter space, but it becomes a nonlinear classifier as a result of the nonlinear mapping of the space of the input patterns into the high-dimensional feature space. The selection of a proper kernel function is very significant for the mapping, and it is dependent on the specific feature data. No proper method exists for the selection of kernel function.

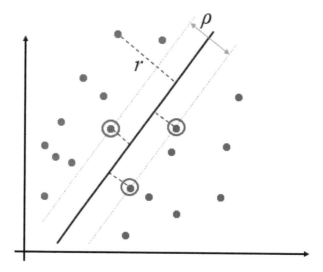

FIGURE 18.1   The figure represents the optimal separating hyperplane along with covered by parallel hyperplanes on both sides. The hyperplanes are shown separating the data points in two groups.

For the present work, the widely used Gaussian radial basis function has been used. The function is given as

$$\varphi(x_i) = exp\left(\frac{-\| x_i - x_j \|^2}{2\sigma^2}\right)$$ (18.3)

where x_j and σ represent the mean and standard deviation of the Gaussian curve, respectively. Using the kernel-induced feature space, the output of the SVM classifier can be computed as

$$Y = sign\left(\sum \alpha_i y_i \varphi(x_i)\right) + b$$ (18.4)

Using the above function, the SVM is trained by a set of training sample $\{x_i y_i\}_{i=1}^N$ with the input vector xi and keeping class labels $\{-1, +1\}$, $\varphi(x_i)$ is the kernel function of SVM, $\lambda_i > 0$ is Lagrangian multipliers calculated by solving a quadratic optimization problem and b is the bias weight.

18.3   RESULT AND DISCUSSION

It has been observed that the maximum classification accuracy using SVM clustering (76.6%) was obtained using the EEG data of 300 samples of two seconds' epochs as shown in Table 18.1. It was seen that the maximum classification accuracy was seen on the C3, Cz, and Pz electrodes, that is, the maximum accuracy is seen at the left central, central, and parietal regions of the brain. Also, the central and parietal regions showed greater classification accuracy than the frontal region of the brain.[5,7] It was also seen that the classification accuracy depended on the number of EEG samples that were used for classification. For the extraction of maximum information from acquired data, the selection of optimal channel locations is very important. Few works have been reported using VEP for the selection of optimal number of channels.[4,9] Conversely, in this work, considering the diminished limbic control in alcoholic subjects;[2] nine channels from cerebral motor cortex region have been selected for the analysis.

The highest classification accuracy obtained with the EEGs extracted from Cz area can be explained in the light of past findings that demonstrated desynchronized firing of neurons in alcoholic subjects,[5-8] which is also reflected as increased amplitude in cortical EEG. Similar observations have been marked in F3 focal area of cerebral cortex. Based on these findings, it can be suggested that the right part of motor cortex area is active or fail to present synchronizing activity in chronic alcoholism.

TABLE 18.1 Classification Accuracy in Percentage for the Alcoholic and Control Subject's EEG Epochs of Two Seconds Tested on Different Number of Samples.

Electrode position	Number of alcoholic and control EEG epochs selected			
	150	160	180	200
F3	60	60	58.33	60
Fz	70	65.25	61.11	62.5
F4	73.3	71.87	72	72.5
C3	76.6	75	75	75
Cz	76.6	71.875	72.22	70
C4	70	68.75	66.66	70
P3	73.33	71.87	72.22	65
Pz	76.67	71.78	72.22	72.5
P4	70	68.75	66.66	70

In this study, the classification accuracy obtained from SVM was compared with another classifier Fuzzy C-Mean Clustering (FCM) and found superior than FCM. The less computational time of the proposed algorithm in classifying EEG signals, arising because of the simplicity of algorithm would make its implementation on a digital chip, a fairly easy task. The future work in this direction is planned on the use of a weighted combination of different band specific features for classification and the use of a global optimization technique for clustering.

The conclusion of the study can be explained as the classifier is a separating hyperplane and most "important" training points are support vectors; they define the hyperplane. The quadratic optimization algorithms can identify which training points xi are support vectors with nonzero Lagrangian multipliers αi. The selection of a proper kernel function plays a very significant role, and no proper method exists for the selection of kernel function.

KEYWORDS

- SVM
- EEG
- HHT
- classification
- alcoholism

REFERENCES

1. Brunelli, R. *Template Matching Techniques in Computer Vision: Theory and Practice*; Wiley: New York, 2009.
2. Chen, Y. W.; Barson, J. R.; Chen, A.; Hoebel, B. G.; Leibowitz, S. F. Opioids in the Perifornical Lateral Hypothalamus Suppress Ethanol Drinking. *Alcohol* **2013**, *47*, 31e38.
3. Guler, I.; Ubeyli, E. Multiclass Support Vector Machines for EEG-Signals Classification. *IEEE Trans. Inform. Technol. Biomed.* **2007**, *11*, 117–126.
4. Kim, D. J.; Jeong, J.; Kim, K. S.; Chae, J. H.; Jin, S. H.; Ahn, K. J.; Myrick, H.; Yoon, S. J.; Kim, H. R.; Kim, S. Y. Complexity Changes of the EEG Induced by Alcohol Cue Exposure in Alcoholics and Social Drinkers. *Alcohol Clin. Exp. Res.* **2003**, *27*, 1955–1961.
5. Kumar, S.; Ghosh, S.; Sinha, R. K. Automated Identification of Chronic Alcoholism from Brain Signals. *Online J. Health Allied Sci.* **2015**, *14*(4), 20. http://www.ojhas.org/issue56/2015-4-20.html.
6. Kumar, S.; Ghosh, S.; Sinha, R. K. Using Computational Classifiers to Detect Chronic Alcoholism. *J. Clin. Eng.* **2016**, *41*(2), 90–94.
7. Kumar, S.; Ghosh, S.; Tetarway, S.; Sinha, R. K. Support Vector Machine and Fuzzy C-mean Clustering-Based Comparative Evaluation of Changes in Motor Cortex Electroencephalogram Under Chronic Alcoholism. *Med. Biol. Eng. Comput.* **2015**, *53*(7), 609–622.
8. Moselhy, H. F.; Georgiou, G.; Kahn, A. Frontal Lobe Changes in Alcoholism: A Review of the Literature. *Alcohol.* **2001**, *36*, 357–368.
9. Porjesz, B.; Begleiter, H. Alcoholism and Human Electrophysiology. *Alcohol Res. Health.* **2003**, *27*, 153–160.
10. Yeh, C. L.; Lee, P. L.; Chen, W. M.; Chang, C. Y.; Wu, Y. T.; Lan, G. Y. Improvement of Classification Accuracy in a Phase-Tagged Steady-State Visual Evoked Potential-Based Brain Computer Interface Using Multiclass Support Vector Machine. *BioMed. Eng. Online* **2013**, *12*, 46. DOI: 10.1186/1475-925X-12-46.
11. Zhang, Z.; Ye, N. Constraint Projections for Discriminative Support Vector Machines. *Int. Joint Conf. Bioinf. Syst. Biol. Intell. Comput.* **2009**, *21*, 21–30.

CHAPTER 19

OPTIMIZATION OF COST EQUITY OF A HYBRID RENEWABLE ENERGY SYSTEM

SARAT KUMAR SAHOO*, R. NIVEDITA*, UDIT MISHRA, and INDRAYUDH GANGULY

School of Electrical Engineering, VIT University, Vellore, India

*Corresponding author. E-mail: sksahoo@vit.ac.in; r_nivedita@ outlook.com

CONTENTS

ABSTRACT

Energy considerations in India as well as the whole world have become more pertinent over the last few decade, with a considerable amount of subsidy and resources being accorded to the systems in India, because of a massive requirement followed by an increasing home and industrial consumption rate. This chapter suggests a modus operandi for optimal sizing of stand-alone PV/WG systems using genetic algorithms. Uniting the necessary working metrics, a full workup of the solar patterns is provided including the irradiation losses, system losses, and the performance ratios of the system. An analysis is done on existing solar power plants and their perceived costs and the impact of varying power features and parameters over time was assessed. Finally, we developed predictive systems for the same using artificial neural networks to derive fitness function that can then be operated upon to get the absolute minimum cost of any stand-alone PV/WG hybrid system.

19.1   INTRODUCTION

Photovoltaic and wind generator energy sources are widely used to supply power in remote areas. This paper deals with the absolutes of cost analysis for solar photovoltaic power plants. The primary reason for which they can be used in a hybrid system is the complementary characteristics which both the sources possess.[1,2] This paper runs an initiation that deals with the absolutes of cost analysis for solar photovoltaic power plants. In the current energy context, it is important to assess the various ways in which the cost optimization approaches can be realized in a real context system. There is a need for thorough analysis of the system, which takes into consideration all the solar parameters, before a cost system algorithm is applied.

The block diagram of a standard stand-alone PV/WG hybrid system is shown in Figure 19.1. Battery chargers shown in the figure are used to charge battery banks from respective photovoltaic and wind generators in case the power output of these components is more than the load demand. The battery bank is used to supply load in case the wind speed is low or due to poor irradiations. A DC/AC converter is used for the purpose of interfacing the DC battery voltage to consumer's AC load requirements. As the solar radiation and the wind speed are continuously varying, which in turn highly influence the resulting energy production, these constraints were to be established and applied in a manner such that it is possible to take care of the load power required for the specific project and since this varies based on environmental conditions, this required a complete environmental analysis on a given location for contextual studies.[3,4]

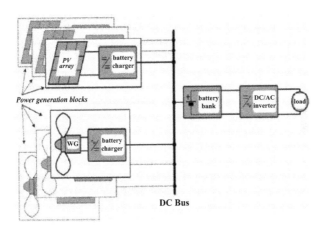

FIGURE 19.1   Hybrid PV/WG system block diagram.

19.2   MATLAB SIMULATION AND COMPARATIVE ANALYSIS

- Carry out a comprehensive sun path and weather analysis of the probable locations of the solar power plant.
- Estimate the solar radiation and wind profile of the given site.
- Calculations of load demands to be met by the hybrid system. Battery capacity based on design loads.
- Estimating the output of a single PV module and wind generator.
- Calculating the size of PV array and the number of wind generators.

The overall analysis deals with the interdependency of the system and its comprehensive development as a factor of the final cost. We have implemented a few methodologies to derive a comprehensive and complete understanding of the same.

19.3   HOMER IMPLEMENTATION OF HYBRID SYSTEM

The HOMER Hybrid Optimization Modelling Software is used for designing and analyzing hybrid power systems, which contains a mix of wind turbines and solar photovoltaics. It allows the user to input an hourly power consumption profile and match renewable energy generation to the required load.

HOMER contains a powerful optimizing function that is useful in determining the cost of various energy project scenarios. It can perform a sensitivity analysis on almost any input by ass igning more than one value to each input of interest. HOMER repeats the optimization process for each value of the input so that one can examine the effect of changes in the value on the results. It helps to understand the level of disparity in cost per unit of electricity, by creating a system that was able to account for changes in environmental factors and use hybrid systems to approximate costs.

Figure 19.2 shows the schematic diagram of the hybrid power plant to be installed in VIT University. Tables 19.1 and 19.2 give us information about the solar irradiation and the wind profile of Vellore, respectively.

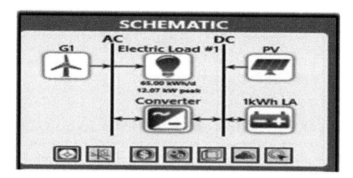

FIGURE 19.2   Schematic hybrid system model of 20-kW power plant.

TABLE 19.1   Solar Resource Inputs.

Month	Daily irradiation (kW h/m²/day)
January	5.232
February	6.166
March	6.348
April	6.923
May	6.271
June	5.153
July	5.005
August	5.072
September	5.325
October	4.998
November	4.815
December	4.694

The net capital cost required to set up a hybrid plant in Vellore would amounts to Rs. 7,264,325.

19.3.1 OPTIMIZING THE COST FUNCTION USING GA

Genetic algorithm (GA) helps to realize the optimization of large power systems. It generally uses the biological properties of crossover, recombination and mutation in the optimization procedure. For every set of parameters, different solutions are grouped together to form chromosomes (Table 19.2).[5]

TABLE 19.2 Wind Source Inputs.

Month	Average speed (m/s)
January	2.830
February	2.880
March	3.270
April	3.620
May	3.950
June	4.280
July	3.900
August	3.800
September	2.910
October	2.430
November	2.650
December	2.920

GA is particularly very helpful for finding optimal solutions in case of nonlinear problems,[6] such as situations where both the supply system as well as the load fluctuates. The flow chart of GA optimization is shown in Figure. 19.3.

Following are set of parameters considered for optimizing the total cost of wind–PV hybrid system:

N_{PV}: number of PV modules
N_{WG}: number of wind generators
N_{BAT}: number of batteries
N_{ch}^{PV}: number of PV chargers
h: height of wind generator

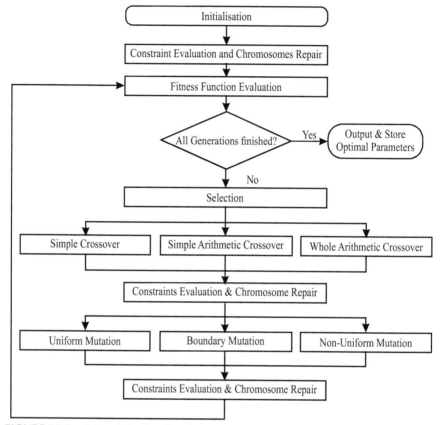

FIGURE 19.3   GA optimization flowchart.

The total system cost function is equal to the sum of total capital cost and total maintenance cost. Hence, the multiobjective function whose cost is to be minimized for a time period of 20 years is defined as

$$J(N_{PV}, N_{WG}, N_{BAT}, N_{ch}^{PV}, h) = N_{PV} \times (C_{PV} + 20 \times M_{PV}) + N_{WG}(C_{WG} + 20 \times M_{WG} + h \times C_h + 20 \times h \times C_{hm}) + N_{BAT}[C_{BAT} + y_{BAT} \times C_{BAT} + (20 - y_{BAT}1) \times M_{BAT}] + N_{ch}^{PV} \times C_{ch}^{PV} \times (y_{ch}^{PV} + 1) + N_{ch}^{PV} \times M_{ch}^{PV}1) + C_{INV} \times (y_{INV} + 1) + M_{INV} \times (20 - y_{INV} - 1)$$

Subject to constraints:

$$N_{PV} \geq 0, N_{PV} \geq 0, N_{BAT} \geq 0, N_{ch}^{PV} \geq 0$$

$$8 \leq h \leq 15, X_4 = 0.183\ X_1\ P = X_1 \times P_{PV} + X_2 \times P_{WG}\ N_{PV} > N_{WG}$$

Tables 19.3 and 19.4 summarize the result we got for the optimal cost of a 20-kW hybrid power plant using genetic algorithm.

TABLE 19.3   Capital and Maintenance Costs of the Hybrid System Devices.

Component	Specifications	Unit cost (INR)	Maintenance cost (INR)
PV module	55 W	16883.57	168.83
	110 W	32991.60	329.91
Wind generator	1000 W	110946	1109.46
	400 W	33792	337.92
Battery	230 A-h	16768.61	167.68
	100 A-h	8003.20	80.03
PV charges	300 W	12703.49	127.03
	240 W	970.64	59.70
Inverter	1500 W	123350.89	1233.50

19.3.2   ARTIFICIAL NEURAL NETWORK TRAINED MODEL FOR PREDICTION OF COST

In machine learning, artificial neural networks (ANNs) are a family of statistical learning algorithms inspired by biological neural networks (the central nervous systems of animals, in particular, the brain) and are used to estimate or approximate functions that can depend on a large number of inputs and are generally unknown.

Following is the step-wise execution of ANN:

- Assign random weights to all linkages to start the algorithm.
- Using the inputs and the hidden linkages, find the activation rate of hidden nodes.
- Using the activation rate of hidden nodes and linkages to output, find the activation rate of output nodes.
- Find the error rate at the output node and recalibrate all the linkages between hidden nodes and the output nodes.
- Using the weights and the error found at output node, cascade down the error to hidden nodes.
- Recalibrate the weights between hidden node and the input nodes.
- Repeat the process till the convergence criterion is met.

TABLE 19.4 Optimized Cost of the Power Plant.

Type					Number of modules				Number of WG	GA cost (INR)
PV module	Battery	Charge	WG	Inverter	Number of PV modules	Number of batteries	Number of charges	Number of inverters		
1	1	1	1	1	75	14	14	1	40	6,711,540
1	2	1	2	1	82	13	15	1	38	6,911,320
2	1	1	1	1	84	15	12	1	37	6,854,364
2	2	1	2	1	81	12	13	1	42	7,278,549

Using the final linkage weights, score the activation rate of the output nodes shown in Figure 19.4.

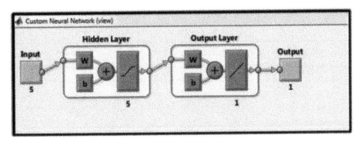

FIGURE 19.4   ANN model schematic.

Figure 19.5 shows the efficiency of ANN model in four parts of the set of values provided, a fraction bock was used first to train the system, the next block of values was then used to validate the system. Ultimately, the graph gave the best fit system as shown in the bottom right graph. The cost prediction using ANN trained model is presented in Table 19.5. Figure 19.6 shows the percentage accuracy of prediction of optimal solution by ANN. Table 19.6 shows the comparative analysis of HOMER, ANN, and GA.

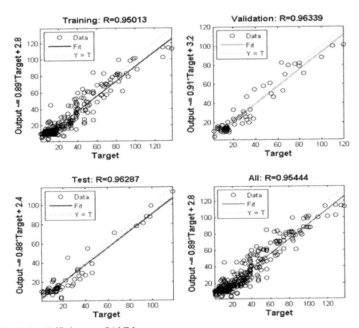

FIGURE 19.5   Efficiency of ANN.

TABLE 19.5   Cost Prediction Using ANN Trained Model.

	No. of PV modules	No. of wind generators	Height of wind generator	No. of batteries	No. of PV chargers	No. of inverters	Total cost (INR)
ANN	71	39	14	14	11	1	6,614,124

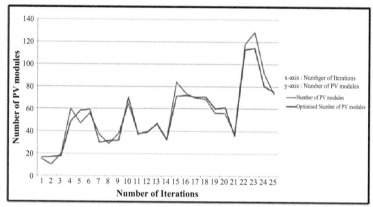

FIGURE 19.6   Assumed no. of PV modules vs ANN predicted no. of PV modules.

TABLE 19.6   Comparative Analysis of HOMER, ANN, and GA.

	No. of PV modules	No. of wind generators	Height of wind generator	No. of batteries	No. of PV chargers	No. of inverters	Total cost (INR)
HOMER	182	1	14	14	34	2	7,264,325
ANN	71	39	14	14	11	1	6,614,124
GA	75	40	14	14	14	1	6,711,540

19.4 CONCLUSION

With the HOMER simulation, we were first able to understand that the minimization factor of cost is such that it corresponds with a wind/PV hybrid system on account of the environmental conditions specifically. We used the ANN algorithm to create a system that is consonant with these findings. The final studies allow the user to compute the cost value of all possible configurations.

KEYWORDS

- solar photovoltaic
- wind generator
- genetic algorithm
- hidden nodes
- PV/WG hybrid system
- cost function

REFERENCES

1. Belfkira, R.; Hajji, O.; Nichita, C.; Barakat, G. In *Optimal Sizing of Stand-Alone Hybrid Wind/PV System with Battery Storage*, Power Electronics and Applications, 2007 European Conference on, 2007, pp 1–10.
2. Kellogg, W. D.; Nehrir, M. H.; Venkataramanan, G.; Gerez, V. Generation Unit Sizing and Cost Analysis for Stand-Alone Wind, Photovoltaic and Hybrid Wind/PV Systems. *IEEE Trans. Energy Convers.* **1998,** *13*(1), 70–75.
3. Coelho, R. F.; Concer, F.; Martins, D. C. In *A proposed Photovoltaic Module and Array Mathematical Modeling Destined To Simulation*, Industrial Electronics, 2009 ISIE 2009 IEEE International Symposium, 2009, pp 1624–1629.
4. Ramirez-Rosado, I. J.; Bernal-Agustin, J. L. Genetic Algorithms Applied to the Design of Large Power Distribution Systems. *IEEE Trans. Power Syst.* **1998,** *13*(2), 696–703.
5. Ganga Prasanna, M.; Mahammed Sameer, S.; Hemavathi, G. Financial Analysis of Solar Photovoltaic Power plant in India. *IOSR J. Econ. Finance* (IOSR-JEF) e-ISSN: 2321-5933, p-ISSN: 2321-5925, pp 09–15.
6. Lazou, A.; Papatsoris, A. The Economics of Photovoltaic Stand-Alone Residential Households: A Case Study for Various European and Mediterranean Locations. *Sol. Energy Mater. Sol. Cells* **2000,** *62*, 411–427.

INDEX